Group-Theoretic Methods in Mechanics and Applied Mathematics

Differential and Integral Equations and their Applications
A series edited by:
A.D. Polyanin
Institute for Problems in Mechanics, Moscow, Russia

Volume 1
Handbook of First Order Partial Differential Equations
A.D. Polyanin, V.F. Zaitsev and A. Moussiaux

Volume 2
Group-Theoretic Methods in Mechanics and Applied Mathematics
D.M. Klimov and V.Ph. Zhuravlev

Volume 3
Quantization Methods in Differential Equations
V.E. Nazaikinskii, B.-W. Schulze and B.Yu. Sternin

Group-Theoretic Methods in Mechanics and Applied Mathematics

D. M. Klimov
and
V. Ph. Zhuravlev

Institute for Problems in Mechanics
Russian Academy of Science
Moscow, Russia

CRC Press
Taylor & Francis Group
Boca Raton London New York

CRC Press is an imprint of the
Taylor & Francis Group, an **informa** business

CRC Press
Taylor & Francis Group
6000 Broken Sound Parkway NW, Suite 300
Boca Raton, FL 33487-2742

ISBN 13: 978-0-415-29863-6 (hbk)

Visit the Taylor & Francis Web site at
http://www.taylorandfrancis.com

and the CRC Press Web site at
http://www.crcpress.com

CONTENTS

FOREWORD

Group theory, which in contemporary science has become a powerful tool for calculus, dates from the middle of the 19th century. The basic notions of group theory were clearly stated by Evariste Galois in connection with the solution of algebraic equations in radicals (1832). Somewhat later, in his fundamental work *Treatise on Permutations*, Camille Jordan not only developed the ideas of Galois, but indicated many other applications of group theory and obtained significant results in the theory itself (1870).

At first the groups studied were finite and discrete. The notion of the continuous group was introduced by Sophus Lie, a pupil of Jordan's, who tried to apply the methods of Galois to the problem of integrating differential equations in quadrature (1891). Lie understood continuous groups as transformation groups expressed by means of analytic (or infinitely differentiable) functions. Later such groups were called Lie groups, while the term *continuous group* has remained, but is now used for abstract groups.

Neither Sophus Lie nor his followers found any constructive methods for integrating differential equations in quadrature, and for a long time the theory was developed to solve purely algebraic problems, regardless of the objective that had initiated its appearance.

The first, and extremely illustrative, applications of Lie group theory pertained to relativistic mechanics (1905) and quantum mechanics (1924).

The group-theoretic analysis of the foundations of relativistic mechanics, carried out by Henri Poincaré, was a brilliant instance of the implementation of Felix Klein's Erlangen program (1872): a meaningful physical theory should be expressed only in terms of invariants of the appropriate group. Poincaré, having postulated that Lorentz transformations provide the connection between different inertial systems, established the group-theoretic character of these transformations, computed the invariants of this group and, using these invariants, found the law of motion of a relativistic particle (the analogue of Newton's second law).

It is also to Poincaré that we owe the first application of group-theoretic ideas to analytic mechanics. He found the generalization of the Lagrange equations by using the operator basis for the transitive group acting on the configuration space of the system (1901).

A special role in mechanics and physics is played by the group SO(3) of orientation-preserving orthogonal transformations, which is a subgroup of the Galilean group and, in addition, the configuration manifold of a rigid body in space with one fixed point. The parametrization of the kinematics of rotations, performed by Hamilton and Klein, based on group variables, allowed this parametrization to be simplified considerably as compared with the parametrization proposed by Euler in 1776. Thus, unlike the Euler equations, the kinematic equations of Poisson are linear. The new kinematics of rigid bodies later found successful applications in the navigation of moving objects.

It is interesting to note that Hamilton, when he "discovered" quaternions in 1843, immediately found applications for them to the kinematics of rigid bodies, long before the notions of the continuous group and of the covering spaces of groups appeared. This shows that not only do group-theoretic ideas serve as the basis for constructing realistic models in mechanics and physics, but the converse is also true: the construction of these models exhibits such a basis.

It is no accident that solutions of autonomous differential equations form a group, whereas those of nonautonomous ones do not. The presence of time on the right-hand side in explicit form indicates that the idealization involved in this model is quite crude. Another example in this field is the equivalence, discovered by J. J. Moreau in 1959, between the dynamics of an ideal liquid in a cavity and the dynamics of an infinite-dimensional rigid body.

It is easy to see that the applications of group theory to mechanics and physics described above were used to create models, derive differential equations, and analyze their properties, and so go far beyond Sophus Lie's narrower goal—integration of differential equations in quadrature. Actually, Lie himself undertook the group-theoretic analysis of partial differential equations (he found the symmetry group of the heat equation), although in such equations the presence of symmetries does not mean that the order of equations can be lowered. However, Lie apparently did not notice that the group invariance of partial differential equations makes it possible to decrease the number of independent variables, to construct particular solutions, and to classify boundary conditions.

In 1950, George Birkhoff pointed out the usefulness of group-theoretic analysis for the equations of hydrodynamics. These ideas were widely developed in the work of L. Ovsiannikov and his pupils.

As to the original goal (the integration of differential equations in quadrature), the first results after Lie were obtained with a certain delay. In 1918, Emmy Noether established the connection between the symmetries of Hamiltonian actions and the first integrals of the corresponding Lagrange equations.

Later on, many authors derived integrability (or the decrease of order) of Hamiltonian systems from the existence of symmetries for the Hamilton equations.

There are only two or three examples of completely integrable systems possessing solvable symmetry groups of the appropriate order, and they are all classical problems whose solutions by other methods were known long before group-theoretic methods appeared.

The only constructive applications of this theory were given by groups of linear transformations, whose existence as symmetries is usually easy to observe, and this makes it possible to lower the order of the system. However, on the one hand, this only requires a small part of the theory and, on the other, it is completely covered by dimensional and similarity considerations.

Here we should note that the problem of exact integration of differential equations, which was regarded as the main problem in the 19th century, is no longer considered so important today. Often the decrease of order by the use of first integrals spoils the system so much that, if complete integrability is not achieved, the procedure is useless. Sometimes it is advantageous to raise the order, whenever this improves the analytical properties of the system, making it more convenient for investigation, e.g., by using a computer.

Not so long ago a new domain of application of group-theoretic methods, in which they turned out to be extremely efficient, arose. This is the formalization of various procedures for finding approximate solutions of differential equations. Well-known asymptotic methods, e.g., the Krylov–Bogolyubov method, Poincaré's method of normal forms, the multiscale method, and others, widely used in celestial mechanics and in the theory of oscillations, have a significant drawback: as the number of approximation steps grows, the amount of computations increases catastrophically, so that the feasibility of high-order approximations lags behind the demand for them.

Still, the procedures involved in such calculations consist of various operations, and if one can discern an algebraic structure in the set of operations, or organize them so that they obey such a structure, then the existing algorithms can be significantly simplified.

Thus the following main domains of application of group-theoretic methods in contemporary science have emerged:

- Methods of exact integration of differential equations.
- Construction of paradigms, or models, possessing some group action.
- Various methods of qualitative analysis involving ideas of group theory in one way or the other.
- Group-theoretic formalizations of approximate solution methods for integrating differential equations.

We have listed these domains roughly in the same order as the appearance of their main results.

Here, when we speak of the applied character of various results, we should clarify what is meant by "applied." Thus, if as our starting point we take, say, the theory of continuous groups as presented by Pontryagin, then the group-theoretic analysis of differential equations as described by Ovsiannikov or Olver is an applied field. Nevertheless, its fundamental role in the theory of differential equations itself is indisputable.

It seems natural to call "applied" any result obtained by group-theoretic analysis if it pertains to a field that lies outside group theory or differential equations (e.g., mechanics or physics).

What are the goals of the present book, how does it differ from other monographs on group-theoretic methods in differential equations and their applications to physics?

First of all, it is obvious that the dominating topics in the literature devoted to applications of groups to physics are quantum mechanics, crystallography, and nuclear physics. The authors of the present monograph would like to shift the emphasis towards mechanics and applied mathematics: classical mechanics, relativistic mechanics, theory of oscillations, perturbation theory, and concrete problems of a purely mechanical type, e.g., problems involving dry friction and the like.

Nevertheless, in this connection we hardly touch on continuum mechanics, since this topic is presented in the literature nearly as fully as quantum mechanics or crystallography.

Secondly, the authors intend to present constructive methods only, i.e., only those where there is no place for statements of the type "if something or other exists, then this or that can be found," or if such a statement appears, then it is immediately followed by an explicit description of the premise.

The contents of the book may be roughly divided into three parts. The first is an introduction to Lie groups sufficient for understanding the specific features of the application of group-theoretic ideas in contemporary mechanics and applied mathematics.

The second part contains a large variety of mechanics problems to which group-theoretic methods can be successfully applied. This includes the analysis of the foundations of classical and relativistic mechanics, perturbation theory for configuration spaces of resonance systems, the kinematics of rigid bodies, elements of Hamiltonian mechanics, several examples of the integration of the equations of motion with the help of symmetry groups.

In the third part we present a group-theoretic formalization of asymptotic methods of applied mathematics and illustrate it by its application to problems of oscillation theory.

The book is intended for research scientists, engineers, and graduate and undergraduate students specializing in applied mathematics, mechanics and physics. Individual chapters may be used as the basis for specialized graduate courses in universities and colleges.

The only prerequisites are the standard introductory courses in advanced calculus and in differential equations.

We would like to express our deep gratitude to Alexei Zhurov who translated this book into English and made a lot of useful comments. We also appreciate the help of Alexei Sossinsky who translated the Foreword.

Dmitry M. Klimov
Victor Ph. Zhuravlev

AUTHORS

Dmitry M. Klimov, Professor, D.Sc., Ph.D., is a noted scientist in the fields of mechanics and applied mathematics. Dmitry Klimov graduated from the Department of Mechanics and Mathematics of the Moscow State University in 1950. He received his Ph.D. degree in 1960 and D.Sc. degree in 1965 at the Moscow State University. Since 1967, Dmitry Klimov has been a member of the staff of the Institute for Problems in Mechanics of the Russian Academy of Sciences. In 1989 he was elected Director of the Institute for Problems in Mechanics.

Professor Klimov is a Full Member of the Russian Academy of Sciences. He is the Head of the Department of Mechanics and Control Processes of the Russian Academy of Sciences and the Head of the Department of Mechanics of Controlled and Gyroscopic Systems at the Moscow Institute of Physics and Technology. He is also the Deputy Editor of the journal *Mechanics of Solids*.

Professor Klimov is an author of five books and more than 100 research papers. He has made important contributions to the fields of navigation systems and gyroscopes, nonlinear oscillations, asymptotic methods, and rheology. For his research in mechanics, Professor Klimov was awarded a U.S.S.R. State Prize in 1976 and a Russian State Prize in 1994.

Victor Ph. Zhuravlev, Professor, D.Sc., Ph.D., is a noted scientist in mechanics and applied mathematics. After having received his first M.S. degree in Mechanics of Gyroscopes at the Bauman Moscow Higher Technical School in 1966, Victor Zhuravlev entered Moscow State University, where he received his second M.S. degree in Differential Equations in 1970.

Since 1970, Victor Zhuravlev has been working at the Institute for Problems in Mechanics, Russian Academy of Sciences, where he received his Ph.D. degree in 1973 and D.Sc. degree in 1978. For over ten years, he has been an Associated Professor with the Moscow Institute of Physics and Technology and is now Head of the Institute's Chair of Mechanics.

Professor Zhuravlev's basic scientific results are in the theory of gyroscopes, analytical mechanics, and application of Lie groups in perturbation theory.

Since 1994, Professor Zhuravlev has been an Associate Member of the Russian Academy of Sciences. He is a member of the Editorial Board of the journals *Mechanics of Solids* and *Applied Mathematics and Mechanics*. In 1994, he was awarded a State Prize for his contribution to mechanics.

Professor Zhuravlev is an author of over 150 scientific publications, four monographs, and a textbook "Fundamentals of Theoretical Mechanics."

Chapter 1

Basic Notions of Lie Group Theory

1.1. Notion of Group

Let the symbol G denote a set of elements of arbitrary nature (e.g., a set of numbers, functions, or some objects of geometric nature, etc.).

The set G is said to be a *group* if:

(1) An operation is defined over the set G which assigns a unique element $C \in G$ to any two elements $A \in G$ and $B \in G$ taken in a certain order. This operation is called multiplication (or composition) and written symbolically as

$$A \circ B = C.$$

In general, this operation is noncommutative, i.e., $A \circ B \neq B \circ A$, which defines the order of taking the elements in the product.

(2) There exists an element $E \in G$ such that

$$A \circ E = E \circ A = A$$

for any $A \in G$. This element is called the *identity element* or *unit* of the group.

(3) For any $A \in G$ there exists an element A^{-1} such that

$$A \circ A^{-1} = A^{-1} \circ A = E.$$

This element is called the *inverse* of A.

(4) The associativity of multiplication holds, i.e.,

$$A \circ (B \circ C) = (A \circ B) \circ C$$

for any A, B, and C of G.

The above four conditions define the notion of an abstract group, i.e., a group in which the nature of its constituent elements is of no importance.

Condition (1) is referred to as the condition of closure of the set G under the operation $\circ$ thus defined on this set.

If there exists a subset $H \subset G$ closed under the same operation, then such a subset is called a *subgroup*.

If the operation introduced in the group G is commutative, i.e., $A \circ B = B \circ A$ for any elements A and B of G, then the group is called *commutative*, or *abelian*.

Let us demonstrate that in any group the unit is unique. We proceed by reductio ad absurdum. Suppose there exists an $E_1 \neq E$ such that

$$A \circ E_1 = E_1 \circ A = A.$$

Multiply this equation by A^{-1}. We have

$$A^{-1} \circ A \circ E_1 = A^{-1} \circ A \quad \text{and} \quad E_1 \circ A \circ A^{-1} = A \circ A^{-1}.$$

Whence, with reference to conditions (3) and (4), we have $E_1 = E$.

The uniqueness of the inverse elements can be proved in much the same simple manner.

Exercise. Show that in the definition of group it suffices to require the existence of a *right* unit and a right inverse, i.e., $A \circ E = A$ and $A \circ A^{-1} = E$, since this will imply the existence of a left unit and its identity with the right unit, as well as the existence of a left inverse and its identity with the right one.

Examples of groups.

1. The set of all real numbers. The group operation is addition. The role of the unit is played by zero, and the inverse of an element is the negative of the element. The set of all rational numbers is a subgroup. Other subgroups include the set of all integers and the set of all even numbers. The set of all odd numbers is not a subgroup.

2. The set of reals is a group under the operation of arithmetical multiplication, provided that the number zero is excluded.

3. Any finite number of elements with the operation defined by the Cayley table (an example of 5 elements is given):

$\circ$	A	B	C	D	E
A	B	C	D	E	A
B	C	D	E	A	B
C	D	E	A	B	C
D	E	A	B	C	D
E	A	B	C	D	E

For example, $C \circ D = B$. The role of the unit is played by E.

4. The set of points lying on a circle. The operation is defined as follows. Let the position of a point A relative to some reference point be defined by the angle φ_A and let the position of a point B be defined by the angle φ_B. To the points A and B the operation assigns the point C whose position is defined by the angle $\varphi_C = \varphi_A + \varphi_B$.

5. The set of $n \times n$ matrices with determinants other than zero. The group operation is matrix multiplication.

Below are some examples of sets which are not groups under the operations introduced over them.

1. The set of integers with multiplication operation. Although the product of two integers is an integer and there exists a unit, none element has its inverse, except for the number one.

2. The set of vectors in the three-dimensional space with the operation of outer product. Except for the first condition in the definition of a group, no other conditions are satisfied.

The groups in examples 1 through 4 are commutative, or abelian, groups. Example 5 presents a noncommutative group.

1.2. Lie Group. Examples

One may easily see that in examples 1, 2, 4, and 5 (see Section 1.1) it is possible to introduce, independently of the group axioms, a notion of closeness between any two elements for the corresponding sets of elements. By virtue of this, the group operations turn out to be continuous functions, which makes it possible to treat such groups in two aspects, from the viewpoint of algebra and from the viewpoint of analysis.

This unification turns out to be quite fruitful. This is what the theory of Lie groups utilizes rather essentially.

Presently, by the term "Lie group" a wider object is understood compared with that introduced by Lie himself. In what follows, we consider this wider object.

By G we mean a set of transformations of an n-dimension real arithmetic space into itself,

$$q' = Q(q, a) \qquad (q \in \mathbb{R}^n),$$

on which an operation satisfying the group axioms is defined. The set of transformations is numbered by a parameter a. If a is real, then the set of transformations is said to be one-parameter. If a is a k-vector, $a \in \mathbb{R}^k$, the set of transformations is k-parameter. In this case, all $a_1, \ldots, a_k$ must be essential, that is, irreducible to fewer parameters by a transformation of the parameters.

The operation introduced over the set G is the composition of two transformations. For example, suppose the transformation $q \to q'$ with a fixed value of a is followed by a transformation $q' \to q''$ with a different value of a, denoted b:

$$q'' = Q(q', b).$$

If the composition

$$q'' = Q\big(Q(q, a), b\big)$$

defining the transformation $q \to q''$ is a transformation from the same set (corresponding to some other value c of the parameter), i.e.,

$$q'' = Q\big(Q(q, a), b\big) = Q(q, c),$$

then this means that an operation is defined on the set of transformations in question (the composition of two transformations from the set does not go beyond the limits of the set).

The transformations are usually assumed to be defined on some open set from $\mathbb{R}^n$ and in a sufficiently small neighborhood of the point a. This assumption is made to avoid discussions about the domain of definition of the functions $q' = Q(q, a)$ in both the variable q and the parameter a.

Thereby the function $q' = Q(q, a)$ determines a local family of local transformations.

DEFINITION. *A set of transformations $q' = Q(q, a)$ is called a local Lie group of transformations (from now on, a Lie group) if:*

(1) *the composition of any two transformations from this set is again a transformation from the same set (i.e., the set of transformations is closed with respect to the composition);*
(2) *the identity transformation, which plays the role of the group unit, belongs to the set of transformations in question;*
(3) *for any transformation from the set there exists its inverse, which belongs to the same set, so that the composition of the two transformations yields the identity transformation;*
(4) *the function $q' = Q(q, a)$ is analytic in q and a in some open set of variation of q and in some neighborhood of the unit element, e, for the variable a.*

Note that the requirement that the operation be associative, which is used in the abstract definition of a group, is unnecessary here, since the composition of transformations obviously satisfies this property.

Group operations. The composition of the transformations $q' = Q(q,a)$ and $q'' = Q(q',b)$ determines the transformation $q'' = Q(q,c)$ in which the parameter c is related to a and b by

$$c = \gamma(a,b).$$

It is this function, analytic in a neighborhood of the unit ($a = e$ and $b = e$), which represents an expression for the group operation.

By using the notion of group operation, one can rewrite the definition of a Lie group more concisely as follows.

A set of transformations $q \to q' : q' = Q(q,a)$ is called a Lie group if:
(1) $Q\big(Q(q,a),b\big) = Q\big(q,\gamma(q,a)\big) \quad \forall\, a,b;$
(2) $\exists\, a = e: \quad \forall\, b \quad \gamma(e,b) = \gamma(b,e) = b;$
(3) $\forall\, a \quad \exists\, b = a^{-1}: \quad \gamma(a,a^{-1}) = \gamma(a^{-1},a) = e;$ and
(4) all the functions involved in the definition are analytic.

Most important examples of Lie groups.

1. *Group of translations*: $q' = q + a$. Here the dimensionality of the parameter a is the same as that of the variable q.

2. *Group of extensions*: $q' = a_i q_i$ $(i = 1,\dots,n)$. If all $a_i = a$, where a is a scalar quantity, then this group is called a *similarity group*.

3. *Group of orthogonal transformations*: $q'_i = \sum_{j=1}^n a_{ij}q_j$ $(i = 1,\dots,n)$, where $A = \{a_{ij}\}$ is an orthogonal matrix, $A^{\mathrm T} = A^{-1}$. This group is conventionally denoted O(n). The most important subgroup of this group is the subset of transformations with $\det A = 1$ (in general, $\det A = \pm 1$). It is conventional to denote this subgroup by SO(n) and call it a *group of rotations*.

4. *Group of linear transformations*: $q'_i = \sum_{j=1}^n a_{ij}q_j$ $(i = 1,\dots,n)$ with $\det A \neq 0$. This group is designated as GL(n), which stands for a *general linear group* in $\mathbb{R}^n$. If we require additionally that $\det A = 1$, then we obtain a volume-preserving subgroup of transformations. This subgroup is called a *special linear group*, or a *unimodular group*, and denoted by SL(n).

5. *Group of motions*: $q' = a + Aq$, where Aq stands for $\sum_{j=1}^n a_{ij}q_j$ $(i = 1,\dots,n)$ for brevity, and A is an orthogonal matrix with $\det A = 1$. In particular, this group contains subgroups such as the group of translations and the group of rotations.

6. *Affine group*: $q' = a + Aq$, where $\det A \neq 0$. The group of motions is a subgroup of the affine group.

7. *Projective group*:

$$q'_i = \frac{\sum_{j=1}^n a_{ij}q_j + b_i}{\sum_{j=1}^n a_j q_j + b} \quad \text{with} \quad \det\begin{pmatrix} a_{ij} & b_i \\ a_j & b \end{pmatrix} \neq 0 \qquad (i,j = 1,\dots,n).$$

In the matrix under the determinant sign, the matrix $A = \{a_{ij}\}$ is extended by a column of b_i and a row of a_j, with the last diagonal entry being b.

8. *Volume-preserving group of transformations*:

$$\det\left\{\frac{\partial q'_i}{\partial q_j}\right\} \equiv 1 \qquad (i,j = 1,\dots,n).$$

Below are two groups which are of particular importance in mechanics and physics.

9. *Galilean group*:

$$t' = t + t_0, \quad r' = r + r_0 + vt + Ar \qquad (A^{\mathrm T} = A^{-1}),$$

where notation customary in mechanics is used. The Galilean group involves ten parameters: t_0, three components of the translation vector r_0 of the origin, three components of the velocity v, and three independent parameters of the rotation matrix A.

10. *Lorentz group*:

$$t' = \frac{t - vx}{\sqrt{1-v^2}}, \quad x' = \frac{x - vt}{\sqrt{1-v^2}},$$

where the x-axis of the fixed reference frame is directed along the velocity v at which the origin of the primed reference frame moves.

Let us find the group operation of the Lorentz group. We have

$$t' = \frac{t - v_1 x}{\sqrt{1-v_1^2}}, \quad x' = \frac{x - v_1 t}{\sqrt{1-v_1^2}};$$

$$t'' = \frac{t' - v_2 x'}{\sqrt{1-v_2^2}} = \frac{t - (v_1+v_2)x/(1+v_1v_2)}{\sqrt{1-[(v_1+v_2)/(1+v_1v_2)]^2}},$$

$$x'' = \frac{x' - v_2 t'}{\sqrt{1-v_2^2}} = \frac{x - (v_1+v_2)t/(1+v_1v_2)}{\sqrt{1-[(v_1+v_2)/(1+v_1v_2)]^2}}.$$

Thus the group operation has the form

$$v_3 = \frac{v_1+v_2}{1+v_1v_2}.$$

It represents the law of addition of velocities in relativistic mechanics.

1.3. Group Generator. Lie Algebra

Let the group parameter a be scalar. We transform the parameter in accordance with the formula $a \to \mu : a = e + \mu$. Then the value $\mu = 0$ corresponds to the identity transformation, i.e., the unit of the group.

Let μ be the parameter occurring in the equation of the group, $q' = Q(q,\mu)$. Expand this expression in a series in powers of μ about the point $\mu = 0$:

$$q' = q + \mu\eta(q) + \cdots.$$

The linear part of the group represented by the first two terms of this expansion is called the *germ of the group*. Let a scalar function $F(q)$ be specified in the space of the variables q. The group transformation $q \to q'$ takes the function $F(q)$ to the form

$$F(q) \to F(q') = F\big(q + \mu\eta(q) + \cdots\big) = F(q) + \mu\eta(q)\frac{dF}{dq} + \cdots.$$

The part of the increment of $F(q')$ linear in μ is given by

$$\Delta F(q') = \mu\eta(q)\frac{dF}{dq} \equiv \mu\sum_{i=1}^{n}\eta_i(q)\frac{\partial F}{\partial q_i} \equiv \mu U F,$$

where U is a linear differential operator of first order,

$$U = \eta_1(q)\frac{\partial}{\partial q_1} + \cdots + \eta_n(q)\frac{\partial}{\partial q_n},$$

This operator is referred to as the *infinitesimal generator of the group*, also called the generator or operator of the group.

If the group is multiparameter, then the above procedure can be carried out with respect to each of the parameters. Thus, a multiparameter group has as many infinitesimal generators as the number of independent parameters.

Below we present expressions of the infinitesimal generators corresponding to the Lie groups listed in the above examples.

1. Group of translations:

$$U_i = \frac{\partial}{\partial q_i} \qquad (i = 1, \dots, n).$$

2. Group of extensions:

$$U_i = q_i \frac{\partial}{\partial q_i} \qquad (i = 1, \dots, n);$$

$$U = q_1 \frac{\partial}{\partial q_1} + \dots + q_n \frac{\partial}{\partial q_n} \qquad \text{(similarity group)}.$$

3. Group of rotations. Consider a small neighborhood of the identity matrix E in the set of orthogonal matrices:

$$A = E + \mu N,$$

where μ is a small scalar parameter. Neglecting the terms of the second order of smallness, we find from the condition of orthogonality of the matrix A,

$$(E + \mu N)^{\mathrm{T}}(E + \mu N) = E,$$

that $N^{\mathrm{T}} = -N$. Thus, a small variation of an orthogonal matrix is a skew-symmetric matrix which is what determines $\frac{1}{2}n(n-1)$ independent parameters of an orthogonal group: $\mu_{ij} = \mu n_{ij}$, where $\{n_{ij}\} = N$ and $n_{ij} = -n_{ji}$. The germ of an orthogonal group has the form

$$q_i' = q_i + \sum_{j=1}^{n} \mu_{ij} q_j;$$

to each parameter μ_{ij} there corresponds the generator

$$U_{ij} = q_j \frac{\partial}{\partial q_i} - q_i \frac{\partial}{\partial q_j} \qquad (i > j = 1, \dots, n).$$

4. Group of linear transformations. The variation of the nondegenerate matrix A in a neighborhood of the identity matrix is given by $A = E + M$, where $M = \{\mu_{ij}\}$ is a matrix with small independent entries μ_{ij}. The number of independent generators is equal to n^2. These are

$$U_{ij} = q_j \frac{\partial}{\partial q_i} \qquad (i, j = 1 \dots, n).$$

5. The generators of the group of motions involve the generators of the translation group and those of the rotation group.

6. The generators of the affine group involve the generators of the groups of translations and linear transformations.

7. Projective group. Introduce notation for the parameters μ so that they vanish for the identity transformation:

$$\{a_{ij}\} = E + \{\mu_{ij}\}, \quad b_i = \mu_i, \quad a_j = \mu_j, \quad b = 1 + \mu, \qquad (i, j = 1, \dots, n).$$

Then the projective group becomes

$$q_i' = \frac{q_i + \sum_{j=1}^n \mu_{ij} q_j + \mu_i}{1 + \mu + \sum_{j=1}^n \mu_j q_j}.$$

The germ of the projective group, i.e., the linear part of the group with respect to all parameters, has the form

$$q_i' = q_i + \sum_{j=1}^n \mu_{ij} q_j + \mu_j - q_i \left(\mu + \sum_{j=1}^n \mu_j q_j \right).$$

This expression permits one to find the generators corresponding to each group of parameters.

For the parameters μ_{ij},

$$U_{ij} = q_j \frac{\partial}{\partial q_i} \qquad (i, j = 1, \ldots, n).$$

For μ_i,

$$U_i = \frac{\partial}{\partial q_i} \qquad (i = 1, \ldots, n).$$

For μ,

$$U = -\sum_{i=1}^n q_i \frac{\partial}{\partial q_i}.$$

For μ_j,

$$U_j = -q_j \sum_{i=1}^n q_i \frac{\partial}{\partial q_i} \qquad (j = 1, \ldots, n).$$

Note that the minus signs in the last two generators can be omitted, since the respective parameters can be changed as $\mu \to -\mu$ and $\mu_j \to -\mu_j$.

Note also that the generator of the similarity subgroup, $U = \sum_{i=1}^n q_i (\partial / \partial q_i)$, corresponding to the parameter μ, is a linear combination of some generators of the extension subgroup:

$$U = \sum_{i=1}^n U_{ii}.$$

Thus, the number of linearly independent generators in a projective group is equal to $n(n+2)$.

8. Group of volume-preserving transformations. Consider an arbitrary nearly-identical transformation,

$$q' = q + \mu \eta(q).$$

As is known, the volume is expressed by the integral $\int \ldots \int dq_1 \ldots dq_n$. The following rule of the change of variables in the integral is valid:

$$\int \ldots \int dq_1' \ldots dq_n' = \int \ldots \int \det \left\{ \frac{\partial q_i'}{\partial q_j} \right\} dq_1 \ldots dq_n.$$

For the volume to remain unchanged under the transformation it is necessary that

$$\det \left\{ \frac{\partial q_i'}{\partial q_j} \right\} = 1,$$

or, in the expanded form,

$$\det \left(E + \mu \left\{ \frac{\partial \eta_i}{\partial q_j} \right\} \right) = 1 + \mu \sum_{i=1}^n \frac{\partial \eta_i}{\partial q_i} + O(\mu^2).$$

The volume is preserved under any transformation for which

$$\sum_{i=1}^{n} \frac{\partial \eta_i}{\partial q_i} \equiv \operatorname{div} \eta \equiv 0.$$

Thus, the generator of a volume-preserving group has the form

$$U = \sum_{i=1}^{n} \eta_i(q) \frac{\partial}{\partial q_i} \qquad (\operatorname{div} \eta = 0).$$

The above example presents an *infinite continuous group*, i.e., a group depending on infinitely many parameters or on arbitrary functions.

9. Galilean group. The generators corresponding to translations in time and coordinates, associated with the parameters t_0 and r_0, respectively, have the form

$$\frac{\partial}{\partial t}, \quad \frac{\partial}{\partial x}, \quad \frac{\partial}{\partial y}, \quad \frac{\partial}{\partial z}$$

The generators determined by the velocity v are

$$t\frac{\partial}{\partial x}, \quad t\frac{\partial}{\partial y}, \quad t\frac{\partial}{\partial z}.$$

The rotation matrix A determines the generators

$$y\frac{\partial}{\partial x} - x\frac{\partial}{\partial y}, \quad z\frac{\partial}{\partial x} - x\frac{\partial}{\partial z}, \quad z\frac{\partial}{\partial y} - y\frac{\partial}{\partial z}.$$

10. Lorentz group. The group generator is given by

$$U = x\frac{\partial}{\partial t} + t\frac{\partial}{\partial x}.$$

Lie algebra. Let an m-parameter group has m generators $U_1, \ldots, U_m$. If all m parameters are essential, i.e., cannot be reduced by any transformations to fewer parameters, then these generators are linearly independent (Lie's first basic theorem). This means that there are numbers $\lambda_1, \ldots, \lambda_m$, not all zero, such that $\lambda_1 U_1 + \cdots + \lambda_m U_m = 0$.

Otherwise the generators are said to be *linearly dependent*. The numbers $\lambda_1, \ldots, \lambda_m$ are assumed to be independent of q.

The linear independence of the generators permits one to take them as a basis of the linear space of generators associated with the group. Any element of this space is a differential operator of the form

$$U = \lambda_1 U_1 + \cdots + \lambda_m U_m,$$

where $\lambda_1, \ldots, \lambda_m$ are the coordinates of the generator U in the linear space.

An operation of multiplication of operators can be introduced in this space. Let U and V be two operators from the space in question. Introduce the product of operators as follows:

$$[U, V] = UV - VU.$$

This product is called the *commutator* of the operators U and V. This means that the action of the operator $[U, V]$ on a function $F(q)$ involves the following: (i) the operator V acts on $F(q)$, (ii) then the operator U acts on the resulting function, and (iii) the result of the action of these operators on $F(q)$ in reverse order is subtracted.

This definition requires that some facts be checked. Firstly, is $[U, V]$ an operator of first order? The straightforward computation shows that the second derivatives that appear in calculating $U(VF)$ are canceled after subtracting $V(UF)$. If the operator U has the form

$$U = \sum_{i=1}^{m} \eta_i \frac{\partial}{\partial q_i}$$

and the operator V is defined as

$$V = \sum_{i=1}^{m} \xi_i \frac{\partial}{\partial q_i},$$

then the resulting operator $[U, V]$ is given by

$$[U, V] = \sum_{i=1}^{n} \alpha_i \frac{\partial}{\partial q_i},$$

where $\alpha_i(q) = U\xi_i(q) - V\eta_i(q)$.

Secondly, does $[U, V]$ belong to the space of operators in question? In other words, can this operator be expressed by a linear combination of the original basis operators U_i? If m arbitrary linearly independent operators are specified, it is not at all necessary that the product of any two of them be expressible as a linear combination of the m operators. It turns out that if these m operators are generators of an m-parameter group, then the product of any two operators linearly expressible in terms of them is also a linear combination of the m operators. The space of generators of an m-parameter group is closed with respect to the introduced operation of operator multiplication (Lie's second basic theorem):

$$[U, V] = k_1 U_1 + \cdots + k_m U_m.$$

Since the introduced product possesses obviously the property of distributivity for addition, i.e.,

$$[U, V + W] = [U, V] + [U, W],$$

then, to find the constants $k_1, \ldots, k_m$ in the case of arbitrary U and V, it suffices to know these constants for the basis operators,

$$[U_i, U_j] = \sum_{s=1}^{m} C_{ij}^{s} U_s.$$

The constants C_{ij}^{s} are called the structural constants of a group and determine the group completely.

Thirdly, it is important to know whether the product introduced is well-formed in the following sense. Dealing with a transformation of the space or a subdomain of it, we express this transformation using the coordinates q. If the space is referred to different coordinates, say, r, then the same transformations, expressed in terms of functions of the coordinates r, will have a different form. Also differently will look the generators (the rule of transformation of generators under a change of variables will be given in Section 1.7).

The question arises whether the operation of generator multiplication, $[U, V]$, depends on the variables in terms of which it is calculated. In other words, is the sequence of operations of significance, first the change of variables is performed in the generators and then the derivatives are computed or vise versa? It is not difficult to verify that this does not matter. The product introduced is invariant under a change of variables.

Thus, we obtain the following object: a linear space of generators with an operation of multiplication introduced in it. Such an object is called a *Lie algebra*. A Lie algebra is generated by a Lie group. Conversely, given m linearly independent differential operators such that the commutator of any two of them is a linear combination of the m operators, they generate an m-parameter Lie group (Lie's second inverse theorem).

Note that the commutator is linear in both arguments; for example,

$$[U, aV_1 + bV_2] = a[U, V_1] + b[U, V_2],$$

where a and b are scalar quantities.

In addition, the commutator is skew-symmetric, $[U, V] = -[V, U]$, and satisfies the Jacobi identity

$$\big[[U, V], W\big] + \big[[V, W], U\big] + \big[[W, U], V\big] = 0.$$

The abstract definition of a Lie algebra is related to neither groups nor generators. A Lie algebra is defined by introducing in a linear space an operation of multiplication which is bilinear and skew-symmetric and satisfies the Jacobi identity. For example, the three-dimensional vector space with the operation of outer product is a Lie algebra.

Example. The generators of the SL(2) group,

$$U_1 = q_2\frac{\partial}{\partial q_1}, \quad U_2 = q_1\frac{\partial}{\partial q_2}, \quad U_3 = q_1\frac{\partial}{\partial q_1} - q_2\frac{\partial}{\partial q_2},$$

form the basis of a three-dimensional Lie algebra with the following rule of multiplication of the basis generators:

$$[U_1, U_2] = -U_3, \quad [U_1, U_3] = 2U_1, \quad [U_2, U_3] = -2U_2.$$

Hence, the structural constants of the SL(2) group are:

$$C_{12}^3 = -1, \quad C_{13}^1 = 2, \quad C_{23}^2 = -2,$$

the other C_{ij}^s being equal to zero.

Example. Consider the linear space of 2×2 real matrices. Introduce the basis

$$e_1 = \begin{pmatrix} 1 & 0 \\ 0 & 1 \end{pmatrix}, \quad e_2 = \begin{pmatrix} 0 & 1 \\ -1 & 0 \end{pmatrix}, \quad e_3 = \begin{pmatrix} 1 & 0 \\ 0 & -1 \end{pmatrix}, \quad e_4 = \begin{pmatrix} 0 & 1 \\ 1 & 0 \end{pmatrix},$$

so that any 2×2 real matrix A can be represented as $A = \lambda_1 e_1 + \lambda_2 e_2 + \lambda_3 e_3 + \lambda_4 e_4$.

Let us transform this linear space into a Lie algebra. We introduce the following rule of multiplication of two arbitrary 2×2 matrices A and B:

$$A \circ B = \tfrac{1}{2}(AB - BA).$$

It is not difficult to verify that the product introduced is bilinear and skew-symmetric and satisfies the Jacobi identity.

The following table of multiplication of basis matrices holds:

$\circ$	e_1	e_2	e_3	e_4
e_1	0	0	0	0
e_2	0	0	$-e_4$	e_3
e_3	0	e_4	0	e_2
e_4	0	$-e_3$	$-e_2$	0

1.4. One-Parameter Groups. Uniqueness Theorem

In the rest of Chapter 1, we mainly deal with one-parameter groups. In the previous section, we already mentioned the second Lie theorem: any m-parameter group generates an m-dimensional Lie algebra, and vise versa, any m-dimensional Lie algebra of differential operators generates a group of the same dimensionality. In the case $m = 1$, the theorem has a nonlocal character.

THEOREM. *Given the infinitesimal generator $U = \sum_{i=1}^{n} \eta_i \frac{\partial}{\partial q_i}$ of a one-parameter group, the group itself, $q' = Q(q, \mu)$, can be reconstructed uniquely (up to a change of the parameter μ).*

Proof. Consider a small variation of the group parameter, $\mu + \delta\mu$, which results in a small variation of the image of the point q,

$$q' + \delta q' = Q(q, \mu + \delta\mu).$$

Taking advantage of the fact that $q' = Q(q, \mu)$ is a group, we represent the inverse function as $q = Q(q', \mu^{-1})$ and substitute this q into the right-hand side of the above equation to obtain

$$q' + \delta q' = Q\big(Q(q', \mu^{-1}),\ \mu + \delta\mu\big).$$

By using the group property once again, we find that

$$q' + \delta q' = Q\big(q',\ \gamma(\mu^{-1}, \mu + \delta\mu)\big).$$

On expanding the group operation $\gamma(\mu^{-1}, \mu+\delta\mu)$ into a series in powers of $\delta\mu$, we have $\gamma(\mu^{-1}, \mu+\delta\mu) = \gamma(\mu^{-1}, \mu) + \Gamma(\mu)\,\delta\mu + \cdots$, where $\gamma(\mu^{-1}, \mu) = 0$ and $\Gamma(\mu) = \partial\gamma(\mu_1, \mu_2)/\partial\mu_2$ at $\mu_1 = \mu^{-1}$ and $\mu_2 = \mu$. Thus, we obtain

$$q' + \delta q' = Q\big(q',\ \Gamma(\mu)\,\delta\mu + \cdots\big).$$

In turn, expanding the function Q into a series in powers of $\delta\mu$ yields

$$q' + \delta q' = q' + \eta(q')\,\Gamma(\mu)\,\delta\mu + \cdots .$$

By proceeding to the limit as $\delta\mu \to 0$, we have

$$\frac{dq'}{d\mu} = \Gamma(\mu)\,\eta(q').$$

The parameter μ in this relation is arbitrary and the substitution $q' = Q(q, \mu)$ must make the relation an identity. This means that the group $q' = Q(q, \mu)$ can be obtained from this relation, viewed as a differential equation subject to the boundary condition $q'\big|_{\mu=0} = q$.

The uniqueness of the group follows from the existence and uniqueness theorem for the solution of a Cauchy problem. The uniqueness is the case here, since the right-hand side is analytic.

If μ is changed to τ in accordance with the formula

$$\tau = \int_0^{\mu} \Gamma(\mu)\,d\mu,$$

then the above differential equation becomes

$$\frac{dq'}{d\tau} = \eta(q'), \qquad q'(0) = q.$$

The right-hand side of the equation is determined by the group generator alone. Solving this equation, we reconstruct the group entirely, up to the change of the parameter as specified above. This proves the theorem.

Remark 1. The theorem means that there is a one-to-one correspondence (up to an unessential change of the parameter) between all one-parameter groups in $\mathbb{R}^n$ and all autonomous ordinary differential equations with analytic right-hand sides.

Remark 2. The parameter $\tau = \int_0^\mu \Gamma(\mu)\,d\mu$ is referred to as the *canonical* parameter. The construction of a group by solving an autonomous differential equation determines this group automatically in terms of the canonical parameter.

The fact that a parameter is canonical implies that the group operation for it has the simplest form, $\tau_3 = \tau_1 + \tau_2$, and the inverse of an element is equal to the negative of the element, $\tau^{-1} = -\tau$.
This follows from the properties of solutions of the initial-value (Cauchy) problem for ordinary differential equations.

Example. Find the group operation and the canonical parameter in the similarity group

$$q' = aq = (1+\mu)q.$$

Let $q' = (1+\mu_1)q$ and $q'' = (1+\mu_2)q'$. Then $q'' = (1+\mu_1+\mu_2+\mu_1\mu_2)q$, and hence, the group operation has the form $\gamma(\mu_1,\mu_2) = \mu_1+\mu_2+\mu_1\mu_2$.
Determine the inverse:

$$\gamma(\mu,\mu^{-1}) = 0 \quad\Longrightarrow\quad \mu+\mu^{-1}+\mu\mu^{-1} = 0 \quad\Longrightarrow\quad \mu^{-1} = -\frac{\mu}{1+\mu}.$$

Compute the derivative:

$$\frac{\partial\gamma}{\partial\mu_2} = 1+\mu_1.$$

We now find the function $\Gamma(\mu)$:

$$\Gamma(\mu) = \frac{\partial\gamma}{\partial\mu_2} \quad\text{at}\quad \mu_1 = \mu^{-1} \text{ and } \mu_2 = \mu.$$

Whence $\Gamma(\mu) = 1/(1+\mu)$. The canonical parameter is given by

$$\tau = \int_0^\mu \frac{d\mu}{1+\mu} = \ln(1+\mu).$$

In terms of the canonical parameter, the group is expressed as $q' = e^\tau q$.

Example. Consider a projective group on the plane:

$$q_1' = \frac{a_{11}q_1+a_{12}q_2+b_1}{a_1q_1+a_2q_2+b}, \qquad q_2' = \frac{a_{21}q_1+a_{22}q_2+b_2}{a_1q_1+a_2q_2+b}.$$

One of the generators of the group is as follows:

$$U = q_1^2\frac{\partial}{\partial q_1} + q_1q_2\frac{\partial}{\partial q_2}.$$

Let us construct the one-parameter subgroup generated by this operator.
The subgroup is determined by the differential equations

$$\frac{dq_1'}{d\tau} = q_1'^2, \quad \frac{dq_2'}{d\tau} = q_1'q_2'; \qquad q_1'(0) = q_1, \quad q_2'(0) = q_2.$$

The solution of this initial-value problem yields the desired subgroup:

$$q_1' = \frac{q_1}{1-\tau q_1}, \quad q_2' = \frac{q_2}{1-\tau q_2}.$$

Choosing any generator from the Lie algebra of group generators, one can thus construct all one-parameter subgroups of the group in question.

1.5. Liouville Equation. Invariants. Eigenfunctions

Let a group be given in terms the canonical parameter τ,

$$q' = q + \eta(q)\,\tau + \cdots$$

and let the system of ordinary differential equations

$$\frac{dq'}{d\tau} = \eta(q')$$

be equivalent to this group, in accordance with the preceding.

Consider once again the issue as to how a function $F(q)$ is transformed under the group action. If the group takes q to q', then

$$F(q) \to F(q') = F\big(q + \eta(q)\,\tau + \cdots\big) \equiv \widetilde{F}(q, \tau).$$

Compute the derivative of the function $F(q')$ with respect to τ:

$$\frac{dF}{d\tau} = \frac{dF}{dq'}\frac{dq'}{d\tau} = \frac{dF}{dq'}\eta(q') = UF.$$

Find the relationship of the transformed function $\widetilde{F}(q, \tau)$ with the original function $F(q)$ and the group generator U.

Expanding $\widetilde{F}(q, \tau)$ into a Taylor series, we have

$$\widetilde{F}(q, \tau) = \widetilde{F}(q, 0) + \frac{\partial \widetilde{F}}{\partial \tau}\bigg|_{\tau=0}\tau + \cdots .$$

Successively,

$$\widetilde{F}(q, 0) = F(q), \quad \frac{\partial \widetilde{F}}{\partial \tau} = \frac{dF}{d\tau} = UF(q'), \quad \frac{\partial \widetilde{F}}{\partial \tau}\bigg|_{\tau=0} = UF(q),$$

$$\frac{\partial^2 \widetilde{F}}{\partial \tau^2} = \frac{d^2 F}{d\tau^2} = U^2 F(q'), \quad \frac{\partial^2 \widetilde{F}}{\partial \tau^2}\bigg|_{\tau=0} = U^2 F(q),$$

and so on. Finally, we arrive at the desired relationship represented as the series

$$F(q') \equiv \widetilde{F}(q, \tau) = F(q) + \tau U F(q) + \frac{\tau^2}{2!}U^2 F(q) + \cdots .$$

This series is referred to as the Lie series. It can be represented concisely as

$$F(q') = e^{\tau U} F(q).$$

The Lie series serves to define the operator exponential function. Differentiating the Lie series with respect to τ, we obtain

$$\frac{\partial \widetilde{F}}{\partial \tau} = UF + \tau U^2 F + \cdots = U(F + \tau UF + \cdots) = U\widetilde{F}.$$

The equation

$$\frac{\partial \widetilde{F}(q, \tau)}{\partial \tau} = U\widetilde{F}(q, \tau)$$

is termed the Liouville equation. Or, in a more detailed representation,

$$\frac{\partial \widetilde{F}}{\partial \tau} = \sum_i \eta_i(q) \frac{\partial \widetilde{F}}{\partial q_i}$$

Supplementing this equation with the initial condition $\widetilde{F}(q, 0) = F(q)$, we obtain an initial-value problem for the linear partial differential equation for the unknown function $\widetilde{F}(q, \tau)$. This problem is equivalent to the system of nonlinear differential equations

$$\frac{dq'}{d\tau} = \eta(q').$$

The equivalence is understood in the following sense. If the general solution of the system of ordinary differential equations is known, $q' = Q(q, \tau)$, where q denotes the initial conditions, then the solution of the Liouville equation is given by

$$\widetilde{F}(q, \tau) = F\big(Q(q, \tau)\big),$$

where the function $F(q)$ determines the initial condition for the Liouville equation.

Conversely, if the solution, $\widetilde{F}(q, t)$, of the initial-value problem for this equation is known, then by taking $F(q) \equiv q$ to be the initial function, we obtain the general solution $q' = Q(q, t)$ of the system $dq'/d\tau = \eta(q')$.

This equivalence makes it possible to change studying a nonlinear system of ordinary differential equations for studying a linear first-order partial differential equation, which is often more convenient.

The Lie series for the particular function $F(q) \equiv q$ permits us to represent the group as a Taylor series in τ:

$$q' = q + \tau U q + \frac{\tau^2}{2!} U^2 q + \cdots .$$

Example. Given the generator $U = q_2 \frac{\partial}{\partial q_1} - q_1 \frac{\partial}{\partial q_2}$ of the rotation group SO(2), reconstruct the group.

Successively we find

$$U q_1 = q_2, \quad U^2 q_1 = U q_2 = -q_1, \quad U^3 q_1 = -U q_1 = -q_2, \quad \ldots,$$
$$U q_2 = -q_1, \quad U^2 q_2 = -U q_1 = -q_2, \quad U^3 q_2 = -U q_2 = q_1, \quad \ldots$$

Hence,

$$q_1' = q_1 + \tau q_2 - \frac{\tau^2}{2!} q_1 - \frac{\tau^3}{3!} q_2 + \cdots = q_1 \cos\tau + q_2 \sin\tau,$$
$$q_2' = q_2 - \tau q_1 - \frac{\tau^2}{2!} q_2 + \frac{\tau^3}{3!} q_1 + \cdots = -q_1 \sin\tau + q_2 \cos\tau.$$

DEFINITION. *A function $G(q)$ is called an invariant of a group if it is left unchanged under the action of the group, i.e.,*

$$G(q') \equiv G(q).$$

The Lie series shows that for an analytic function of its variables to be an invariant of the group it necessary and sufficient that the function be a root of the group generator,

$$UG(q) = 0 \quad \text{for any} \quad q.$$

Property of invariant. Let $G(q)$ be an invariant of a group with generator U and let $F(q)$ be an arbitrary function. Then

$$U[G(q)\,F(q)] = F(q)\,UG(q) + G(q)\,UF(q) = G(q)\,UF(q).$$

Hence, an invariant of a group can be factored out just as a usual constant.

DEFINITION. *A function $P(q)$ is called an eigenfunction of a generator U if*

$$UP(q) = \lambda(q)\,P(q),$$

where $\lambda(q)$ is an invariant. In this case, the invariant $\lambda(q)$ is called the eigenvalue corresponding to the eigenfunction $P(q)$.

If $P(q)$ is an eigenfunction of a generator U, then this function is transformed by the group corresponding to this generator as follows:

$$\widetilde{P}(q,\tau) = G(q,\tau)\,P(q),$$

where $G(q,\tau)$ is an invariant of the group for any fixed τ,

$$G(q,\tau) = 1 + \tau\lambda(q) + \frac{\tau^2}{2!}\lambda^2(q) + \cdots.$$

Evidently, the following property of an eigenfunction of a generator is valid. If $P_1(q)$, $P_2(q)$ are eigenfunctions and $\lambda_1(q)$, $\lambda_2(q)$ are their respective eigenvalues, then

$$U(P_1P_2) = (\lambda_1 + \lambda_2)P_1P_2.$$

Hence, P_1P_2 is also an eigenfunction, which corresponds to the eigenvalue $\lambda_1 + \lambda_2$.

Invariant family of manifolds. Let a function $G(q)$ be an invariant of a group. Then by setting $G(q) = C$, where C is an arbitrary constant, we obtain a family of hypersurfaces each of which is transformed into itself by the group. In other words, each of these hypersurfaces (manifolds) is invariant.

The following situation, different from that above, is of interest. Let a family of hypersurfaces, $\omega(q) = C$, be specified. Suppose $\omega(q)$ is not an invariant of the group. In this case, the group transformation modifies each of the hypersurfaces of the family. We are interested in the case where these hypersurfaces change to other hypersurfaces of the same family under the group transformation. Such a family is referred to as an invariant family.

Example. The family of straight lines $q_1/q_2 = C$ passing through the origin of the plane (q_1, q_2) is an invariant family of the rotation group SO(2).

Let us establish the condition for the function $\omega(q)$ to define an invariant family.

Let $\omega(q) = C$ and $\sigma(q) = k$ be two representations of a single family. Each hypersurface of one family is congruent to a hypersurface of the other family.

This means that for each C of the first representation there exists a k of the second family such that both representations determine the same hypersurface. That is, k is a function of C: $k = f(C)$. But then also $\sigma(q) = f\big(\omega(q)\big)$. Thus, the condition of invariance of the family is the following:

$$\omega(q') \equiv \widetilde{\omega}(q,\tau) = f\big(\omega(q),\tau\big).$$

Expanding this condition into a series in τ, we obtain

$$\widetilde{\omega}(q,\tau) = f_0\big(\omega(q)\big) + \tau f_1\big(\omega(q)\big) + \frac{\tau^2}{2!} f_2\big(\omega(q)\big) + \cdots.$$

Moreover, since $\tau = 0$ determines the identity transformation, we have $f_0\big(\omega(q)\big) \equiv \omega(q)$.

Compare this series with the Lie series for $\widetilde{\omega}(q,\tau)$,

$$\widetilde{\omega}(q,\tau) = \omega(q) + \tau U\omega + \frac{\tau^2}{2!}U^2\omega + \cdots.$$

It is apparent that the invariance condition can be expressed in terms of the group generator as

$$U\omega = f_1(\omega).$$

Thus, for the function $\omega(q)$ to determine an invariant family of the group it is necessary that the group generator transform this function into some function of $\omega(q)$.

This condition can be represented in a form more convenient for applications. To this end, we get rid of the arbitrariness in selecting $f_1(\omega)$. We will use the above condition to seek a function of $\omega(q)$, $\Omega\big(\omega(q)\big)$, rather than the function $\omega(q)$ itself. Since $\Omega\big(\omega(q)\big)$ defines the same family, we have

$$U\Omega\big(\omega(q)\big) = h\big(\omega(q)\big),$$

where $h(\omega)$ is some other arbitrary function.

Since $U\Omega\big(\omega\big) = \frac{d\Omega}{d\omega}U\omega(q)$, then

$$\frac{d\Omega}{d\omega}U\omega(q) = h\big(\omega(q)\big).$$

Since there is an arbitrariness in selecting $\Omega(\omega)$, we dispose of this arbitrariness so that $\frac{d\Omega}{d\omega} = h(\omega)$. Then we obtain an invariance criterion for the family of hypersurfaces in the form

$$U\omega(q) = 1.$$

We will consider this condition the basic criterion for determining invariant families.

Example. The function $\omega(q_1, q_2) = q_1/q_2$ determines an invariant family of the rotation group SO(2). Let us verify all above conditions.

In terms of the group $q_1' = q_1 \cos\tau + q_2 \sin\tau$, $q_2' = -q_1 \sin\tau + q_2 \cos\tau$, we have

$$\omega(q_1', q_2') \equiv \widetilde{\omega}(q_1, q_2, \tau) = \frac{q_1 \cos\tau + q_2 \sin\tau}{-q_1 \sin\tau + q_2 \cos\tau} = \frac{\omega \cos\tau + \sin\tau}{-\omega \sin\tau + \cos\tau},$$

that is,

$$\widetilde{\omega}(q_1, q_2, \tau) = f\big(\omega(q_1, q_2), \tau\big), \quad \text{where} \quad f(\omega, \tau) = \frac{\omega \cos\tau + \sin\tau}{-\omega \sin\tau + \cos\tau}.$$

In terms of the group generator $U = q_2 \frac{\partial}{\partial q_1} - q_1 \frac{\partial}{\partial q_2}$:

$$U\omega = U\left(\frac{q_1}{q_2}\right) = 1 + \frac{q_1^2}{q_2^2} = 1 + \omega^2,$$

that is,

$$U\omega = f_1(\omega), \quad \text{where} \quad f_1(\omega) = 1 + \omega^2.$$

Basic criterion. If we take the function determining the invariant family in the form

$$\omega(q_1, q_2) = \arctan\frac{q_1}{q_2},$$

then we have

$$U \arctan\frac{q_1}{q_2} = 1.$$

1.6. Linear Partial Differential Equations

As one can see from the preceding, the problem of finding invariants and invariant families, as well as eigenfunctions, necessitates solving linear partial differential equations of the from

$$Q_1(q_1,\dots,q_n)\frac{\partial\omega}{\partial q_1}+\dots+Q_n(q_1,\dots,q_n)\frac{\partial\omega}{\partial q_n}=Q_{n+1}(q_1,\dots,q_n,\omega).$$

In the case of searching for an invariant, the equation is homogeneous, i.e., $Q_{n+1}=0$. If an invariant family or an eigenfunction is sought, then the equation is nonhomogeneous.

Consider first the homogeneous equation. To this equation there corresponds the system of ordinary differential equations

$$\frac{dq_1}{Q_1}=\frac{dq_2}{Q_2}=\dots=\frac{dq_n}{Q_n}.$$

THEOREM. *A function $\omega(q_1,\dots,q_m)$ is a solution of the equation*

$$\sum_{i=1}^{n}Q_i\frac{\partial\omega}{\partial q_i}=0$$

if and only if ω is a first integral of the above system.

The proof of the theorem follows from the equivalence of the Liouville equation to the corresponding system of ordinary differential equations (see the previous section).

The solution of the nonhomogeneous equation is reduced to the solution of a homogeneous equation if the solution is sought in implicit form,

$$\Phi(\omega,q)=0.$$

Then

$$\frac{\partial\omega}{\partial q_k}=\frac{\partial\Phi}{\partial q_k}\bigg/\frac{\partial\Phi}{\partial\omega}.$$

On substituting this ratio into the nonhomogeneous equation, we obtain

$$Q_1\frac{\partial\Phi}{\partial q_1}+\dots+Q_n\frac{\partial\Phi}{\partial q_n}+Q_{n+1}\frac{\partial\Phi}{\partial\omega}=0.$$

The general solution of this equation is an arbitrary function of n independent first integrals, $\alpha_1(q,\omega),\dots,\alpha_n(q,\omega)$, of the system of ordinary differential equations

$$\frac{dq_1}{Q_1}=\frac{dq_2}{Q_2}=\dots=\frac{dq_n}{Q_n}=\frac{d\omega}{Q_{n+1}},$$

that is,

$$\Phi=\Phi\big(\alpha_1(q,\omega),\dots,\alpha_n(q,\omega)\big).$$

For a system of ordinary differential equations of order $n+1$, there always exist (which is proved in the general theory of ordinary differential equations) exactly n independent first integrals in a sufficiently small neighborhood of any nonsingular point, i.e., any point at which at least one of the functions Q_i is nonzero.

By setting $\Phi=0$ and solving the obtained implicit relation for ω, one arrives at the solution of the original nonhomogeneous equation.

Example. Find all invariant families of the rotation group SO(2). The group generator, $U = q_2\frac{\partial}{\partial q_1} - q_1\frac{\partial}{\partial q_2}$, allows us to represent the equation for determining the desired families in the form

$$q_2\frac{\partial\omega}{\partial q_1} - q_1\frac{\partial\omega}{\partial q_2} = 1.$$

To this equation there corresponds the homogeneous equation

$$q_2\frac{\partial\Phi}{\partial q_1} - q_1\frac{\partial\Phi}{\partial q_2} + \frac{\partial\Phi}{\partial\omega} = 0$$

The equivalent system of ordinary differential equations,

$$\frac{dq_1}{q_2} = \frac{dq_2}{-q_1} = \frac{d\omega}{1},$$

has the following first integrals. From the equation

$$\frac{dq_1}{q_2} = \frac{dq_2}{-q_1},$$

we obtain

$$\alpha_1 = \sqrt{q_1^2 + q_2^2}.$$

Solving the equation

$$\frac{d\omega}{1} = \frac{dq_1}{\sqrt{\alpha_1^2 - q_1^2}}$$

yields

$$\alpha_2 = \omega - \arcsin\frac{q_1}{\alpha_1} = \omega - \arcsin\frac{q_1}{\sqrt{q_1^2 + q_2^2}} = \omega - \arctan\frac{q_1}{q_2}.$$

The general solution of the homogeneous equation has the form

$$\Phi = \Phi(\alpha_1, \alpha_2) = \Phi\left(\sqrt{q_1^2 + q_2^2},\ \omega - \arctan\frac{q_1}{q_2}\right).$$

Equating this solution to zero and solving the resulting equation for ω, we obtain

$$\omega = \arctan\frac{q_1}{q_2} + F(q_1^2 + q_2^2),$$

where F is an arbitrary function.

If $F \equiv 0$, then we arrive at the bundle of straight lines studied in previous examples. If $F = \sqrt{q_1^2 + q_2^2}$, then the equation $\arctan(q_1/q_2) + \sqrt{q_1^2 + q_2^2} = c$, where c is a constant, determines a family of Archimedean spirals.

1.7. Change of Variables. Canonical Coordinates of a Group

What specific form the generator $U = \sum_{i=1}^{n} \eta_i(q)\frac{\partial}{\partial q_i}$ of a group acting in the space of the variables q has depends on the choice of q. Let us find out how this form changes if we pass from the coordinates q to some coordinates r defined by

$$r = R(q),$$

where R is some function. Let the generator have the following form in the new coordinates:

$$\widetilde{U} = \sum_{i=1}^{n} \widetilde{\eta}_i(r) \frac{\partial}{\partial r_i}.$$

The corresponding differential equations determining the group in the new coordinates are expressed as

$$\frac{dr'}{dr} = \widetilde{\eta}(r'), \qquad r'(0) = r.$$

Let us write the Liouville equation for each component of the function $R(q')$:

$$\frac{dR_i}{dr} = UR_i.$$

Hence,

$$\widetilde{\eta}_i(r') = UR_i\big|_{q'=R^{-1}(r')}.$$

Thus, the variables in the generator are changed according to the formula

$$\widetilde{U} = \sum_{i=1}^{n} (UR_i) \frac{\partial}{\partial r_i},$$

in which the variables q must be expressed, after applying the generator in terms of the old variables to $R(q)$, in terms of r by using the inverse of the change of variables, $q = R^{-1}(r)$.

Example. Express the generator

$$U = q_1^2 \frac{\partial}{\partial q_1} + q_1 q_2 \frac{\partial}{\partial q_2}$$

of a subgroup of the projective group in the plane (q_1, q_2) in terms of the polar coordinates (ρ, φ),

$$\rho = \sqrt{q_1^2 + q_2^2}, \quad \varphi = \arctan \frac{q_2}{q_1}; \qquad q_1 = \rho \cos\varphi, \quad q_2 = \rho \sin\varphi.$$

We find successively

$$U\rho = \frac{q_1^3}{\sqrt{q_1^2 + q_2^2}} + \frac{q_1 q_2^3}{\sqrt{q_1^2 + q_2^2}} = q_1 \sqrt{q_1^2 + q_2^2} = \rho^2 \cos\varphi,$$
$$U\varphi = 0.$$

Thus, in the polar coordinates the generator becomes

$$\widetilde{U} = \rho^2 \cos^2\varphi \frac{\partial}{\partial r}.$$

Canonical coordinates of group. It is quite natural to raise the question of finding coordinates such that the group expressed in terms of them would have the simplest form. One may require that the maximum number of components of the transformed generator $\widetilde{U} = \sum_{i=1}^{n} (UR_i) \frac{\partial}{\partial r_i}$ becomes equal to zero:

$$UR_i = 0.$$

But this means that the functions R_i, determining the transition to the new coordinates, must be invariants of the group. The invariants of a group are nothing but first integrals of the differential equations determining the group and, by virtue of the above-mentioned theorem, the number of independent first integrals of an n-dimensional group is $n-1$. Therefore, all components of the generator but one can be made equal to zero. One can take the remaining component to be any function of q, in particular, identically equal to unity.

The coordinates in which the generator $\widetilde{U}$ has such a form are called canonical.

Thus, to pass in the generator $U = \sum_{i=1}^{n} \eta_i(q)\frac{\partial}{\partial q_i}$ to canonical coordinates, it is necessary to perform a change of variables $q \to r : q = R(r)$ such that $UR_1 = 1$ and $UR_i = 0$ $(i > 1)$. Then the generator becomes

$$\widetilde{U} = \frac{\partial}{\partial r_1}.$$

This is a generator of a group of translations.

The above result that any one-parameter group is similar to a group of translations along one of the coordinate axes is equivalent to the theorem of straightening out the vector field.

Note that the function determining an invariant family and satisfying the equation $UR_1 = 1$ is the logarithm of an eigenfunction corresponding to the eigenvalue $\lambda = 1$. Indeed,

$$Ue^{R_1} = e^R UR = e^R,$$

and hence, $\exp[R_1(q)]$ is an eigenfunction with eigenvalue $\lambda(q) \equiv 1$.

Thus, the role of canonical coordinates of a group is played by the logarithm of an eigenfunction of the group generator and $n-1$ independent invariants.

1.8. Hausdorff's Formula. Symmetry Groups

In Section 1.5, it was found out how a function defined in some domain of a coordinate space is transformed by a one-parameter group of transformations. The result was represented in two forms, the Liouville equation and the Lie series. Both forms related the expression of the function in the old coordinates to that in the new coordinates, as well as to the infinitesimal generator of the group defining the transformation from the old coordinates to the new ones.

In this section, we consider a similar problem, but here the object of transformation is a system of ordinary differential equations,

$$\frac{dq}{dt} = K(q), \qquad q \in \mathbb{R}^n,$$

rather than a function.

Consider a group of transformations

$$q \to q' : q' = Q(q, \tau).$$

It is required to find out how the above system of ordinary differential equations is modified under this change of variables.

In terms of group generators, the problem can be restated as follows. The generator of the group corresponding to the differential system has the form

$$A = \sum_{i=1}^{n} K_i(q)\frac{\partial}{\partial q_i}.$$

The generator of the group of transformations is expressed as

$$U = \sum_{i=1}^{n} \eta_i(q)\frac{\partial}{\partial q_i}.$$

In the new variables, the differential system acquires the form

$$\frac{dq'}{dt} = \widetilde{K}(q', \tau),$$

with operator

$$\widetilde{A} = \sum_{i=1}^{n} \widetilde{K}_i(q', \tau)\frac{\partial}{\partial q'_i}.$$

It is necessary to establish the relationship of A with $\widetilde{A}$ and U.

The solution of this problem is given by Hausdorff's formula. We now proceed to the derivation of this formula.

Let us represent the transformation determined by the group and its inverse in the exponential form (see Section 1.5):

$$q' = e^{\tau U}q, \quad q = e^{-\tau U}q'.$$

Recall that $U = \sum_{i=1}^{n} \eta_i(q)\frac{\partial}{\partial q_i}$ in these expressions; for the inverse transformation, the generator has the same form, with the old variables formally replaced by the new ones,

$$U = \sum_{i=1}^{n} \eta_i(q')\frac{\partial}{\partial q'_i}.$$

By using the formulas of the change of variables in a generator (see the previous section), we pass in $\widetilde{A}$ back from the new variables to the old ones to obtain

$$A = \sum_{i=1}^{n} (\widetilde{A}q_i)\frac{\partial}{\partial q_i} = \sum_{i=1}^{n} (\widetilde{A}e^{-\tau U}q'_i)\frac{\partial}{\partial q_i}.$$

In this formula, the components of A are functions of the new variables. By changing them for the old variables, we arrive at the components of the original operator A:

$$\widetilde{A}e^{-\tau U}q'_i\big|_{q'_i = e^{\tau U}q_i} = K_i(q).$$

These expressions are independent of τ, and consequently,

$$\frac{d}{d\tau}(\widetilde{A}e^{-\tau U}q'_i) = 0.$$

Whence,

$$\frac{\partial \widetilde{A}}{\partial \tau}e^{-\tau U}q'_i - \widetilde{A}Ue^{-\tau U}q'_i + U\widetilde{A}e^{-\tau U}q'_i = 0.$$

The first and second terms of this relation are obtained by the differentiation with respect to the group parameter τ that occurs explicitly. The third term stems from differentiating a function depending on q' with respect to τ; recall that q' depends on τ, $q' = e^{\tau U}q$. Such a differentiation of a function $F(q')$ along the trajectories of the group was carried out in Section 1.5: $\frac{d}{d\tau}F(q') = UF(q')$.

As a result, we obtain

$$\frac{\partial \widetilde{A}}{\partial \tau} = \widetilde{A}U - U\widetilde{A} = [\widetilde{A}, U].$$

This equation must be subject to the initial condition

$$\widetilde{A}(q', \tau)\big|_{\tau=0} = A(q')$$

to obtain the equation relating A, $\widetilde{A}$, and U.

The equation $\frac{\partial}{\partial\tau}\widetilde{A} = [\widetilde{A}, U]$, governing the transformed operator $\widetilde{A}$, is an analogue of the Liouville equation, governing the transformed function. The equation obtained reveals the meaning of the second name of the commutator—the derivative operator. The commutator is literally the derivative of the operator $\widetilde{A}$ with respect to the parameter of the group determined by the generator U.

The above initial-value (Cauchy) problem for the operator equation can be solved by expanding $\widetilde{A}$ into a Taylor series in τ:

$$\widetilde{A}(q', \tau) = A(q') + \tau \frac{\partial \widetilde{A}}{\partial \tau}\bigg|_{\tau=0} + \frac{\tau^2}{2!}\frac{\partial^2 \widetilde{A}}{\partial \tau^2}\bigg|_{\tau=0} + \cdots .$$

Let us calculate first $\frac{\partial}{\partial\tau}\widetilde{A}\big|_{\tau=0}$:

$$\frac{\partial \widetilde{A}}{\partial \tau}\bigg|_{\tau=0} = \left[\widetilde{A}, U\right]_{\tau=0} = [A, U].$$

Calculating the second derivative yields

$$\frac{\partial^2 \widetilde{A}}{\partial \tau^2} = \frac{\partial}{\partial \tau}\left[\widetilde{A}, U\right].$$

If the operator $\widetilde{B} = \left[\widetilde{A}, U\right]$ results from transforming an operator B by a group with generator U, then we are entitled to write

$$\frac{\partial \widetilde{B}}{\partial \tau} = \left[\widetilde{B}, U\right] = \left[[\widetilde{A}, U], U\right].$$

Let us demonstrate that this is the case indeed. The operator B can be taken in the from $B = [A, U]$. Since the commutator is invariant under changes of variables (see Section 1.3), we have $\widetilde{B} = \left[\widetilde{A}, \widetilde{U}\right]$.

We now show that $\widetilde{U} \equiv U$. Indeed, the change of variables defined by the generator U is $q' = Q(q, \tau)$. This function solves the system $\frac{d}{d\tau}q' = \eta(q')$. The generator U is transformed by the group it determines as follows:

$$\widetilde{U} = (UQ)\frac{\partial}{\partial q'}.$$

However, according to the Liouville formula, we have $UQ = \frac{dq'}{d\tau} = \eta$ and $\widetilde{U} = \eta\frac{\partial}{\partial q'} = U$.

Hence,

$$\frac{\partial^2 \widetilde{A}}{\partial \tau^2}\bigg|_{\tau=0} = \frac{\partial}{\partial \tau}\left[\widetilde{A}, U\right]_{\tau=0} = \left[[\widetilde{A}, U], U\right]_{\tau=0} = \left[[A, U], U\right].$$

The other derivatives can be calculated in a similar manner.

Thus, we arrive at *Hausdorff's formula*

$$\widetilde{A} = A + \tau[A, U] + \frac{\tau^2}{2!}\left[[A, U], U\right] + \cdots .$$

It follows that if $[A, U] = 0$, then $\widetilde{A} = A$. This means that a group with generator U leaves the operator A (or the corresponding differential equations) unchanged.

In this case, the group $q' = Q(q, \tau)$ is called a *symmetry group* of the differential system $\frac{dq}{dt} = K(q)$. In other words, this differential system admits the group. If equations remain unchanged under group transformations, then solutions of these equations are transformed by the symmetry group into solutions of these equations again. This fact can serve as another definition of a symmetry group. In this case, for the Liouville equation $\frac{\partial}{\partial t}\widetilde{F} = A\widetilde{F}$ equivalent to the differential system

$\frac{dq}{dt} = K(q)$, the following statement is valid. If $\widetilde{F}(q,t)$ is a solution of the Liouville equation, then $U\widetilde{F}(q,t)$ is also a solution of it. That is, a solution of the Liouville equation is transformed by the generator of the symmetry group into a solution of this equation as well. Indeed, multiply the Liouville equation by U: $U\frac{\partial}{\partial t}\widetilde{F} = UA\widetilde{F}$. Taking into account the fact that $U\frac{\partial}{\partial t}\widetilde{F} = \frac{\partial}{\partial t}U\widetilde{F}$ and $UA\widetilde{F} = AU\widetilde{F}$, we have

$$\frac{\partial}{\partial t}U\widetilde{F} = AU\widetilde{F}.$$

This is the above Liouville equation with $\widetilde{F}$ substituted by $U\widetilde{F}$.

The utility of establishing symmetries of differential equations is demonstrated by the following theorem.

THEOREM. *Let a system*

$$\frac{dq}{dt} = K(q)$$

be specified. If a symmetry group of this system with generator $U = \sum_{i=1}^{n} \eta_i(q)\frac{\partial}{\partial q_i}$ *is known, then the order of the system can be reduced.*

Proof. We indicate an algorithm for reducing the order. The group generated by the generator U is assumed to be known to an extent such that canonical coordinates of the group, $r = R(q)$, are known and in these coordinates the group generator U has the simplest form $\widetilde{U} = \frac{\partial}{\partial r_1}$. Then the condition $[\widetilde{A},\widetilde{U}] = 0$ acquires the form

$$[\widetilde{A},\widetilde{U}] = \left[\widetilde{A}, \frac{\partial}{\partial r_1}\right] = 0.$$

The computation of the latter commutator is reduced to the differentiation of the components of $\widetilde{A}$ with respect to r_1. The fact that the commutator is equal to zero means that the operator A in the canonical coordinates of the group U is independent of r_1 (it is $\widetilde{A}$ which is the operator A rewritten in the canonical coordinates of the group U). Here and in what follows, we often identify a group and its generator.

Example. Reduce the order of the equation

$$\frac{d^2y}{dt^2} + M(t)\frac{dy}{dt} + N(t)y = 0.$$

We reduce first this equation to a system of first order equations:

$$\frac{dx}{dt} = 1, \quad \frac{dy}{dt} = z, \quad \frac{dz}{dt} = -M(x)z - N(x)y.$$

One of the symmetry groups of this system is apparent. It is related to scaling of the variable y. This is a group of extensions (a similarity group):

$$x' = x, \quad y' = ky = (1+\mu)y, \quad z' = kz = (1+\mu)z.$$

Hence, the operators A and U are expressed as

$$A = \frac{\partial}{\partial x} + z\frac{\partial}{\partial y} - [M(x)z + N(x)y]\frac{\partial}{\partial z},$$

$$U = y\frac{\partial}{\partial y} + z\frac{\partial}{\partial z}.$$

One can verify that $[A,U] = 0$.

As follows from the preceding, the order of the original equation can be reduced by passing to canonical coordinates of the similarity group. We seek canonical coordinates of the group U in the from

$$p = p(x, y, z), \quad q = q(x, y, z), \quad r = r(x, y, z)$$

by imposing the conditions

$$Up = 1, \quad Uq = 0, \quad Ur = 0.$$

To this end, in accordance with the procedure described in Section 1.6, we search for first integrals of the system

$$\frac{dx}{0} = \frac{dy}{y} = \frac{dz}{z} = \frac{dp}{1}.$$

We find that $p = \ln y$, $q = z/y$, and $r = x$. In the new variables, the operator A has the from

$$\widetilde{A} = (Ap)\frac{\partial}{\partial p} + (Aq)\frac{\partial}{\partial q} + (Ar)\frac{\partial}{\partial r} = \frac{z}{y}\frac{\partial}{\partial p} - \left\{\frac{1}{y}\left[M(x)z + N(x)y\right] + \frac{z^2}{y^2}\right\}\frac{\partial}{\partial q} + \frac{\partial}{\partial r}.$$

Expressing x, y, and z in terms of p, q, and r, we obtain

$$\widetilde{A} = q\frac{\partial}{\partial p} - \left[M(r)q + N(r) + q^2\right]\frac{\partial}{\partial q} + \frac{\partial}{\partial r}.$$

In other words, in the new variables the original differential system acquires the form

$$\dot{p} = q, \quad \dot{q} = N(r) + M(r)q + q^2, \quad \dot{r} = 1,$$

where the dot stands for $\frac{d}{dt}$. Thus, the problem is reduced to a Riccati equation.

Note that the above technique of reducing the order requires canonical coordinates of a symmetry group to be known. However, sometimes it suffices to know the mere group generator. For example, such a situation is possible in the cases where the system is of the second order. Consider this case. Let us have a system

$$\frac{dx}{dt} = X(x, y), \quad \frac{dy}{dt} = Y(x, y)$$

with operator

$$A = X(x, y)\frac{\partial}{\partial x} + Y(x, y)\frac{\partial}{\partial y}$$

and let the generator of its symmetry group be known,

$$U = \xi(x, y)\frac{\partial}{\partial x} + \eta(x, y)\frac{\partial}{\partial y}.$$

Additionally, we assume that U is linearly unrelated to A. (The condition of linear unrelatedness is more severe than that of linear independence. The former condition is stated as follows: generators $U_1, \ldots, U_k$ are linearly unrelated if there are no $\lambda_1(q), \ldots, \lambda_k(q)$ such that

$$\lambda_1 U_1 + \cdots + \lambda_k U_k = 0,$$

provided that the λ_i are not all zero. Unlike the definition of linear independence, here the coefficients λ can depend on the variables q.)

To integrate the system in question, it suffices to find a first integral of it, $\omega(x, y)$, which, by definition, satisfies the condition

$$A\omega(x, y) = 0.$$

Since a symmetry group transforms a solution into a solution, the family of integral curves $\omega(x,y)=C$ must be an invariant family of the group, i.e.,

$$U\omega(x,y)=1.$$

The two relations form the system

$$X(x,y)\frac{\partial\omega}{\partial x}+Y(x,y)\frac{\partial\omega}{\partial y}=0,$$
$$\xi(x,y)\frac{\partial\omega}{\partial x}+\eta(x,y)\frac{\partial\omega}{\partial y}=1.$$

Solving this system for $\frac{\partial\omega}{\partial x}$ and $\frac{\partial\omega}{\partial y}$, we obtain

$$\frac{\partial\omega}{\partial x}=-\frac{Y(x,y)}{X(x,y)\,\eta(x,y)-Y(x,y)\,\xi(x,y)},$$
$$\frac{\partial\omega}{\partial y}=\frac{X(x,y)}{X(x,y)\,\eta(x,y)-Y(x,y)\,\xi(x,y)}.$$

These equations permit us to find the first integral in quadrature:

$$\omega(x,y)=\int\frac{X(x,y)\,dy-Y(x,y)\,dx}{X(x,y)\,\eta(x,y)-Y(x,y)\,\xi(x,y)}=\int_0^1\frac{X(tx,ty)\,y-Y(tx,ty)\,x}{X(tx,ty)\,\eta(tx,ty)-Y(tx,ty)\,\xi(tx,ty)}\,dt.$$

A similar result holds for the case of arbitrary order of the system. Chebotarev[1] formulated this result as follows.

THEOREM. *If a system* $\frac{dq}{dt}=K(q)$ *of order n admits an* $(n-1)$*-parameter solvable group whose generators,* $U_1,\ \ldots,\ U_{n-1}$*, together with the operator of the system,* $A=K(q)\frac{\partial}{\partial q}$*, form a linearly unrelated system, then the differential system can be integrated in quadrature.*

The term *solvable group* just stems from the possibility to use such groups for integrating systems of ordinary differential equations in quadrature. We do not present a rigorous definition of the term *solvable group*. Instead, we give a criterion that allows one to establish the presence of this property: An n-parameter group is solvable if and only if its structural constants (see Section 1.3) satisfy the relation

$$C^{\mu}_{i\lambda}C^{\nu}_{j\mu}C^{\lambda}_{k\nu}=C^{\lambda}_{i\mu}C^{\mu}_{j\nu}C^{\nu}_{k\lambda}\qquad(i,j,k=1,\ldots,n),$$

where summation is assumed over the indices i, j, and k. This criterion was first proved by Cartan.

Extension of the notion of symmetry group. Consider once again the system

$$\frac{dq}{dt}=K(q)$$

with operator A.

Let a group be given with a generator U such that the relation

$$[A,U]=\lambda_1(a)A$$

holds. Using Hausdorff's formula, we will find out how the differential system in question changes in this case. Successively, we have

$$\begin{aligned}\big[[A,U],U\big]&=[\lambda_1A,U]=\lambda_1AU-U(\lambda_1A)=\lambda_1AU-\lambda_1UA-(U\lambda_1)A\\&=\lambda_1[A,U]-(U\lambda_1)A=\big(\lambda_1^2-(U\lambda_1)\big)A=\lambda_2A,\end{aligned}$$

where $\lambda_2=\lambda_1^2-(U\lambda_1)$.

[1] N. G. Chebotarev, *Theory of Lie Groups*, GITTL, Moscow, 1940 [in Russian].

Similarly, we can obtain expressions of the other terms of Hausdorff's series. Eventually, this series permits us to write

$$\widetilde{A} = \lambda(q)A.$$

Thus, in this case the group $q' = Q(q, \tau)$ brings the system $\dot{q} = K(q)$ to the form

$$\frac{dq'}{dt} = \lambda(q')\,K(q').$$

It is apparent that although the equations have changed, the phase trajectories have remained the same, since the factor $\lambda(q')$ common for all right-hand sides cancels out when dividing all the equations by one of them.

A group satisfying the condition $[A, U] = \lambda A$ is also called a symmetry group. Such a group transforms the phase trajectories into phase trajectories.

THEOREM. *If a differential system admits a symmetry group in the extended sense, then the order of this system can be reduced.*

Proof. Let us proceed to canonical coordinates of the group U: $q \to r$. Then $\widetilde{U} = \frac{\partial}{\partial r}$ and the extended symmetry condition becomes

$$\left[\widetilde{A}, \frac{\partial}{\partial r_1}\right] = \widetilde{\lambda}\widetilde{A} \quad \text{or} \quad -\frac{\partial \widetilde{A}}{\partial r_1} = \widetilde{\lambda}\widetilde{A}.$$

In terms of the components of the operator $\widetilde{A}$, this can be represented as

$$-\frac{\partial \widetilde{K}_i}{\partial r_1} = \widetilde{\lambda}\widetilde{K}_i \qquad (i = 1, \ldots, n).$$

On dividing all these equations by the first one, we obtain

$$\frac{\partial \widetilde{K}_i}{\partial \widetilde{K}_1} = \frac{\widetilde{K}_i}{\widetilde{K}_1} \qquad (i = 2, \ldots, n).$$

Integrating yields

$$\widetilde{K}_i = C_i \widetilde{K}_1 \qquad (i = 2, \ldots, n).$$

The constants C_i depend on the other variables except for r_1, i.e., $C_i = C_i(r_2, \ldots, r_n)$.

But this means that, in the canonical coordinates of the group U, the system acquires the form

$$\dot{r}_1 = \widetilde{K}_1, \quad \dot{r}_i = C_i(r_2, \ldots, r_n)\,\widetilde{K}_1 \qquad (i = 2, \ldots, n).$$

On dividing all equations by the first one, we arrive at a system of order $n - 1$, which proves the theorem.

Example. Consider the Blasius equation

$$y\frac{d^2y}{dt^2} + \left(\frac{dy}{dt}\right)^2 + \frac{t}{2}\frac{dy}{dt} = 0.$$

It is required to lower the order.

As a preliminary, we reduce this equation to the Cauchy normal form by representing it as an autonomous system:

$$\frac{dx}{dt} = 1, \quad \frac{dy}{dt} = z, \quad \frac{dz}{dt} = -\frac{z^2}{y} - \frac{xz}{2y}$$

with operator

$$A = \frac{\partial}{\partial x} + z\frac{\partial}{\partial y} - \left(\frac{z^2}{y} + \frac{xz}{2y}\right)\frac{\partial}{\partial z}.$$

We seek a symmetry group by scaling the variables (group of extensions):

$$x = kx', \quad y = my', \quad z = sz'.$$

Substituting these expressions into the system yields

$$\frac{dx'}{dt} = \frac{1}{k}, \quad \frac{dy'}{dt} = \frac{s}{m}z', \quad \frac{dz'}{dt} = -\frac{s}{m}\frac{z'^2}{y'} - \frac{k}{m}\frac{x'z'}{2y'}.$$

It can be seen that the right-hand sides of these equations have a common scalar factor if $s = k$ and $1/k = s/m$.

Hence, we have the one-parameter symmetry group (in the extended sense)

$$x = kx', \quad y = k^2y', \quad z = kz',$$

or, as usual, by changing the parameter k for $1+\mu$, we write out the germ of the obtained symmetry group as follows:

$$x' = (1-\mu), \quad y' = (1-2\mu+\cdots)y, \quad z' = (1-\mu+\cdots)z.$$

The generator of this group is expressed as

$$U = x\frac{\partial}{\partial x} + 2y\frac{\partial}{\partial y} + z\frac{\partial}{\partial z}.$$

The commutator of the operators A and U is equal here to A,

$$[A, U] = A.$$

To reduce the order of the system, it suffices to pass from the variables in terms of which it is written to variables that are canonical coordinates of its symmetry group, $(x, y, z) \to (p, q, r)$. We have

$$x\frac{\partial p}{\partial x} + 2y\frac{\partial p}{\partial y} + z\frac{\partial p}{\partial z} = 1,$$
$$x\frac{\partial q}{\partial x} + 2y\frac{\partial q}{\partial y} + z\frac{\partial q}{\partial z} = 0,$$
$$x\frac{\partial r}{\partial x} + 2y\frac{\partial r}{\partial y} + z\frac{\partial r}{\partial z} = 0.$$

This system completely represents the conditions that must be obeyed by the canonical coordinates. However, in practical calculations, there is no need to write out the last two equations, since all necessary integrals can be obtained from the first equation:

$$\frac{dx}{x} = \frac{dy}{2y} = \frac{dz}{z} = \frac{dp}{1}.$$

Whence, $p = \ln x$, $q = x^2/y$, and $r = x/z$. The inverse change of variables is expressed as

$$x = e^p, \quad y = \frac{1}{q}e^{2p}, \quad z = \frac{1}{r}e^p.$$

It remains to find the representation of the operator A in terms of the new variables. We have

$$\widetilde{A} = e^{-p}\left[\frac{\partial}{\partial p} + \left(2q - \frac{q^2}{r}\right)\frac{\partial}{\partial q} + \left(r + q + \frac{rq}{2}\right)\frac{\partial}{\partial r}\right].$$

Thus, in the new variables the original equations become

$$\frac{dp}{dt} = e^{-p}, \quad \frac{dq}{dt} = e^{-p}\left(2q - \frac{q^2}{r}\right), \quad \frac{dr}{dt} = e^{-p}\left(r + q + \frac{rq}{2}\right).$$

This example demonstrates the difference between the cases $[A, U] = 0$ and $[A, U] = \lambda A$. In the former case, after passing to the canonical coordinates, the system becomes independent of one of the variables. In the latter case, the dependence on one of the variables resides only in the common scalar factor on the right-hand sides (e^{-p} in this example), which does not prevent us from reducing the order of the system. We have

$$\frac{dq}{dp} = 2q - \frac{q^2}{r}, \quad \frac{dr}{dp} = q + r + \frac{qr}{2}.$$

The condition $[A, U] = \lambda(q)A$, which determines a symmetry group in the extended sense, permits us to pose the following problem: Given a differential system with operator A, find a symmetry group of the system. In this case, the above condition is a system of equations for determining the coefficients $\eta(q)$ of the unknown generator U. This system has the form

$$\sum_{i=1}^{n}\left(K_i(q)\frac{\partial \eta_j}{\partial q_i} - \eta_i(q)\frac{\partial K_j}{\partial q_i}\right) = \lambda(q)\,K_j, \qquad j = 1, \ldots, n.$$

This system of equations for the unknowns η_i and λ is referred to as a *determining system*. It is a system of the form

$$\sum_{i=1}^{n} a_i(x_1, \ldots, x_n, y_1, \ldots, y_m)\frac{\partial y_\mu}{\partial x_i} = b_\mu(x_1, \ldots, x_n, y_1, \ldots, y_m), \qquad \mu = 1, \ldots, m.$$

Such systems are equivalent to the single equation

$$\sum_{i=1}^{n} a_i \frac{\partial F}{\partial x_i} + \sum_{\mu=1}^{m} b_\mu \frac{\partial F}{\partial y_\mu} = 0,$$

which is of the same form as that discussed in Section 1.6. Let $F_\lambda(x_1, \ldots, x_n, y_1, \ldots, y_m) = C_\lambda$ ($\lambda = 1, \ldots, m$) be m particular solutions of this equations such that

$$\frac{\partial(F_1, \ldots, F_m)}{\partial(y_1, \ldots, y_m)} \neq 0.$$

Then $y_1, \ldots, y_m$ obtained from the equations $F_\lambda = C_\lambda$ are solutions of the original system.

We will point out an important property of the determining system. Specifically, a set of generators satisfying this system form an algebra. This means the following. Let U be the generator of a symmetry group of a system $\frac{dq}{dt} = K(q)$ with operator $A = K(q)\frac{\partial}{\partial q}$ and let V be the generator of another symmetry group of this system; then the commutator $[U, V]$ is also the generator of a symmetry group of the system.

Let us demonstrate this. According to the problem data, $[A, U] = \lambda A$ and $[A, V] = \nu A$. Then we have

$$[[A, U], V] = [\lambda A, V] = \lambda[A, V] - (V\lambda)A = \big(\lambda\nu - (V\lambda)\big)A,$$
$$[[A, V], U] = [\nu A, U] = \nu[A, U] - (U\nu)A = \big(\lambda\nu - (U\nu)\big)A.$$

Subtracting the second equation from the first one yields

$$\big[[A, U], V\big] - \big[[A, V], U\big] = \big((U\nu) - (V\lambda)\big)A = \mu(q)A.$$

By the Jacobi identity,

$$\big[[A, U], V\big] - \big[[A, V], U\big] = \big[[U, V], A\big].$$

Hence,

$$\big[[U, V], A\big] = \mu(q)A.$$

But this just means that $[U, V]$ is a symmetry generator.

Some examples of the solution of determining equations with the aim to find symmetry algebras will be considered in Sections 1.12, 2.2, 2.3, and 2.5.

1.9. Principle of Superposition of Solutions and Separation of Motions in Nonlinear Mechanics

Consider two systems of ordinary differential equations,

$$\begin{cases} \dfrac{dq_1}{dt} = f_1(q_1, \ldots, q_n), \\ \ldots\ldots\ldots\ldots\ldots\ldots \\ \dfrac{dq_n}{dt} = f_n(q_1, \ldots, q_n), \end{cases} \quad \text{and} \quad \begin{cases} \dfrac{dq_1}{d\tau} = F_1(q_1, \ldots, q_n), \\ \ldots\ldots\ldots\ldots\ldots\ldots \\ \dfrac{dq_n}{d\tau} = F_n(q_1, \ldots, q_n), \end{cases}$$

or, in brief notation,

$$\frac{dq}{dt} = f(q) \quad \text{and} \quad \frac{dq}{d\tau} = F(q).$$

To these systems there correspond the groups

$$q' = u(q, t), \qquad q' = v(q, \tau)$$

which are the solutions of the respective systems subject to the boundary conditions $q'(0) = q$.

The generators of the groups have the from

$$A = f_1\frac{\partial}{\partial q_1} + \cdots + f_n\frac{\partial}{\partial q_n} \equiv f(q)\frac{\partial}{\partial q},$$
$$B = F_1\frac{\partial}{\partial q_1} + \cdots + F_n\frac{\partial}{\partial q_n} \equiv F(q)\frac{\partial}{\partial q}.$$

Consider the composition of the transformations defined by the two different groups:

$$q' = u(q, t), \quad q'' = v(q', \tau) = v\big(u(q, t), \tau\big).$$

Let us find out when the composition does not depend on the order of applying the transformations. In such a case, the two groups are said to commute,

$$v\big(u(q, t), \tau\big) \equiv u\big(v(q, \tau), t\big).$$

PROPOSITIONS 1. *Two groups commute if and only if their generators commute.*

Proof. We take advantage of the exponential representation of groups (see Section 1.5):

$$u\big(v(q,\tau),t\big) = e^{At}v(q,\tau) = e^{At}e^{B\tau}q,$$
$$v\big(u(q,t),\tau\big) = e^{B\tau}u(q,t) = e^{B\tau}e^{At}q.$$

It follows that the groups commute if and only if

$$e^{At}e^{B\tau} = e^{B\tau}e^{At}.$$

For this to hold it is necessary and sufficient that

$$AB = BA, \quad \text{or} \quad [A,B] = 0.$$

PROPOSITIONS 2. *If two groups commute, then their composition is a group, provided that the group parameters are the same, $t = \tau$, i.e., $u\big(v(q,t),t\big)$ is a group.*

Proof. For commuting groups we have

$$u\big(v(q,t),t\big) = e^{At}e^{Bt}q = e^{Bt}e^{At}q = e^{(A+B)t}q.$$

The rightmost expression is a group with generator $A + B$.

THEOREM (PRINCIPLE OF SUPERPOSITION IN NONLINEAR SYSTEMS). *Let the system of differential equations*

$$\dot{q} = \frac{dq}{dt} = f(q) + F(q)$$

be such that the operators A and B of the systems $\dot{q} = f(q)$ and $\dot{q} = F(q)$ commute, $[A,B] = 0$. Then the solution of this system is the superposition of the solutions of the system $\dot{q} = f(q)$ and $\dot{q} = F(q)$, specifically,

$$q = v\big(u(q_0,t),t\big) = u\big(v(q_0,t),t\big),$$

where q_0 is the initial condition.

The proof of this theorem follows from the above propositions.

Example. Consider the system

$$\dot{x} = kx + \lambda y, \quad \dot{y} = ky - \lambda x.$$

Let us divide it into two subsystems as follows:

$$(1) \quad \dot{x} = kx, \quad \dot{y} = ky,$$
$$(2) \quad \dot{x} = \lambda y, \quad \dot{y} = -\lambda x.$$

The operators of the subsystems,

$$A = k\left(x\frac{\partial}{\partial x} + y\frac{\partial}{\partial y}\right), \quad B = \lambda\left(y\frac{\partial}{\partial x} - x\frac{\partial}{\partial y}\right),$$

commute, $[A,B] = 0$.

Solution of subsystem (1):

$$x = x_0 \exp(kt), \quad y = y_0 \exp(kt).$$

Solution of subsystem (2):

$$x = x_0 \cos \lambda t + y_0 \sin \lambda t, \quad y = -x_0 \sin \lambda t + y_0 \cos \lambda t.$$

The solution of the full system is the superposition of these solutions in either order (the solutions of one subsystem must be substituted for the initial conditions of the other subsystem), specifically,

$$x = e^{kt}(x_0 \cos \lambda t + y_0 \sin \lambda t), \quad y = e^{kt}(-x_0 \sin \lambda t + y_0 \cos \lambda t).$$

In other words, the proved property of superposition means separation of motions, i.e., the motion of the full system is the superposition of the motions of the constituent subsystems. This circumstance is directly related to methods of separation of motions (usually called fast and slow motions) in nonlinear mechanics, where the problem is formulated as follows (see Chapter 5):

A system

$$\frac{dx}{dt} = f(x, t) + \varepsilon F(t, x)$$

is given in which the parameter ε is considered small. For the degenerate system ($\varepsilon = 0$),

$$\frac{dx}{dt} = f(x, t),$$

its general solution,

$$x = u(C, t),$$

is known (C is an arbitrary constant). While preserving this form for the solution of the full system and assuming that $C = C(t)$, it is required to find an equation for $C(t)$.

If the operators of the subsystems

$$\frac{dx}{dt} = f(t, x) \quad \text{and} \quad \frac{dx}{dt} = \varepsilon F(t, x)$$

commute, then the result follows immediately from the principle of superposition. Consequently, the task of the theory of perturbations is to find a change of variables $x \to y$ which would reduce the system $\dot{x} = f(t, x) + \varepsilon F(t, x)$ to a form in which the constituent subsystems commute. It is just this point of view which will be effectively used in Chapter 5.

1.10. Prolongation of Groups. Differential and Integral Invariants

The problem of transformation of differential equations can be viewed somewhat differently as compared with the approach discussed in Section 1.8. For example, let it be required to find out how a system of differential equations represented in the form

$$F(t, q, \dot{q}) = 0$$

is modified by a group. Here, t is the independent variable (time), and q and F are n-vectors.

Let a group

$$t' = t + \xi(t, q)\tau + \cdots,$$
$$q' = q + \eta(t, q)\tau + \cdots$$

act in the space (t, q). We introduce the notation $\dot{q} = p$ and treat the system $F(t, q, p) = 0$ as the equations of a manifold in the space of the variables (t, q, p). By their meaning, these variables are related to each other. Therefore, the transformation generated by this group in the space (t, q, p) is completely determined by the coefficients $\xi(t, q)$ and $\eta(t, q)$. We have

$$\begin{aligned} t' &= t + \xi(t, q)\tau + \cdots, \\ q' &= q + \eta(t, q)\tau + \cdots, \\ p' &= p + \zeta(t, q, p)\tau + \cdots, \end{aligned}$$

where $\zeta(t, q, p)$ must be uniquely determined by $\xi(t, q)$ and $\eta(t, q)$. By definition,

$$p' = \frac{dq'}{dt'} = \frac{dq'}{dt} \Big/ \frac{dt'}{dt} = \frac{\dot{q} + \dot{\eta}\tau + \cdots}{1 + \dot{\xi}\tau + \cdots} = \dot{q} + (\dot{\eta} - \dot{q}\dot{\xi})\tau + \cdots,$$

where $\dot{\eta}$ and $\dot{\xi}$ are the total derivatives of η and ξ with respect to t:

$$\dot{\eta} = \frac{\partial \eta}{\partial t} + \frac{\partial \eta}{\partial q}\dot{q}, \quad \dot{\xi} = \frac{\partial \xi}{\partial t} + \frac{\partial \xi}{\partial q}\dot{q}.$$

Hence, $\zeta(t, q, p) = \dot{\eta} - p\dot{\xi}$. A group prolonged in such a manner to the space (t, q, p) is referred to as a group of *first prolongation*. Accordingly, its generator is referred to as a generator of first prolongation and denoted by

$$\overset{(1)}{U} = \xi(t, q)\frac{\partial}{\partial t} + \eta(t, q)\frac{\partial}{\partial q} + \zeta(t, q, p)\frac{\partial}{\partial p}.$$

The equations containing derivatives of any order can be treated in a similar manner. For example, consider an equation of the form $F(t, q, \dot{q}, \ddot{q}) = 0$. This is the equation of a manifold in the space of the variables t, q, $p = \dot{q}$, and $w = \ddot{q}$. The group produced by the original group has the form

$$\begin{aligned} t' &= t + \xi(t, q)\tau + \cdots, \\ q' &= q + \eta(t, q)\tau + \cdots, \\ p' &= p + \zeta(t, q, p)\tau + \cdots, \\ w' &= w + \delta(t, q, p, w)\tau + \cdots. \end{aligned}$$

Here, what remains to be found is the coefficient $\delta(t, q, p, w)$. By analogy with the preceding, we have

$$w' = \frac{dp'}{dt'} = \frac{\dot{p} + \dot{\zeta}\tau + \cdots}{1 + \dot{\xi}\tau + \cdots} = \dot{p} + (\dot{p}\dot{\xi} - \dot{\zeta})\tau + \cdots = w - (w\dot{\xi} - \dot{\zeta})\tau + \cdots.$$

Hence, $\delta(t, q, p, w) = -w\dot{\xi} + \dot{\zeta}$ and the generator of second prolongation is expressed as

$$\overset{(2)}{U} = \xi(t, q)\frac{\partial}{\partial t} + \eta(t, q)\frac{\partial}{\partial q} + \zeta(t, q, p)\frac{\partial}{\partial p} + \delta(t, q, p, w)\frac{\partial}{\partial w}.$$

By using the concept of a prolonged generator, one can transform differential systems with the aid of Lie series rather than Hausdorff's formula. In this case, the variable t can be viewed as a variable of the extended phase space.

The invariants of a prolonged group are called *differential invariants of the group*.

Example. Find the differential invariants of the rotation group

$$U = q\frac{\partial}{\partial t} - t\frac{\partial}{\partial q} \qquad (q \text{ is a scalar})$$

up to the second order inclusive.

Since $\xi = q$ and $\eta = -t$, we have

$$\zeta = \dot{\eta} - \dot{q}\dot{\xi} = -1 - \dot{q}^2.$$

Hence, the generator of first prolongation is given by

$$\overset{(1)}{U} = q\frac{\partial}{\partial t} - t\frac{\partial}{\partial q} - (1 + \dot{q}^2)\frac{\partial}{\partial \dot{q}}.$$

To find the generator of second prolongation, we carry out the following:

$$\ddot{q}' = \frac{d\dot{q}'}{dt'} = \frac{d\dot{q}'}{dt}\Big/\frac{dt'}{dt} = \ddot{q} + (\dot{\zeta} - \ddot{q}\dot{\xi})\tau + \cdots.$$

Whence,

$$\delta(t, q, \dot{q}, \ddot{q}) = \dot{\zeta} - \ddot{q}\dot{\xi} = -2\dot{q}\ddot{q} - \dot{q}\ddot{q} = -3\dot{q}\ddot{q}.$$

Thus, the generator of second prolongation has the from

$$\overset{(2)}{U} = q\frac{\partial}{\partial t} - t\frac{\partial}{\partial q} - (1 + \dot{q}^2)\frac{\partial}{\partial \dot{q}} - 3\dot{q}\ddot{q}\frac{\partial}{\partial \ddot{q}}.$$

Finding differential invariants is reduced to finding first integrals of the system

$$\frac{dt}{-q} = \frac{dq}{t} = \frac{d\dot{q}}{1 + \dot{q}^2} = \frac{d\ddot{q}}{3\dot{q}\ddot{q}}.$$

Solving the first equations yields $C_1 = \sqrt{t^2 + q^2}$. The equation

$$\frac{dq}{\sqrt{C_1^2 - q^2}} = \frac{d\dot{q}}{1 + \dot{q}^2}$$

implies

$$C_2 = \arctan\frac{q}{t} - \arctan\dot{q}, \quad \text{or} \quad \tan C_2 = \frac{\dot{q} - q/t}{1 + q\dot{q}/t}.$$

At last, on solving the equation

$$\frac{d\dot{q}}{1 + \dot{q}^2} = \frac{d\ddot{q}}{3\dot{q}\ddot{q}},$$

we obtain

$$C_3 = \frac{(1 + \dot{q}^2)^{3/2}}{\ddot{q}}.$$

This invariant has the meaning of the curvature of a curve in the (t, q) plane.

Integral invariants. Consider the following three types of integrals:

$$J_1 = \int_{t_1}^{t_2} \Phi(t, q, \dot{q}, \ldots)\, dt, \quad J_2 = \oint_\gamma \Phi_k(q)\, dq_k, \quad J_3 = \int \cdots \int_V \Phi(q)\, dq_1 \ldots dq_n,$$

where $\Phi(t, q, \dot{q}, \ldots)$, $\Phi_k(q)$, and $\Phi(q)$ are some functions, t_1 and t_2 are some instants of time, and γ and V are, respectively, a closed contour and a domain in the phase space $(q_1, \ldots, q_n)$. The integral J_1 is a functional along the trajectories $q(t)$; it depends on m first derivatives of q with respect to time. In the integral J_2, the summation over k is implied.

Let a group of transformations,

$$t' = t + \xi(t, q)\,\tau + \cdots,$$
$$q' = q + \eta(t, q)\,\tau + \cdots,$$

be specified. Compute the above integrals in terms of the new variables with the same integrand functions:

$$J_1' = \int_{t_1'}^{t_2'} \Phi(t', q', \dot{q}', \ldots)\, dt', \quad J_2' = \oint_{\gamma'} \Phi_k(q')\, dq_k', \quad J_3' = \int \cdots \int_{V'} \Phi(q')\, dq_1' \ldots dq_n'.$$

In general, these integrals become functions of the group parameter τ: $J_1'(\tau)$, $J_2'(\tau)$, and $J_3'(\tau)$.

DEFINITION. *The last three integrals are said to be integral invariants of the group in question if*

$$J_1'(\tau) \equiv J_1, \quad J_2'(\tau) \equiv J_2, \quad J_3'(\tau) \equiv J_3.$$

For this to be the case it is necessary and sufficient that

$$\frac{dJ_1'}{d\tau} \equiv 0, \quad \frac{dJ_2'}{d\tau} \equiv 0, \quad \frac{dJ_3'}{d\tau} \equiv 0.$$

Find what conditions the functions $\Phi(t, q, \dot{q}, \ldots)$, $\Phi_k(q)$, and $\Phi(q)$ must satisfy for the respective integrals to be invariants of the group.

We begin with the first integral and consider a small increment of the group parameter, $\tau + \delta\tau$. We have

$$J_1'(\tau + \delta\tau) = \int_{t_1'(\tau+\delta\tau)}^{t_2'(\tau+\delta\tau)} \Phi\big(t'(\tau + \delta\tau),\, q'(\tau + \delta\tau),\, \ldots\big)\, dt'(\tau + d\tau).$$

In this integral, we carry out the change of variables

$$t'(\tau + \delta\tau) \to t'(\tau), \quad q'(\tau + \delta\tau) \to q'(\tau)$$

in accordance with the formulas

$$t'(\tau + \delta\tau) = t'(\tau) + \frac{dt'}{d\tau}\,\delta\tau + \cdots,$$
$$q'(\tau + \delta\tau) = q'(\tau) + \frac{dq'}{d\tau}\,\delta\tau + \cdots.$$

By virtue of the uniqueness theorem (see Section 1.4), we have

$$\frac{dt'}{d\tau} = \xi(t', q'), \quad \frac{dq'}{d\tau} = \eta(t', q').$$

Then the above integral acquires the form

$$J_1'(\tau+\delta\tau) = \int_{t_1'}^{t_2'} \left[\Phi(t', q', \dot{q}', \dots) + \frac{d\Phi}{d\tau}\delta\tau + \cdots \right] d[t' + \xi(t', q')\,\delta\tau + \cdots].$$

Taking into account the fact that $d\Phi/d\tau = \overset{(m)}{U}\Phi$, we obtain

$$J_1'(\tau+\delta\tau) = \int_{t_1'}^{t_2'} \Phi(t', q', \dot{q}', \dots)\, dt' + \delta\tau \int_{t_1'}^{t_2'} \left(\overset{(m)}{U}\Phi + \frac{d\xi}{dt'}\Phi \right) dt' + \cdots .$$

It follows that

$$\lim_{\delta\tau\to 0} \frac{J_1'(\tau+\delta\tau) - J_1'(\tau)}{\delta\tau} \equiv \frac{dJ_1'}{d\tau} = \int_{t_1'}^{t_2'} \left(\overset{(m)}{U}\Phi + \dot{\xi}\Phi \right) dt'.$$

In order that $dJ_1'/d\tau \equiv 0$ along every trajectory $q'(t')$, it is necessary and sufficient that

$$\overset{(m)}{U}\Phi(t', q', \dot{q}', \dots) + \dot{\xi}(t', q')\,\Phi(t', q', \dot{q}', \dots) = 0.$$

Inasmuch as this expression does not depend on τ explicitly, t' and q' have the meaning of dummy variables, and hence, the primes can be omitted:

$$\overset{(m)}{U}\Phi + \dot{\xi}\Phi = 0.$$

This equation expresses the desired criterion for J_1 to be an integral invariant.

When calculating the derivatives of J_2' and J_3' with respect to the group parameter, we proceed quite similarly. Since time does not occur in the integrands, one can set $\xi(t, q) \equiv 0$ in the group of transformations as well. We choose to give the final results for the derivatives:

$$\frac{dJ_2'}{d\tau} = \oint_{\gamma'} \left(\eta_l \frac{\partial\Phi_k}{\partial q_l'} + \Phi_i \frac{\partial\eta_i}{\partial q_k'} \right) dq_k' \qquad \text{(summation over } i, k, l \text{ from 1 to } n\text{)},$$

$$\frac{dJ_3'}{d\rho} = \int \cdots \int_{V'} (U\Phi + \Phi\,\mathrm{div}\,\eta)\, dq_1' \dots dq_n'.$$

Whence we obtain the invariance criteria for J_2' and J_3':

$$\frac{d}{dq_s}\left(\eta_l \frac{\partial\Phi_k}{\partial q_l} + \Phi_i \frac{\partial\eta_i}{\partial q_k} \right) = \frac{d}{dq_k}\left(\eta_l \frac{\partial\Phi_s}{\partial q_l} + \Phi_i \frac{\partial\eta_i}{\partial q_s} \right) \qquad (\text{for } J_2'),$$

$$U\Phi + \Phi\,\mathrm{div}\,\eta = 0 \qquad (\text{for } J_3').$$

Example. Find integral invariants of the J_1 type for the rotation group SO(2). The prolonged generator of the group is given by

$$\overset{(m)}{U} = q\frac{\partial}{\partial t} - t\frac{\partial}{\partial q} - (1+\dot{q}^2)\frac{\partial}{\partial \dot{q}} - 3\dot{q}\ddot{q}\frac{\partial}{\partial \ddot{q}} + \cdots .$$

The invariance condition for J_1 is expressed as

$$\overset{(m)}{U}\Phi + \dot{\xi}\Phi = q\frac{\partial\Phi}{\partial t} - t\frac{\partial\Phi}{\partial q} - (1+\dot{q}^2)\frac{\partial}{\partial \dot{q}} - 3\dot{q}\ddot{q}\frac{\partial\Phi}{\partial \ddot{q}} + \cdots + \dot{q}\Phi = 0.$$

The system of ordinary differential equations corresponding to this partial differential equation has the form

$$\frac{dt}{-q} = \frac{dq}{t} = \frac{d\dot{q}}{1+\dot{q}^2} = \frac{d\ddot{q}}{3\dot{q}\ddot{q}} = \cdots = \frac{d\Phi}{\dot{q}\Phi}.$$

The equations that do not contain Φ permit one to calculate differential invariants up to the mth order inclusive,

$$C_1 = \sqrt{t^2+q^2}, \quad C_2 = \frac{t\dot{q}-q}{t+q\dot{q}}, \quad C_3 = \frac{(1+\dot{q}^2)^{3/2}}{\ddot{q}}, \quad \ldots$$

From the equation $d\dot{q}/(1+\dot{q}^2) = d\Phi/(\dot{q}\Phi)$ it follows that $\Phi = C\sqrt{1+\dot{q}^2}$, or

$$C = \frac{\Phi}{\sqrt{1+\dot{q}^2}}.$$

The general solution of the equation expressing the invariance of J_1' is the function Φ. This function should be determined from the condition $G(C_1, C_2, \ldots, C) = 0$, where G is an arbitrary function of the first integrals obtained. Solving this equation for Φ yields

$$\Phi = \sqrt{1+\dot{q}^2}\, F(C_1, C_2, \ldots),$$

where F is an arbitrary function of the first m differential invariants.

Thus, the general expression of the integral invariant J_1 of the rotation group SO(2) is as follows:

$$\int_{t_1}^{t_2} \sqrt{1+\dot{q}^2}\, F\left(t^2+q^2, \frac{t\dot{q}-q}{1+\dot{q}^2}, \ldots\right) dt.$$

In particular, for $F \equiv 1$, this functional expresses the length of the curve $q(t)$.

1.11. Equations Admitting a Given Group

The theory of prolongation of group generators permits one to formulate the conditions for differential equations to be invariant under a one-parameter group of transformations in a different manner.

For instance, let us deal with a differential equation specified in the xy-plane and represented implicitly,

$$F\left(x, y, \frac{dy}{dx}\right) = 0.$$

As follows from the results of the previous section, for this equation to be invariant under the group with generator

$$U = \xi(x,y)\frac{\partial}{\partial x} + \eta(x,y)\frac{\partial}{\partial y}$$

it is necessary and sufficient that

$$\overset{(1)}{U} F(x, y, \overset{1}{y}) = 0 \quad \text{for} \quad F(x, y, \overset{1}{y}) = 0 \qquad \left(\overset{1}{y} \equiv \frac{dy}{dx}\right).$$

That is, the equation $F(x, y, \overset{1}{y}) = 0$ must define an invariant surface in the $(x, y, \overset{1}{y})$ space of the group prolonged one time.

If this first order equation is solvable for $\overset{1}{y}$, it can be represented by two autonomous first order equations as

$$\frac{dx}{dt} = X(x,y), \quad \frac{dy}{dt} = Y(x,y) \qquad \left(A = X\frac{\partial}{\partial x} + Y\frac{\partial}{\partial y}\right).$$

The invariance condition (see Section 1.8) for these equations has the form $[A, U] = \lambda A$.

Let us demonstrate that the two criteria are equivalent in this case.

Rewrite the system as a single equation,

$$X(x, y)\overset{1}{y} - Y(x, y) = 0,$$

and apply to it the first prolongation generator

$$\xi\left(\frac{\partial X}{\partial x}\overset{1}{y} - \frac{\partial Y}{\partial x}\right) + \eta\left(\frac{\partial X}{\partial y}\overset{1}{y} - \frac{\partial Y}{\partial y}\right) + \zeta(x, y, \overset{1}{y})X = 0 \quad (\text{with } X\overset{1}{y} - Y = 0),$$

where $\zeta(x, y, \overset{1}{y}) = \eta_x + (\eta_y - \xi_x)\overset{1}{y} - \xi_y \overset{1}{y}{}^2$.

On eliminating $\overset{1}{y}$ and ζ from these relations, we obtain

$$Y\left(\xi\frac{\partial X}{\partial x} + \eta\frac{\partial X}{\partial y}\right) - X\left(\xi\frac{\partial Y}{\partial x} + \eta\frac{\partial Y}{\partial y}\right) + X\left(X\frac{\partial \eta}{\partial x} + Y\frac{\partial \eta}{\partial y}\right) - Y\left(X\frac{\partial \xi}{\partial x} + Y\frac{\partial \xi}{\partial y}\right) = 0.$$

In terms of the operators A and U, this equation can be rewritten as

$$YUX - XUY + XA\eta - YA\xi = 0.$$

Whence,

$$\frac{UX - A\xi}{X} = \frac{UY - A\eta}{Y}.$$

But this means that the coordinates of the commutator $[A, U]$ are proportional to those of the operator A, i.e., $[A, U] = \lambda A$.

For higher order equations the proved equivalence is not the case. To explain this fact and to discuss its consequences, we consider the following nth order equation solved for the highest derivative:

$$\overset{n}{y} = f(x, y, \overset{1}{y}, \ldots, \overset{n-1}{y}) \qquad \left(n > 1, \quad \overset{k}{y} \equiv \frac{d^k y}{dx^k}\right).$$

The condition for the invariance of this equation under some group can be stated in various forms, similar to those derived for the first order equation but not equivalent now to each other for $n > 1$.

First form. Consider a group acting in the (x, y) plane with generator $U = \xi(x, y)\frac{\partial}{\partial x} + \eta(x, y)\frac{\partial}{\partial y}$. The invariance criterion for the equation is the criterion for the invariance of the manifold $\overset{n}{y} - f = 0$ in the $(x, y, \overset{1}{y}, \ldots, \overset{n}{y})$ space under the group prolonged n times,

$$\overset{(n)}{U}(\overset{n}{y} - f) = 0 \quad \text{for} \quad \overset{n}{y} = f.$$

As applied to finding a symmetry group for a specified differential equation, this criterion is a partial differential equation for two unknowns, $\xi(x, y)$ and $\eta(x, y)$. Since these functions do not depend on the $\overset{k}{y}$, the coefficients of like powers and mixed products of $\overset{k}{y}$ must be equated to zero. This yields an overdetermined system for $\xi(x, y)$ and $\eta(x, y)$. The cases where this overdetermined system is solvable is of great importance in mechanics and physics. However, the system can have no solutions.

Second form. We introduce the notation

$$x = x_1, \quad y = x_2, \quad \overset{1}{y} = x_3, \quad \ldots, \quad \overset{n}{y} = x_{n+2}$$

and rewrite the differential equation in question as a system of ordinary differential equations to obtain

$$\frac{dx_1}{d\tau} = 1, \quad \frac{dx_2}{d\tau} = x_3, \quad \ldots, \quad \frac{dx_n}{d\tau} = x_{n+1}, \quad \frac{dx_{n+1}}{d\tau} = f(x_1, \ldots, x_{n+1}).$$

Then, in order to find symmetries, one can use the criterion $[A, U] = \lambda A$. In this case, the desired generator, $U = \sum_{i=1}^{n+1} \eta_i(x)\frac{\partial}{\partial x_i}$, has $n + 1$ independent components, which are not related by any conditions of prolongation. For this reason, the problem is always solvable.

The invariance concept in the form $[A, U] = \lambda A$ was discussed in Section 1.8. In the current section, we consider invariance in the form $\overset{(n)}{U} F(x, y, \overset{1}{y}, \ldots, \overset{n}{y}) = 0$, $F = 0$. In respect of this condition, the following two statements of the problem are of interest. First problem: the generator $U = \zeta\frac{\partial}{\partial x} + \eta\frac{\partial}{\partial y}$ is assumed to be known and what is required to find are all differential equations invariant under the group determined by this generator. The second problem was already mentioned above: given a differential equation, find the algebra of its symmetry generators.

Examples of solving the second problem will be given in Section 1.12 and Chapter 2. In this section, we dwell on the first problem.

We start the solving of this problem with the first order differential equation

$$F(x, y, \overset{1}{y}) = 0, \qquad \overset{1}{y} = \frac{\partial y}{\partial x}.$$

Inasmuch as every such differential equation invariant under the group U must be an invariant surface of the group prolonged one time in the $(x, y, \overset{1}{y})$ space, finding the general form of this surface solves the problem. Since the generator

$$\overset{(1)}{U} = \xi(x, y)\frac{\partial}{\partial x} + \eta(x, y)\frac{\partial}{\partial y} + \zeta(x, y, \overset{1}{y})\frac{\partial}{\partial \overset{1}{y}}$$

has only two independent invariants (denote them by u and v), any invariant is a function of the two invariants, $M(u, v)$. Hence, the general form of the invariant surface is $M(u, v) = 0$ (M is an arbitrary function).

As u we take an invariant of U; hence, it does not depend on $\overset{1}{y}$. Denote this invariant by G_0. As v we take an invariant of $\overset{(1)}{U}$, i.e., a first order differential invariant of the generator U. Denote this invariant by G_1. Then the general form of differential equations invariant under the group with generator U is as follows:

$$G_1(x, y, \overset{1}{y}) = N(G_0(x, y)),$$

where N is an arbitrary function.

Example. Previously (see Section 1.10), invariants of the rotation group SO(2) with $U = y\frac{\partial}{\partial x} - x\frac{\partial}{\partial y}$ were found. These are: the basic invariant, $u = x^2 + y^2$, and the differential invariant, $v = (x\overset{1}{y} - y)/(x + y\overset{1}{y})$. Hence, the general form of the first order differential equations that are invariant under this group is given by

$$\frac{x\overset{1}{y} - y}{x + y\overset{1}{y}} = N(x^2 + y^2), \qquad \overset{1}{y} = \frac{dy}{dx}.$$

In particular, if $N(x^2 + y^2) \equiv x^2 + y^2$, then solving this equation for the derivative yields

$$\frac{dy}{dx} = \frac{y + (x^2 + y^2)x}{x - (x^2 + y^2)y},$$

or, in the form of an autonomous system,

$$\frac{dx}{dt} = x - (x^2 + y^2)y, \quad \frac{dy}{dt} = y + (x^2 + y^2)x.$$

Consider now the second order equation

$$F\big(x, y, \overset{1}{y}, \overset{2}{y}\big) = 0, \qquad \overset{k}{y} = \frac{d^k y}{dx^k}.$$

We search for the general form of the equations that are invariant under a prescribed group with generator $U = \xi(x,y)\frac{\partial}{\partial x} + \eta(x,y)\frac{\partial}{\partial y}$. Considerations quite similar to those for the first order equation lead us to the expression

$$G_2\big(x, y, \overset{1}{y}, \overset{2}{y}\big) = M\big(G_0(x,y),\, G_1(x, y, \overset{1}{y})\big),$$

where G_0, G_1, and G_2 are invariants of the zeroth, first, and second order, respectively, and M is an arbitrary function.

In the example of the rotation group (see Section 1.10), a second order differential invariant was found,

$$w = \overset{2}{y}\big(1 + \overset{1}{y}{}^2\big)^{-3/2}.$$

Therefore, the general form of the differential equations that are invariant under the rotation group is the following:

$$\overset{2}{y} = \big(1 + \overset{1}{y}{}^2\big)^{3/2} M\left(x^2 + y^2,\ \frac{x\overset{1}{y} - y}{x + y\overset{1}{y}}\right).$$

It is easy to infer that the general form of the nth order differential equations $F(x, y, \overset{1}{y}, \dots, \overset{n}{y}) = 0$ that are invariant under the group $U = \xi(x,y)\frac{\partial}{\partial x} + \eta(x,y)\frac{\partial}{\partial y}$ is given by

$$G_n = M(G_0, \dots, G_{n-1}),$$

where $G_0, G_1, \dots, G_n$ are invariants of the zeroth through nth order, respectively.

To construct any nth order equation which is invariant under a prescribed group, one needs to know invariants of this group up to the nth order inclusive.

Let the generator of the nth prolongation have the form

$$\overset{(n)}{U} = \xi(x,y)\frac{\partial}{\partial x} + \eta(x,y)\frac{\partial}{\partial y} + \eta_1(x, y, \overset{1}{y})\frac{\partial}{\partial \overset{1}{y}} + \dots + \eta_n(x, y, \overset{1}{y}, \dots, \overset{n}{y})\frac{\partial}{\partial y^n}.$$

To find invariants $G_0, \dots, G_n$ of this generator, one has to calculate $n+1$ first integrals of the system of ordinary differential equations

$$\frac{dx}{\xi(z,y)} = \frac{dy}{\eta(x,y)} = \frac{d\overset{1}{y}}{\eta_1(x, y, \overset{1}{y})} = \dots = \frac{d\overset{n}{y}}{\eta_n(x, y, \overset{1}{y}, \dots, \overset{n}{y})}.$$

THEOREM. *To determine all independent first integrals G_0, G_1, ..., G_n from the above system, it suffices know only G_0.*

Proof. We show first how G_1 can be constructed for a given G_0. To this end, consider the system

$$\frac{dx}{\xi(z,y)} = \frac{dy}{\eta(x,y)} = \frac{d\overset{1}{y}}{\eta_1(x, y, \overset{1}{y})}.$$

In Section 1.10 we obtained the following formula for the component η_1 of a generator prolonged one time:

$$\eta_1(x, y, \overset{1}{y}) = \eta_x + (\eta_y - \xi_x)\overset{1}{y} - \xi_y \overset{1}{y}{}^2.$$

Suppose that a first integral, $G_0 = G_0(x, y)$, is found from the equation

$$\frac{dx}{\xi(z, y)} = \frac{dy}{\eta(x, y)}.$$

Then, to find G_1 from the equation

$$\frac{dx}{\xi\big(x, y(x)\big)} = \frac{d\overset{1}{y}}{\eta_x + (\eta_y - \xi_x)\overset{1}{y} - \xi_y \overset{1}{y}{}^2},$$

one has to solve the Riccati equation

$$\frac{d\overset{1}{y}}{dx} = \frac{1}{\xi\big(x, y(x)\big)}\big[\eta_x + (\eta_y - \xi_x)\overset{1}{y} - \xi_y \overset{1}{y}{}^2\big] = L(x, G_0) + L_1(x, G_0)\overset{1}{y} + L_2(x, G_0)\overset{1}{y}{}^2.$$

This equation is solvable in quadrature, since a particular solution of it is known:

$$\overset{1}{y} = \frac{\eta\big(x, y(x)\big)}{\xi\big(x, y(x)\big)}.$$

Thus, given the invariant $G_0(x, y)$, the first order differential invariant $G_1(x, y, \overset{1}{y})$ can be found in the fashion just described.

Finding higher order differential invariants now creates no difficulties. Indeed, consider the general form of the first order differential equations that are invariant under the group U,

$$G_1(x, y, \overset{1}{y}) = N\big(G_0(x, y)\big).$$

Differentiating this condition with respect to x yields

$$\frac{\partial G_1}{\partial x} + \frac{\partial G_1}{\partial y}\overset{1}{y} + \frac{\partial G_1}{\partial \overset{1}{y}}\overset{2}{y} = \frac{dN}{dG_0}\left(\frac{\partial G_0}{\partial x} + \frac{\partial G_0}{\partial y}\overset{1}{y}\right).$$

Whence,

$$\frac{dN}{dG_0} = \frac{\dfrac{\partial G_1}{\partial x} + \dfrac{\partial G_1}{\partial y}\overset{1}{y} + \dfrac{\partial G_1}{\partial \overset{1}{y}}\overset{2}{y}}{\dfrac{\partial G_0}{\partial x} + \dfrac{\partial G_0}{\partial y}\overset{1}{y}}.$$

The left-hand side of this relation is an arbitrary function of the invariant G_0 and, hence, is an invariant as well. It follows that the expression on the right-hand side is also an invariant. This expression depends on $\overset{2}{y}$ and therefore is a second order differential invariant, i.e.,

$$G_2 = \frac{dG_1}{dG_0} \equiv \left(\frac{\partial G_1}{\partial x} + \frac{\partial G_1}{\partial y}\overset{1}{y} + \frac{\partial G_1}{\partial \overset{1}{y}}\overset{2}{y}\right) \Big/ \left(\frac{\partial G_0}{\partial x} + \frac{\partial G_0}{\partial y}\overset{1}{y}\right).$$

Analogously,

$$G_k = \frac{dG_{k-1}}{dG_0},$$

that is, each subsequent invariant can be obtained by differentiating its preceding with respect to the zeroth invariant.

Example. Find differential invariants of the similarity group ($U = x\frac{\partial}{\partial x} + y\frac{\partial}{\partial y}$)

$$x' = x + \tau x, \quad y' = y + \tau y$$

to the second order inclusive.

Calculate two prolongations of the group:

$$\overset{1}{y}{}' = \frac{dy'}{dx'} = \frac{\overset{1}{y} + \tau \overset{1}{y}}{1+\tau} = \overset{1}{y} + \cdots,$$

$$\overset{2}{y}{}' = \frac{d\overset{1}{y}{}'}{dx'} = \frac{\overset{2}{y} + \cdots}{1+\tau} = \overset{2}{y} - \tau \overset{2}{y} + \cdots,$$

$$\overset{2}{U} = x\frac{\partial}{\partial x} + y\frac{\partial}{\partial y} + 0 \times \frac{\partial}{\partial \overset{1}{y}} - \overset{2}{y}\frac{\partial}{\partial \overset{2}{y}}.$$

To find differential invariants to the second order inclusive, $\overset{(2)}{U} G = 0$, we have to calculate first integrals of the system

$$\frac{dx}{x} = \frac{dy}{y} = \frac{d\overset{1}{y}}{0} = \frac{d\overset{2}{y}}{-\overset{2}{y}},$$

thus obtaining $G_0 = y/x$, $G_1 = \overset{1}{y}$, and $G_2 = x\overset{2}{y}$.

As follows from the above theorem, given invariants G_0 and G_1, a second order differential invariant can be found by computing dG_1/dG_0 instead of deriving it from the above system; specifically,

$$\frac{dG_1}{dG_0} = \frac{d\overset{1}{y}}{dx} \Big/ \left(\frac{1}{x}\overset{1}{y} - \frac{1}{x^2}y\right) = \frac{x\overset{2}{y}}{\overset{1}{y} - \dfrac{y}{x}}.$$

The thus obtained invariant is a function of those found earlier:

$$\frac{x\overset{2}{y}}{\overset{1}{y} - \dfrac{y}{x}} = \frac{G_2}{G_1 - G_0}.$$

In addition, the last theorem suggests a way of reducing the order of a differential equation admitting a group U. The general form of the nth order equations that are invariant under this group can be represented as

$$G_n = M(G_0, \ldots, G_{n-1}) = \frac{dG_{n-1}}{dG_0},$$

where M is an arbitrary function. By passing to new variables, $(x, y) \to (u, v)$, in accordance with the formulas $u = G_0(x, y)$ and $v = G_1(x, y, \overset{1}{y})$, we obtain the $(n-1)$st order equation

$$\frac{d^{n-1}v}{du^{n-1}} = M\left(u, v, \frac{dv}{du}, \ldots, \frac{d^{n-2}v}{du^{n-2}}\right).$$

Example. Consider once again the linear equation

$$\frac{d^2y}{dt^2} + M(t)\frac{dy}{dt} + N(t)y = 0,$$

which was used to illustrate the method of reducing the order outlined in Section 1.8. Now we can do this with the technique described above. Unlike Section 1.8, time also must be subjected to transformation here:

$$t' = t + \xi(t, y)\tau + \cdots, \quad y' = y + \eta(t, y)\tau + \cdots .$$

The equation in question is invariant under the extension group $t' = t$, $y' = y + \tau y$, $U = y\frac{\partial}{\partial y}$. The group prolonged one time has the form

$$t' = t,, \quad y' = y + \tau y, \quad \overset{1}{y}{}' = \overset{1}{y} + \tau\overset{1}{y}; \qquad \overset{(1)}{U} = y\frac{\partial}{\partial y} + \overset{1}{y}\frac{\partial}{\partial \overset{1}{y}}.$$

Solving the system

$$\frac{dt}{0} = \frac{dy}{y} = \frac{d\overset{1}{y}}{\overset{1}{y}},$$

we find two first invariants, specifically, $G_0 = t$ and $G_1 = \overset{1}{y}/y$.

We carry out the change of variables $(t, y) \to (u, v)$,

$$u = t, \quad v = \frac{\dot{y}}{y},$$

in the original equation, thus arriving, just as in Section 1.8, at the Riccati equation

$$\frac{dv}{du} = \frac{\ddot{y}}{y} - \frac{\dot{y}^2}{y^2} = -M(u)v - N(u) - v^2.$$

1.12. Symmetries of Partial Differential Equations

Let us deal with a partial differential equation with two independent variables,

$$F\left(x, y, z, \frac{\partial z}{\partial x}, \frac{\partial z}{\partial y}, \frac{\partial^2 z}{\partial x^2}, \frac{\partial^2 z}{\partial x \partial y}, \frac{\partial^2 z}{\partial y^2}\right) = 0.$$

Let a group with generator

$$U = \xi(x, y)\frac{\partial}{\partial x} + \eta(x, y)\frac{\partial}{\partial y} + \zeta(x, y)\frac{\partial}{\partial z}$$

act in the (x, y, z) space. We introduce the notation

$$\frac{\partial z}{\partial x} = p, \quad \frac{\partial z}{\partial y} = q, \quad \frac{\partial^2 z}{\partial x^2} = r, \quad \frac{\partial^2 z}{\partial x \partial y} = s, \quad \frac{\partial^2 z}{\partial y^2} = l$$

and rewrite the partial differential equation as the equation of a surface in the eight-dimensional space:

$$F(x, y, z, p, q, r, s, l) = 0.$$

The original group generates a prolonged group in this space. The construction of the prolonged generator involves the following.

We write out the germ of the prolonged group as follows:

$$\begin{aligned}
x' &= x + \xi(x, y, z)\tau + \cdots, \\
y' &= y + \eta(x, y, z)\tau + \cdots, \\
z' &= z + \zeta(x, y, z)\tau + \cdots, \\
p' &= p + \pi(x, y, z, p, q)\tau + \cdots, \\
q' &= q + \rho(x, y, z, p, q)\tau + \cdots, \\
&\cdots\cdots\cdots\cdots\cdots\cdots\cdots\cdots
\end{aligned}$$

It is required to determine the coefficients π, ρ, etc. The variables x, y, z, p, and q are not independent—they are related by the total differential condition

$$dz = p\,dx + q\,dy.$$

Similarly, the variables p' and q' are related by

$$dz' = p'\,dx' + q'\,dy'.$$

Substituting the group germ into the latter constraint yields

$$\begin{aligned}
dz + \tau(\zeta_x\,dx + \zeta_y\,dy + \zeta_z\,dz) + \cdots &= (p + \pi\tau + \cdots)[dx + \tau(\xi_x\,dx + \xi_y\,dy + \xi_z\,dz) + \cdots] \\
&\quad + (q + \rho\tau + \cdots)[dy + \tau(\eta_x\,dx + \eta_y\,dy + \eta_z\,dz) + \cdots].
\end{aligned}$$

Changing dz for $p\,dx + q\,dy$ and matching the coefficients of dx and dy, we find that

$$\begin{aligned}
\pi &= \zeta_x - p\xi_x - q\eta_x - p(-\zeta_z + p\xi_z + q\eta_z), \\
\rho &= \zeta_y - p\xi_y - q\eta_y - q(-\zeta_z + p\xi_z + q\eta_z).
\end{aligned}$$

These relations are sufficient for constructing the first prolongation generator. To construct the second prolongation generator, one should invoke the total differential conditions

$$d\frac{\partial z}{\partial x} = \frac{\partial^2 z}{\partial x^2}\,dx + \frac{\partial^2 z}{\partial x \partial y}\,dy, \quad d\frac{\partial z}{\partial y} = \frac{\partial^2 z}{\partial x \partial y}\,dx + \frac{\partial^2 z}{\partial y^2}\,dy$$

or, equivalently,

$$dp = r\,dx + s\,dy, \quad dq = s\,dx + l\,dy.$$

The extension of the group germ to r, s, and l is determined by the coefficients α, β, and γ such that

$$r' = r + \alpha\tau + \cdots, \quad s' = s + \beta\tau + \cdots, \quad l' = l + \gamma\tau + \cdots.$$

Searching for α, β, and γ in just the same manner as π and ρ, we obtain

$$\begin{aligned}
\alpha &= \pi_x + p\pi_z + r\pi_p + s\pi_q - r(\xi_x + p\xi_z) - s(\eta_x + p\eta_z), \\
\beta &= \pi_y + q\pi_z + s\pi_p + l\pi_q - r(\xi_y + q\xi_z) - s(\eta_y + q\eta_z), \\
\gamma &= \rho_y + q\rho_z + s\rho_p + l\rho_q - s(\xi_y + q\xi_z) - l(\eta_y + q\eta_z).
\end{aligned}$$

The second prolongation generator has the form

$$\overset{(2)}{U} = \xi\frac{\partial}{\partial x} + \eta\frac{\partial}{\partial y} + \zeta\frac{\partial}{\partial z} + \pi\frac{\partial}{\partial p} + \rho\frac{\partial}{\partial q} + \alpha\frac{\partial}{\partial r} + \beta\frac{\partial}{\partial s} + \gamma\frac{\partial}{\partial l}.$$

The invariance condition for the surface $F(x,y,z,p,q,r,s,l)=0$, or, what is the same, the invariance condition for the corresponding second order partial differential equation acquires the form

$$\overset{(2)}{U}F=0, \qquad F=0.$$

Just as was the case previously, these relations can be viewed from two standpoints: (i) given the generator U of a group, find an invariant surface or (ii) given a surface, find a surface-preserving group. For ordinary differential equations, the two problems are equivalent in complexity, since the ordinary differential equation and the group transforming it are objects of the same nature. As far as partial differential equations are concerned, this is not the case. The equations determining the group are ordinary differential equations, whereas the equation being transformed is a partial differential equation.

The former object is simpler. For this reason, the algorithms of searching for symmetry groups of partial differential equations turns out to be constructive.

If the equation $F(x,y,z,p,q,r,s,l)=0$ is linear, then, just as in the case of ordinary differential equations, the invariance conditions can be formulated without invoking prolongation of generators, in terms of commutators.

Let us rewrite this equation as $Az=0$, where A is a linear differential operator,

$$A = X_{11}(x,y)\frac{\partial^2}{\partial x^2} + X_{12}(x,y)\frac{\partial^2}{\partial x\partial y} + X_{22}(x,y)\frac{\partial^2}{\partial y^2} + \cdots .$$

By virtue of the linearity of the equation, transforming the variable z makes no sense. The fact that a linear homogeneous equation admits a group of extensions along z is obvious and of no interest. Therefore, we will consider a group acting in the (x,y) plane with generator

$$U = \xi(x,y)\frac{\partial}{\partial x} + \eta(x,y)\frac{\partial}{\partial y}.$$

The commutator of A and U, $[A,U]=AU-UA$ is a second order differential operator.

DEFINITION. *A generator U is called a symmetry generator of the equation $Az=0$ if $[A,U]=\lambda(x,y)A$.*

THEOREM. *A symmetry generator takes the solutions of the equation $Az=0$ to other solutions of the same equation.*

Proof. Suppose $z=\varphi(x,y)$ is a solution of the equation. Then

$$A(U\varphi) = U(A\varphi) + \lambda A\varphi = 0,$$

that is, $U\varphi$ is also a solution of the equation.

Example. Consider the Helmholtz equation

$$\frac{\partial^2 z}{\partial x^2} + \frac{\partial^2 z}{\partial y^2} + \omega^2 z = 0, \qquad A = \frac{\partial}{\partial x^2} + \frac{\partial}{\partial y^2} + \omega^2.$$

It is required to find the symmetry algebra of this equation. We proceed from the condition $[A,U]=\lambda A$, in which the scalar coefficient $\lambda(x,y)$ is to be determined, along with the generator U.

Calculating the commutator yields

$$AU-UA = 2\xi_x\frac{\partial^2}{\partial x^2} + 2(\eta_x+\xi_y)\frac{\partial^2}{\partial x\partial y} + 2\eta_y\frac{\partial^2}{\partial y^2} + (\xi_{xx}+\xi_{yy})\frac{\partial}{\partial x} + (\eta_{xx}+\eta_{yy})\frac{\partial}{\partial y}.$$

Subtracting the operator λA from it and equating the coefficients of the resulting operator with zero, we obtain

$$2\xi_x = \lambda, \quad 2\eta_y = \lambda, \quad 2(\eta_x + \xi_y) = 0, \quad \xi_{xx} + \xi_{yy} = 0, \quad \eta_{xx} + \eta_{yy} = 0, \quad \omega^2\lambda = 0.$$

It follows from the last relation that $\lambda \equiv 0$. In this case, the first relation implies that ξ depends on y alone and the fourth relation implies that ξ is a linear function of y: $\xi = a + by$. Similarly, we have $\eta = c + dx$. The condition $\eta_x + \xi_y = 0$ allows us to establish that $d = -b$, i.e., $\xi = a + by$ and $\eta = c - bx$, where a, b, and c are arbitrary constants.

Hence, the general form of U is given by

$$U = (a + by)\frac{\partial}{\partial x} + (c - bx)\frac{\partial}{\partial y}.$$

This operator is an arbitrary element of the three-dimensional algebra of generators

$$U = aU_1 + bU_2 + cU_3$$

with the basis

$$U_1 = \frac{\partial}{\partial x}, \quad U_2 = y\frac{\partial}{\partial x} - x\frac{\partial}{\partial y}, \quad U_3 = \frac{\partial}{\partial y}.$$

The operators U_1 and U_3 are generators of the groups of translations along the x- and y-axes, and U_2 is the generator of the rotation group SO(2).

Chapter 2

Group Analysis of Foundations of Classical and Relativistic Mechanics

2.1. Axiomatization Problem of Mechanics

Classical mechanics is an axiomatic system. The concept of an axiomatic system is common in mathematics and involves the following.

The construction of a theory must begin with the introduction of basic *categories* that are not defined subsequently. Denote these categories by A, B, C, etc. Usually, the categories are given names, for example, point, force, and so on. At the present stage, however, the names do not make much sense. Individual properties of the objects named appear when the objects get related to each other, rather than when they are introduced. The relationships between categories are referred to as *axioms*. By completing the introduced system of categories and axioms by rules of inference, one can develop the theory as deeply as desired, without recourse to any practice or evidence. A theory thus constructed may appear to be a product of pure reason which bears no relation to nature. If, however, in this nature or some other science or a sphere of human activity there are objects that, when identified with the introduced categories A, B, C, etc., turn out to have the same relationships as those prescribed by the axioms, then all conclusions of the theory will hold true for these objects as well, regardless of their other properties. The relationships established between the categories of an axiomatic system and objects of this sort are referred to as *realizations* of the axiomatic system.

Nevertheless, prior to constructing such a theory, one should make sure that the foundation of the theory—i.e., the axiomatic basis selected—is firm.

In this connection, the following three basic axiomatic problems are considered: (i) consistency problem, (ii) minimality problem, and (iii) completeness problem.

The consistency of a system of axioms implies that one cannot obtain by logical reasoning a positive and a negative statement about the same fact on the basis of this system.

To solve the minimality problem is to prove that each statement of the axiomatic system is independent of the other statements, i.e., cannot be obtained from them by inference.

At last, the completeness of an axiomatic system implies the possibility to prove that any nonabsurd statement involving the objects being considered is either true or false.

The requirement to construct physico-mathematical disciplines on an axiomatic basis was put forward by Hilbert at the end of the 19th century (Hilbert's sixth problem). He regarded this requirement as a necessary condition for a discipline to be absolutely rigorous.

Newton was the first to formulate the basic axioms of mechanics in a systematic form in 1687. It was not until 20th century that studies of these axioms were undertook to solve the above three problems. This required that Newton's axioms of mechanics be modified so as to make it possible to apply methods of mathematical logic.

At present, there are several modifications of the axioms of mechanics. Some axiomatic systems are based on considering discrete populations of point particles. There are systems that employ the idea of continuum. Some systems were inspired by Mach's ideas; Mach stated that notions which do not admit a constructive practical check cannot be introduced into a science. In systems of this sort, the requirement of correspondence between original notions on the formal axiomatic level and observable notions on the empirical level is obligatory.

Moreover, an axiomatization in the pure spirit of Hilbert is known; in this case, the original categories are simply listed, and their content is determined by the axioms introduced.

Note that a presentation of any formal axiomatization of classical mechanics is irrelevant in a course of mechanics, since this is, in fact, a chapter of mathematical logic rather than mechanics proper.

Similarly, for instance, an axiomatization of arithmetic is not a subject matter of arithmetic.

By contrast, the group-theoretic analysis of foundations of classical mechanics and relativistic mechanics reflects the most principal features of these physical theories. Without understanding these features, one cannot comprehend these paradigms adequately.

2.2. Postulates of Classical Mechanics

1. The first group of axioms is entirely borrowed from geometry and defines the notion of Euclidean space $\mathbb{E}^3$ and geometric objects in it (point, straight line, and plane).

2. The objects in $\mathbb{E}^3$ are assumed to depend on a scalar parameter t called *time*. This means that mechanics deals with a transformation $\mathbb{R}^1 \to \mathbb{E}^3$ called *motion*.

3. A point particle $(\mathbf{r}, m)$—also called particle or mass point—is a geometric point determined by its position vector $\mathbf{r}$ and associated with a scalar m called the *mass* of the point particle; $\mathbf{r}$ is a vector in Euclidean space related to a Cartesian reference frame. The mass is assumed to be constant and dependent on neither the position of the particle in space nor time.

4. To each couple of point particles $(\mathbf{r}_1, m_1)$ and $(\mathbf{r}_2, m_2)$ a couple of vectors $\mathbf{F}_1$ and $\mathbf{F}_2$ such that $\mathbf{F}_1 = -\mathbf{F}_2 \parallel (\mathbf{r}_1 - \mathbf{r}_2)$ can be assigned.

In this case, it is said that a force $\mathbf{F}$ is applied to a point particle or a force acts on a point particle. Two particles are said to interact with each other if forces satisfying the above condition are assigned to them. In a population of interacting point particles, several forces can be applied to an individual particle. The vector sum of these forces is called the *resultant* or the *resultant force*. (Simultaneously with the category of *force*, this axiom introduces also Newton's third law.)

5. In Euclidean space a Cartesian reference frame can be found and a way of t-parametrization can be selected such that

$$m\ddot{\mathbf{r}} = \mathbf{F}$$

(by a tradition that goes back to Newton, the derivative of a quantity with respect to time is denoted by a dot over the quantity, e.g., $\dot{\mathbf{r}} \equiv \frac{d\mathbf{r}}{dt}$). Such reference frames in $\mathbb{E}^3$ together with a parameter t for which the above equation (Newton's second law) holds are called *inertial reference frames*.

The listed postulates form the axiomatic system of classical mechanics.

Pay attention to the dual character of all basic categories of mechanics. On the one hand, the concepts of force, mass, and acceleration are abstract categories of an axiomatic system, for which Newton's second and third laws serve to define them. On the other hand, these notions appeal to real objects with which human practice deals and on which the axiomatic system of mechanics is realized. "Force" in this case is the measure of physical interaction of bodies; this interaction can be measured by any physical means. "Point particle" is a body of sufficiently small size. A realization of the concept of "Euclidean space" is the space of fixed stars. Light rays are realizations of the category of "straight line." Cartesian coordinate trihedrals, which serve to define the position of any bodies in space, can be associated with fixed stars by means of optical devices. A realization of the category of "time" can be effected by comparing observed processes with one usually involving repeated phases, e.g., the revolution of the Earth about the Sun.

The terms "homogeneity" and "isotropy" of space and "isotropy of time" are beyond the axiomatic scheme, since these are not needed there. These notions belong to the sphere of realization and determine properties of the real physical space and specific ways of measuring time. A space is said to be homogeneous and isotropic if the physical processes depend on neither the site in the space where they take place nor the directions in the space. The homogeneity of time means that it is possible to select a clock by which any moving point particle, unless acted upon by a force, travels equal distances in equal time intervals. Pay attention that the worldly phrases like "time passes" and "time passes at a point" lie outside the axiomatic system of classical mechanics. Time is an argument of a function and nothing else.

Sometimes the statement of Newton's second law is preceded by the statement of Newton's first law: in Euclidean space a Cartesian reference frame can always be found and a way of t-parametrization can be selected such that any point particle continues in the state of rest or uniform motion in a straight line unless acted upon by a force ($\mathbf{F} \equiv \mathbf{0}$). This way of introducing axioms of mechanics contains contradictions.

Indeed, let, in accordance with Newton's first law, an inertial reference frame $\{t, x, y, z\}$ be found. Then any particle which is not acted upon by a force moves uniformly in a straight line. Consider the reference frame $\{t_1, x_2, y_1, z_1\}$ defined as

$$t_1 = t + \omega x, \quad x_1 = x, \quad y_1 = y, \quad z_1 = z.$$

In this system of coordinates also the particle moves uniformly in a straight line, since a linear transformation takes a straight line to a straight line. According to Newton's first law, this reference frame should also be called an inertial system. But this comes into conflict with Newton's second law, which does not hold in this reference frame. Newton's equations of motion in this system become

$$\frac{d^2x_1}{dt_1^2} = \left(1 - \omega\frac{dx_1}{dt_1}\right)^3 F_{x_1},$$

$$\frac{d^2y_1}{dt_1^2} + \omega\left(\frac{dy_1}{dt_1}\frac{d^2x_1}{dt_1^2} - \frac{d^2y_1}{dt_1^2}\frac{dx_1}{dt_1}\right) = \left(1 - \omega\frac{dx_1}{dt_1}\right)^3 F_{y_1},$$

$$\frac{d^2z_1}{dt_1^2} + \omega\left(\frac{dz_1}{dt_1}\frac{d^2x_1}{dt_1^2} - \frac{d^2z_1}{dt_1^2}\frac{dx_1}{dt_1}\right) = \left(1 - \omega\frac{dx_1}{dt_1}\right)^3 F_{z_1}.$$

It can be immediately seen from this set of equations that the particle moves indeed in a straight line for $\mathbf{F} \equiv \mathbf{0}$.

The above reference frame is not artificial and has a clear physical meaning. Imagine a train going eastward. A passenger in the train keeps track of the mile-posts and see the clocks fitted to them and showing local time. The miles at the posts and the time of their clocks form the above reference frame (of course, in regions whose dimensions are small compared with the Earth's radius, so that the Earth's shape could be disregarded). Note that the apparent artificiality in the way of introducing time in this example is in fact a matter of particular internal design of the clocks. For example, one can imagine a clock with a built-in navigation system which constantly identifies the location of the clock and introduces necessary corrections into the readings. If people did not have any other clocks, they would not possibly suspect that some other way of measuring time could exist. In this case, if the passenger in the example with a train had a clock of this sort, then he or she would not need to put the clock forward when getting to a new time zone.

To remove this contradiction a synchronization of clocks at all points of space should be specified. However, the allowance for the synchronization procedure would create serious complications in the formal axiomatic structure of classical mechanics.

It is much simpler to leave the option to choose the way of measuring time at an arbitrary point of space and regard Newton's first law as a mere *necessary* condition for the selected reference frame to be inertial.

Under such an approach, the validity of Newton's second law in a system of coordinates is a necessary and sufficient condition for the inertiality of this system. Then, Newton's first law loses its independence and turn out to be a corollary of Newton's second law.

This is not the case in relativistic mechanics, which is discussed in Section 2.5. Here the prerogative to form the inertiality criterion for a reference frame is transferred to Maxwell's equations of electrodynamics. In relativistic mechanics, Newton's first law turns into an independent condition of inertiality. For this reason, the question how the reference frames which satisfy this condition of inertiality are related to each other is of considerable interest.

2.3. Projective Symmetries of Newton's First Law

Let us discuss in detail the above question for the case of one spatial dimension. The extension of our consideration to three dimensions does not present much difficulties.

A straight line in the (t, x) plane defined by the equation

$$\frac{d^2x}{dt^2} = 0$$

is taken to a straight line by a transformation $(t, x) \to (t', x')$ if the equation of the former is invariant under this transformation.

We seek the unknown group in the form

$$t' = t + \xi(x, t)\tau + \cdots, \quad x' = x + \eta(x, t)\tau + \cdots .$$

Then the above equation must be invariant under a twice prolonged group with generator

$$\overset{(2)}{U} = \xi(t, x)\frac{\partial}{\partial t} + \eta(t, x)\frac{\partial}{\partial x} + \zeta(t, x, v)\frac{\partial}{\partial v} + \delta(t, x, v, w)\frac{\partial}{\partial w},$$

where $v = \dot{x}$ (the velocity) and $w = \ddot{x}$ (the acceleration). The above differential equation determines in the (t, x, v, w) space the hypersurface $w = 0$. The invariance condition of this hypersurface is expressed as

$$\overset{(2)}{U} = 0 \quad \text{on} \quad w = 0,$$

which yields

$$\delta(t, x, v, 0) = 0.$$

In accordance with the theory of prolongation (see Section 1.10), we have

$$\delta = \zeta_t + v\zeta_x + w(\zeta_v - \xi_t) - vw\xi_x,$$
$$\zeta = \eta_t + v(\eta_x - \xi_t) - v^2\xi_x.$$

Hence, the invariance condition becomes

$$\zeta_t + v\zeta_x = \eta_{tt} + v(\eta_{xt} - \xi_{tt}) - v^2\xi_{xt} + v\eta_{tx} + v^2(\eta_{xx} - \xi_{tx}) - v^3\xi_{xx} = 0.$$

Since the functions ξ and η do not depend on v, this relation splits into the following:

$$\eta_{tt} = 0, \quad 2\eta_{xt} - \xi_{tt} = 0, \quad 2\xi_{xt} - \eta_{xx} = 0, \quad \xi_{xx} = 0.$$

Differentiating the second relation with respect to x and the third with respect to t, we obtain

$$2\eta_{xxt} - \xi_{xtt} = 0, \quad 2\xi_{xtt} - \eta_{xxt} = 0.$$

This system of two linear algebraic equations for ξ_{xtt} and η_{xxt} with nonzero determinant has only the trivial solution

$$\xi_{xtt} = 0, \quad \eta_{xxt} = 0.$$

Whence,

$$\xi = a + bt + cx + dtx + et^2,$$
$$\eta = f + gt + hx + ltx + mx^2,$$

where a, b, c, d, e, f, g, h, l, and m are arbitrary constants.

Substituting this solution into the equations $2\eta_{xt} - \xi_{tt} = 0$ and $2\xi_{tx} - \eta_{xx} = 0$ yields $2d - 2m = 0$ and $2l - 2e = 0$.

Thus, the generator of the desired group transforming straight lines in the (t, x) plane into straight lines has the form

$$U = aU_1 + bU_2 + cU_3 + dU_4 + eU_5 + fU_6 + gU_7 + hU_8,$$

where

$$U_1 = \frac{\partial}{\partial t}, \quad U_2 = t\frac{\partial}{\partial t}, \quad U_3 = x\frac{\partial}{\partial t}, \quad U_4 = tx\frac{\partial}{\partial t} + x^2\frac{\partial}{\partial x},$$
$$U_5 = t^2\frac{\partial}{\partial t} + tx\frac{\partial}{\partial x}, \quad U_6 = \frac{\partial}{\partial x}, \quad U_7 = t\frac{\partial}{\partial x}, \quad U_8 = x\frac{\partial}{\partial x},$$

and a, b, c, d, e, f, g, and h are arbitrary constants. The generators $U_1, \ldots, U_8$ form a basis of the projective group (see Section 1.3).

Thus, *all reference frames in which Newton's first law (a necessary condition of inertiality) holds are related to each other by a projective group.*

The same result is valid for the three-dimensional case.

2.4. Newton's Second Law. Galilean Symmetries

Consider now the question of symmetry groups for Newton's equations, i.e., groups which transform inertial reference frames into inertial reference frames. In this case, it is necessary to distinguish between the notions of *invariance* and *covariance* of equations with respect to one or another change of variables and time. The invariance means that the equations have the same form in the original and transformed variables, that is, the transformations under which the equations are invariant transfer their solutions into their solutions again (see Section 1.8). The covariance means that the rules of construction of the equations remain the same. This new notion is specific to mechanics and physics, where various *laws* are used for writing differential equations. An example of such laws is Newton's second law. When we speak about the invariance of Newton's second law under the transition from one inertial reference frame to another, this means that in the new reference frame we must equate the acceleration of the particle times its mass to the same force acting on the particle. In the new variables, however, the physical force can have a different analytical expression.

Thus, the invariance of the rule of construction of differential equations under the transition to the new variables is called the covariance of the differential equations themselves:

covariance of differential equations	≡	invariance of the law according to which they are constructed

Similarly, they speak about the covariance of the differential equations of mechanical systems which are constructed in the form of Lagrange's equations of the second kind under any change of the generalized variables, and about the covariance of equations in the Hamiltonian form under so-called canonical transformations.

In these cases also, they speak about the invariance of the rule of construction of differential equations rather than the invariance of the constructed equations themselves.[2]

It is clear that covariance of equations is not already related to the property of taking solutions of the equations to other solutions of these equations or to the property of having first integrals (see Section 4.4).

Meanwhile, for Newton's equations the question of their covariance can be reduced to the question of their invariance by treating the force as an independent variable of the space of the variables (t, x, F) subjected to transformation, rather than as a function of time, the coordinates and velocities.

Just as in the previous section, we consider this issue for the case of one spatial dimension.

Let the desired group of symmetries of Newton's equation

$$m\ddot{x} = F$$

act in the space of variables (t, x, F) and has the group germ

$$\begin{aligned} t' &= t + \xi(t, x)\tau + \cdots, \\ x' &= x + \eta(t, x)\tau + \cdots, \\ F' &= F + \gamma(F)\tau + \cdots. \end{aligned}$$

The twice prolonged generator of the group is expressed as

$$\overset{(2)}{U} = \xi\frac{\partial}{\partial t} + \eta\frac{\partial}{\partial x} + \zeta\frac{\partial}{\partial v} + \delta\frac{\partial}{\partial w} + \gamma\frac{\partial}{\partial F}.$$

The invariance of the equation $m\ddot{x} = F$ under a group of this sort is the covariance under the group $(t, x) \to (t', x')$. Pay attention to the fact that the function $\gamma(F)$ does not depend on t and x *by definition.*

The condition of invariance of the hypersurface $\omega = mw - F = 0$ is expressed as

$$\overset{(2)}{U}\omega = 0 \quad \text{on} \quad \omega = 0,$$

and hence,

$$\delta\left(t, x, v, \frac{F}{m}\right) - \gamma(F) = 0.$$

Just as previously,

$$\begin{aligned} \delta &= \zeta_t + \zeta_x + w(\zeta_v - \xi_t) - vw\xi_x, \\ \zeta &= \eta_t + v(\eta_x - \xi_t) - v^2\xi_x. \end{aligned}$$

Therefore the invariance condition can be rewritten in the form

$$\zeta_t + v\zeta_x + \frac{F}{m}(\zeta_v - \xi_t) - \frac{F}{m}v\xi_x - \gamma(F) = 0.$$

---

[2] In the textbook V. I. Arnold, *Mathematical Methods of Classical Mechanics*, Springer-Verlag, New York, 1980, Galileo's principle of relativity is treated wrongly, which results from confusion between the notions of invariance and covariance.

Since ζ and ξ do not depend on F, the function $\gamma(F)$ can depend on F only linearly,

$$\gamma = k_1 + k_2 \frac{F}{m}.$$

Hence the above invariance condition splits into two relations

$$\zeta_t + v\zeta_x - k_1 = 0, \quad \zeta_v - \xi_t - v\xi_x - k_2 = 0.$$

Substituting the expression of ζ in terms of ξ and η into these relations yields

$$\eta_{tt} + v(\eta_{xt} - \xi_{tt}) - v^2\xi_{xt} + v\eta_{tx} + v^2(\eta_{xx} - \xi_{tx}) - v^3\xi_{xx} - k_1 = 0,$$
$$\eta_x - \xi_t - 2v\xi_x - \xi_t - v\xi_x - k_2 = 0.$$

Again, taking into account the independence of ξ and η of v, we obtain

$$\eta_{tt} - k_1 = 0, \quad 2\eta_{xt} - \xi_{tt} = 0, \quad 2\xi_{xt} - \eta_{xx} = 0,$$
$$\xi_{xx} = 0, \quad \eta_x - 2\xi_t - k_2 = 0, \quad \xi_x = 0.$$

Just as in the previous section, from the second and third equations it follows that

$$\xi = a + bt + cx + dtx + et^2,$$
$$\eta = f + gt + hx + etx + dx^2.$$

Then the first equation yields $k_1 = 0$ and the last equation implies $c = d = 0$. From the penultimate equation it follows that

$$k_2 = h + et - 2b - 4et.$$

Since k_2 is independent of t, we have $e = 0$. Finally, the desired generator of the group of symmetries of Newton's equation in the one-dimensional case has the form

$$U = aU_1 + bU_2 + fU_3 + gU_4 + hU_5,$$

where a, b, f, g, and h are arbitrary constants and the basis generators $U_1, \dots, U_5$ are expressed as

$$U_1 = \frac{\partial}{\partial t}, \quad U_2 = t\frac{\partial}{\partial t} - 2\frac{F}{m}\frac{\partial}{\partial F}, \quad U_3 = \frac{\partial}{\partial x}, \quad U_4 = t\frac{\partial}{\partial x}, \quad U_5 = x\frac{\partial}{\partial x} + \frac{F}{m}\frac{\partial}{\partial F}.$$

The generator U_1 determines the transformation $t' = t + \tau$, which means a shift of the initial time (a subgroup of translations in time). The generator U_3 corresponds to shifting the origin of the coordinate x: $x' = x + \tau$ (a subgroup of translations in the coordinate).

The generator U_4 determines the motion of the new origin of coordinates with respect to the initial origin of coordinates at a constant velocity of τ: $x' = x + \tau t$. The generator U_2 determines simultaneous scaling of the force F and time, and the generator U_5 simultaneous scaling of the force and the coordinate.

The scaling of the variables is of no interest. The three-parameter subgroup with generators U_1, U_3, and U_4 is referred to as the Galilean group (one-dimensional case).

The determination of the group of symmetries of Newton's equations in the three-dimensional case has no principal distinctions from the above procedure. Omitting the scaling transformation of time, the coordinates, and the force, we list below all the other subgroups of the symmetry group for the equations of classical mechanics:

1. Translation in time: $t' = t + \tau$, $\mathbf{r}' = \mathbf{r}$, $\mathbf{F}' = \mathbf{F}$.
2. Translation in the spatial variables: $t' = t$, $\mathbf{r}' = \mathbf{r} + \mathbf{r}_0$, $\mathbf{F}' = \mathbf{F}$.
3. Translational motion of the coordinate trihedral at a constant velocity: $t' = t$, $\mathbf{r}' = \mathbf{r} + \mathbf{v}t$, $\mathbf{F}' = \mathbf{F}$.
4. Transition to a new basis: $t' = t$, $\mathbf{r}' = A\mathbf{r}$, $\mathbf{F}' = A\mathbf{F}$ ($\det A \neq 0$). If we confine ourselves to the case of orthogonal bases, then the matrix A is also orthogonal, and the transformation means the transition to a turned coordinate system.

What is conventionally called the Galilean group is the subgroup of symmetries of Newton's equations which involves neither scaling of variables nor transitions to nonorthogonal systems of coordinates. This subgroup is obviously a ten-parameter group. The general group of symmetries contains twenty independent parameters.

2.5. Postulates of Relativistic Mechanics

Just as in classical mechanics, the basic axioms of relativistic mechanics must be preceded by the introduction of the categories of space, time, point particle, and force. Unlike the classical case, here the concepts of space and time can be introduced (although not necessarily) as a unified four-dimensional spacetime with the indefinite metric $dl^2 = dt^2 - dx^2 - dy^2 - dz^2$.

In what follows, the basic axioms are:

A. AXIOM OF INERTIAL REFERENCE FRAME (SYSTEM OF COORDINATES). *There exist a reference frame, called an inertial reference frame, in which*

(1) any point particle which is not acted upon by any forces moves in a straight line;

(2) Maxwell's equations in vacuum have the form

$$\operatorname{div}\mathbf{E} = 0, \quad \operatorname{div}\mathbf{B} = 0, \quad \operatorname{curl}\mathbf{E} + \frac{\partial \mathbf{B}}{\partial t} = 0, \quad \operatorname{curl}\mathbf{B} - \frac{1}{c^2}\frac{\partial \mathbf{E}}{\partial t} = 0,$$

where $\mathbf{E}$ *and* $\mathbf{B}$ *are the electric and magnetic components of the electromagnetic field, respectively, and* c *is the speed of light.*

B. AXIOM OF DYNAMICS. *In any inertial systems, the law of motion of a point particle is expressed by a second order differential equation which leads to Newton's second law equations in the limit* $c \to \infty$*:*

$$\lim_{c\to\infty} \Phi(t, \mathbf{r}, \dot{\mathbf{r}}, \ddot{\mathbf{r}}, m, \mathbf{F}, c) = m\ddot{\mathbf{r}} - \mathbf{F},$$

where m *and* $\mathbf{F}$ *are the parameters of the limit equation* $m\ddot{\mathbf{r}} - \mathbf{F} = 0$*, i.e., mass and force.*

In classical mechanics the equations of dynamics (Newton's equations) were postulated and regarded as a criterion of inertiality of the reference frame.

In relativistic mechanics the role of the inertiality criterion is played by the postulated equations of electrodynamics, whereas the equations of dynamics are to be determined.

To this end, it is necessary to find out how the inertial reference frames are related to each other. Since Maxwell's equations must have the same form in any inertial system of coordinates, the transformations relating these systems must be symmetries of these equations.

From Maxwell's equations we successively obtain

$$\operatorname{curl}\dot{\mathbf{B}} = \ddot{\mathbf{E}} \quad \Longrightarrow \quad \operatorname{curl}\operatorname{curl}\mathbf{E} = -\ddot{\mathbf{E}} \quad \Longrightarrow \quad \ddot{\mathbf{E}} - \nabla^2\mathbf{E} = 0.$$

Here, without loss of generality, we set $c = 1$.

Consider for simplicity the one-dimensional case,

$$\frac{\partial^2 \mathbf{E}}{\partial t^2} - \frac{\partial^2 \mathbf{E}}{\partial x^2} = 0.$$

Wherever the non-one-dimensionality is significant, this will be specified.

2.6. Group of Symmetries of Maxwell's Equations

To establish a symmetry group for the above equation, we use the condition $[A, U] = \lambda(t, x)A$ (see Section 1.12), where the operators A and U are expressed as

$$A = \frac{\partial^2}{\partial t^2} - \frac{\partial^2}{\partial x^2}, \quad U = \xi(t,x)\frac{\partial}{\partial t} + \eta(t,x)\frac{\partial}{\partial x}.$$

Substituting these expressions into the above condition yields

$$[A,U] = 2\xi_t\frac{\partial^2}{\partial t^2} + 2(\eta_t - \xi_x)\frac{\partial^2}{\partial t\partial x} - 2\eta_x\frac{\partial^2}{\partial x^2} + (\xi_{tt} - \xi_{xx})\frac{\partial}{\partial t} + (\eta_{tt} - \eta_{xx})\frac{\partial}{\partial x} = \lambda(t,x)\left(\frac{\partial^2}{\partial t^2} - \frac{\partial^2}{\partial x^2}\right).$$

Whence follows the equation for determining the coefficients of U:

$$2\xi_t = \lambda, \quad 2\eta_x = \lambda, \quad \eta_t = \xi_x, \quad \xi_{tt} - \xi_{xx} = 0, \quad \eta_{tt} - \eta_{xx} = 0.$$

The first two equations yield $\xi_t = \eta_x$. From this relation and the third equation it follows that $\xi_{tt} = \eta_{xt}$ and $\xi_{xx} = \eta_{tx}$, which implies the fourth equation. Similarly, from $\xi_t = \eta_x$ and $\eta_t = \xi_x$ we have $\eta_{xx} = \xi_{tx}$ and $\eta_{tt} = \xi_{xt}$, which implies the last equation. Hence, only the first three of the five equations are independent:

$$2\xi_t = \lambda(t,x), \quad 2\eta_x = \lambda(t,x), \quad \eta_t = \xi_x.$$

They form a system of three equations for three unknowns. This system can be reduced to two equations for the two coefficients of the desired generator U:

$$\xi_t = \eta_x, \quad \xi_x = \eta_t.$$

The general solution of this system has the form

$$\xi(t,x) = \varphi(t-x) + \psi(t+x),$$
$$\eta(t,x) = -\varphi(t-x) + \psi(t+x),$$

where $\varphi(\alpha)$ and $\psi(\beta)$ are arbitrary smooth functions.

The presence of arbitrary functions in the general solution of the determining system indicates that the algebra of symmetries of Maxwell's equations is infinite-dimensional. However, we are only interested in symmetries which transform straight lines into straight lines, and hence these symmetries should be sought in the projective group constructed for the (t,x) space in Section 2.3. In other words, the algebra of a group that takes inertial reference frames to inertial reference frames is the intersection of the algebra of symmetries of Maxwell's equations with the algebra of the projective group.

Since the coefficients $\xi(t,x)$ and $\eta(t,x)$ of the projective group are quadratic functions of t and x, the functions $\varphi(\alpha)$ and $\psi(\beta)$ should be restricted to quadratic polynomials in α and β:

$$\varphi(\alpha) = p + q\alpha + r\alpha^2, \quad \psi(\beta) = m + n\beta + l\beta^2.$$

This yields

$$\xi(t,x) = p + m + (q+n)t + (n-q)x + (r+l)t^2 + 2(l-r)tx + (r+l)x^2,$$
$$\eta(t,x) = m - p + (n-q)t + (q+n)x + (l-r)t^2 + 2(l+r)tx + (l-r)x^2.$$

By comparing these relations with those obtained in Section 2.3 for the projective group, one can see that the coefficient of x^2 in the expression for ξ and the coefficient of t^2 in the expression for η must vanish. This leads to the relations $r + l = 0$ and $l - r = 0$, which imply $r = l = 0$.

Finally, the coefficients ξ and η become

$$\xi = C_1 + C_3 t + C_4 x, \quad \eta = C_2 + C_3 x + C_4 t.$$

Hence, the desired generator of the general symmetry group for Maxwell's equations has the form

$$U = C_1 U_1 + C_2 U_2 + C_3 U_3 + C_4 U_4,$$

where C_1, C_2, C_3, and C_4 are arbitrary constants and the generators U_1, U_2, U_3, and U_4 form a basis of a four-dimensional group of symmetries of Maxwell's equations; specifically,

$$U_1 = \frac{\partial}{\partial t} \qquad \text{the generator of translations in time,}$$

$$U_2 = \frac{\partial}{\partial x} \qquad \text{the generator of translations in the spatial variable,}$$

$$U_3 = t\frac{\partial}{\partial t} + x\frac{\partial}{\partial x} \qquad \text{the generator of the similarity subgroup,}$$

$$U_4 = x\frac{\partial}{\partial t} + t\frac{\partial}{\partial x} \qquad \text{the generator of the Lorentz group.}$$

Thus, the only group which differs from translations and uniform extensions and exhausts all inertial reference frames is the Lorentz group.

2.7. Twice Prolonged Lorentz Group

The germ of the Lorentz group in the one-dimensional case has the form

$$t' = t + x\tau, \quad x' = x + \tau t.$$

Calculate the first prolongation:

$$\dot{x}' \equiv \frac{dx'}{dt'} = \frac{\dot{x} + \tau + \cdots}{1 + \dot{x}\tau + \cdots} = \dot{x} + (1 - \dot{x}^2)\tau + \cdots .$$

The second prolongation:

$$\ddot{x}' = \frac{d\dot{x}'}{dt'} = \frac{\ddot{x} - 2\dot{x}\ddot{x}\tau + \cdots}{1 + \dot{x}\tau} = \ddot{x} - 3\dot{x}\ddot{x}\tau + \cdots .$$

Thus, the second prolongation generator is expressed as

$$\overset{(2)}{U}_4 = x\frac{\partial}{\partial t} + t\frac{\partial}{\partial x} + (1 - \dot{x}^2)\frac{\partial}{\partial \dot{x}} - 3\dot{x}\ddot{x}\frac{\partial}{\partial \ddot{x}}.$$

In accordance with the uniqueness theorem for a group (see Section 1.4), this generator can be used to uniquely reconstruct the group; we have the differential equations

$$\frac{dt'}{d\tau} = x', \quad \frac{dx'}{d\tau} = t', \quad \frac{d\dot{x}'}{d\tau} = 1 - \dot{x}'^2, \quad \frac{d\ddot{x}'}{d\tau} = -3\dot{x}'\ddot{x}',$$

whose right-hand sides are the respective components of $\overset{(2)}{U}_4$.

The first two equations yield

$$t' = t\cosh\tau + x\sinh\tau, \quad x' = t\sinh\tau + x\cosh\tau.$$

The third equation implies

$$\ln\sqrt{\frac{1 + \dot{x}'}{1 - \dot{x}'}} = \tau + \ln\sqrt{\frac{1 + \dot{x}}{1 - \dot{x}}}.$$

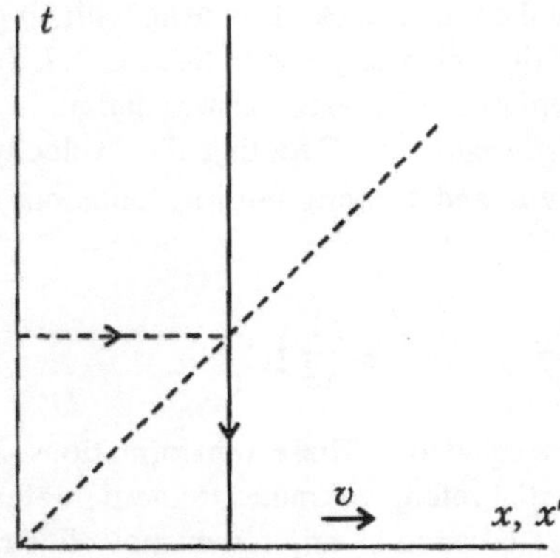

Figure 2.1

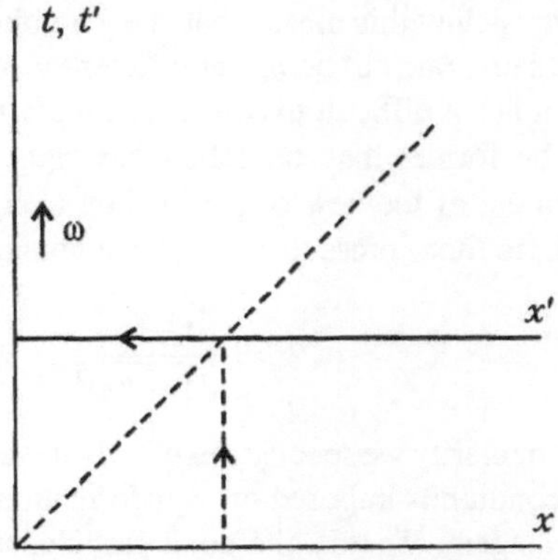

Figure 2.2

This relation is convenient to reveal the mechanical meaning of the parameter τ. Let the reference frame (t', x') move relative to the fixed reference frame (t, x) in the x-direction at a velocity v. Then the velocity $\dot{x}'$ of the origin of the reference frame (t', x') is zero, whereas $\dot{x} = v$. Hence,

$$\tau = \ln\sqrt{\frac{1-v}{1+v}}.$$

Substituting this expression into $\sinh\tau$ and $\cosh\tau$, we arrive at the following equations of the Lorentz group in terms of the velocity v treated as a new group parameter:

$$t' = \frac{t - vx}{\sqrt{1-v^2}}, \quad x' = \frac{x - vt}{\sqrt{1-v^2}}.$$

The remaining two differential equations lead to the relations

$$\dot{x}' = \frac{\dot{x} - v}{1 - \dot{x}v}, \quad \ddot{x}' = \left(\frac{\sqrt{1-v^2}}{1-\dot{x}v}\right)^3 \ddot{x}.$$

These are just the complete expressions for the twice prolonged Lorentz group. The first two relations represent the Lorentz group proper, the third is the law of transformation of velocities, and the fourth is that of accelerations.

Remark 1. The velocity transformation law $\dot{x}' = (\dot{x}-v)/(1-\dot{x}v)$ is often referred to as the law of addition of velocities, which is not quite accurate. The velocity transformation law is actually a mere special case of the velocity addition law where one of the velocities being added (v) does not depend on time. The construction of the velocity addition law where both velocities are time-dependent falls beyond the scope of the Lorentz group, since this requires the consideration of a noninertial reference frame.

Remark 2. In the original Lorentz generator $U_4 = x\frac{\partial}{\partial t} + t\frac{\partial}{\partial x}$, time and the coordinate appear symmetrically. In the above expressions of the group in terms of v, the symmetry is lost. This means that a nonuniqueness situation arose somewhere in the course of the construction of the twice prolonged group and one of the symmetric branches of the theory one chosen.

This occurred in prolonging the group generator and revealing the meaning of the group parameter.

The transition to the reference frame moving along the x-axis means that a new origin of the coordinate x is assigned to each value of time t (see Fig. 2.1).

There is nothing to prevent us from reasoning in a different way. A new origin of time t can be assigned to each value of x (see Fig. 2.2).

Physically this means that at each point of the x-axis there is a clock of its own, with all going at the same rate but being set differently. An example of this sort was given in Section 2.2. In this case, it is not difficult to construct transformations that replace the Lorentz transformation.

The former may be called *conjugate Lorentz transformations.* Denoting the "velocity" of movement of the new origin of time along the t-axis by ω and carrying out manipulations quite similar to those presented above, we obtain

$$t' = \frac{t-\omega x}{\sqrt{1-c^2\omega^2}}, \quad x' = \frac{x-c^2\omega t}{\sqrt{1-c^2\omega^2}} \qquad \left(\omega^2 \ll \frac{1}{c^2}\right).$$

Here for clarity we specify c explicitly not assuming it to be equal to 1. These transformations satisfy both conditions imposed on transformations that take inertial reference frames to inertial reference frames again. That is, the point particles which are not acted upon by any forces move in straight lines in the variables $(t,' x')$ as well, and Maxwell's equations in these variables have the same form as in the original variables (t, x).

The equations of relativistic mechanics, which are the prime objective of our considerations, are also invariant under the conjugate Lorentz transformations.

Remark 3. In principle, the spatial case does not differ much from the one-dimensional case considered. However, the formulas expressing the group transformations, as well as the expressions of the generators, are substantially more cumbersome.

To construct such transformations we can proceed as follows. Suppose the origin of the frame (t', x', y', z') moves relative to the frame (t, x, y, z) at a velocity $\mathbf{v} = \{v_x, v_y, v_z\}$. By virtue of the linearity in the group parameters, the germ of a three-parameter group can be represented as the composition of three one-parameter subgroups (taken in any order). The first subgroup corresponds to the motion along the x-axis:

$$(t, x, y, z) \to (t_1, x_1, y_1, z_1): \quad t_1 = t + x\tau_x, \quad x_1 = x + t\tau_x, \quad y_1 = y, \quad z_1 = z.$$

The second subgroup determines the motion along the y-axis:

$$(t_1, x_1, y_1, z_1) \to (t_2, x_2, y_2, z_2): \quad t_2 = t_1 + y\tau_y, \quad x_2 = x_1, \quad y_2 = y_1 + t\tau_y, \quad z_2 = z_1.$$

The final, third subgroup specifies the motion along the z-axis:

$$(t_2, x_2, y_2, z_2) \to (t', x', y', z'): \quad t' = t_2 + z\tau_z, \quad x' = x_2, \quad y' = y_2, \quad z' = z_2 + t\tau_z.$$

Their composition has the form

$$(t, x, y, z) \to (t', x', y', z'): \quad t' = t + x\tau_x + y\tau_y + z\tau_z, \quad x' = x + t\tau_x, \quad y' = y + t\tau_y, \quad z' = z + t\tau_z.$$

For the sake of convenience, we introduce the vectors $\mathbf{r} = \{x, y, z\}$ and $\boldsymbol{\tau} = \{\tau_x, \tau_y, \tau_z\}$. Then the germ of the group becomes

$$t' = t + \mathbf{r} \cdot \boldsymbol{\tau} + \cdots, \quad \mathbf{r}' = \mathbf{r} + t\boldsymbol{\tau} + \cdots,$$

and the corresponding generator of the Lorentz group in the spatial case acquires the form

$$U = \mathbf{r}\frac{\partial}{\partial t} + t\frac{\partial}{\partial \mathbf{r}}.$$

Calculate the first prolongation of the Lorentz group. We have

$$\dot{\mathbf{r}}' \equiv \frac{d\mathbf{r}'}{dt'} = \frac{d\mathbf{r}'}{dt} \bigg/ \frac{dt'}{dt} = \frac{\dot{\mathbf{r}} + \boldsymbol{\tau} + \cdots}{1 + \dot{\mathbf{r}} \cdot \boldsymbol{\tau} + \cdots} = \dot{\mathbf{r}} + \boldsymbol{\tau} - (\dot{\mathbf{r}} \cdot \boldsymbol{\tau})\dot{\mathbf{r}} + \cdots.$$

For the second prolongation we obtain

$$\ddot{\mathbf{r}}' \equiv \frac{d\dot{\mathbf{r}}'}{dt'} = \frac{d\dot{\mathbf{r}}'}{dt} \Big/ \frac{dt'}{dt} = \frac{\ddot{\mathbf{r}} - (\ddot{\mathbf{r}}\cdot\boldsymbol{\tau})\dot{\mathbf{r}} - (\dot{\mathbf{r}}\cdot\boldsymbol{\tau})\ddot{\mathbf{r}} + \cdots}{1 + \dot{\mathbf{r}}\cdot\boldsymbol{\tau} + \cdots} = \ddot{\mathbf{r}} - (\ddot{\mathbf{r}}\cdot\boldsymbol{\tau})\dot{\mathbf{r}} - 2(\dot{\mathbf{r}}\cdot\boldsymbol{\tau})\ddot{\mathbf{r}} + \cdots .$$

Notice that

$$(\dot{\mathbf{r}}\cdot\boldsymbol{\tau})\dot{\mathbf{r}} = \dot{\mathbf{r}}\dot{\mathbf{r}}^{\mathrm{T}}\boldsymbol{\tau}, \quad (\ddot{\mathbf{r}}\cdot\boldsymbol{\tau})\dot{\mathbf{r}} = \dot{\mathbf{r}}\ddot{\mathbf{r}}^{\mathrm{T}}\boldsymbol{\tau}, \quad (\dot{\mathbf{r}}\cdot\boldsymbol{\tau})\ddot{\mathbf{r}} = \ddot{\mathbf{r}}\dot{\mathbf{r}}^{\mathrm{T}}\boldsymbol{\tau},$$

where, for example,

$$\dot{\mathbf{r}}\dot{\mathbf{r}}^{\mathrm{T}} = \begin{pmatrix} \dot{x} \\ \dot{y} \\ \dot{z} \end{pmatrix} (\dot{x}, \dot{y}, \dot{z}) = \begin{pmatrix} \dot{x}^2 & \dot{x}\dot{y} & \dot{x}\dot{z} \\ \dot{y}\dot{x} & \dot{y}^2 & \dot{y}\dot{z} \\ \dot{z}\dot{x} & \dot{z}\dot{y} & \dot{z}^2 \end{pmatrix}.$$

Hence, the second prolongation generator acquires the form

$$\overset{(2)}{U} = \mathbf{r}\frac{\partial}{\partial t} + t\frac{\partial}{\partial \mathbf{r}} + (E - \dot{\mathbf{r}}\dot{\mathbf{r}}^{\mathrm{T}})\frac{\partial}{\partial \dot{\mathbf{r}}} - (\dot{\mathbf{r}}\ddot{\mathbf{r}}^{\mathrm{T}} + 2\ddot{\mathbf{r}}\dot{\mathbf{r}}^{\mathrm{T}})\frac{\partial}{\partial \ddot{\mathbf{r}}}, \qquad E = \begin{pmatrix} 1 & 0 & 0 \\ 0 & 1 & 0 \\ 0 & 0 & 1 \end{pmatrix}.$$

In order to obtain the expression of the Lorentz group itself in the spatial case, it is simpler to proceed as follows. First, we associate a reference frame (x_*, y_*, z_*) with the velocity $\mathbf{v} = \{v_x, v_y, v_z\}$ so that the x_*-axis points in the direction of $\mathbf{v}$. In this frame, we can take advantage of the results of the one-dimensional case; we have

$$t' = \frac{t - vx_*}{\sqrt{1-v^2}}, \quad x'_* = \frac{x_* - vt}{\sqrt{1-v^2}}, \quad y'_* = y_*, \quad z'_* = z_*.$$

After this we return to the old variables x, y, and z by means of the orthogonal transformation

$$\begin{pmatrix} x \\ y \\ z \end{pmatrix} = \begin{pmatrix} a_{11} & a_{12} & a_{13} \\ a_{21} & a_{22} & a_{23} \\ a_{31} & a_{32} & a_{33} \end{pmatrix} \begin{pmatrix} x_* \\ y_* \\ z_* \end{pmatrix}, \qquad \begin{pmatrix} x_* \\ y_* \\ z_* \end{pmatrix} = \begin{pmatrix} a_{11} & a_{21} & a_{31} \\ a_{12} & a_{22} & a_{32} \\ a_{13} & a_{23} & a_{33} \end{pmatrix} \begin{pmatrix} x \\ y \\ z \end{pmatrix},$$

where

$$a_{11} = \frac{v_1}{v}, \quad a_{21} = \frac{v_2}{v}, \quad a_{31} = \frac{v_3}{v}, \quad v = \sqrt{v_1^2 + v_2^2 + v_3^2}.$$

Substituting

$$x_* = \frac{1}{v}(v_1 x + v_2 y + v_3 z)$$

into $t' = (t - vx_*)/\sqrt{1-v^2}$ yields

$$t' = \frac{t - v_1 x - v_2 y - v_3 z}{\sqrt{1-v^2}} = \frac{t - \mathbf{v}\cdot\mathbf{r}}{\sqrt{1-v^2}}.$$

Performing the orthogonal transformation of the spatial part of the group, we obtain

$$x' = \frac{x_* - vt}{\sqrt{1-v^2}}\frac{v_1}{v} + a_{12}y_* + a_{13}z_*,$$
$$y' = \frac{x_* - vt}{\sqrt{1-v^2}}\frac{v_2}{v} + a_{22}y_* + a_{23}z_*,$$
$$z' = \frac{x_* - vt}{\sqrt{1-v^2}}\frac{v_3}{v} + a_{32}y_* + a_{33}z_*.$$

By expressing x_*, y_*, and y_* on the right-hand sides in terms of x, y, and z, we find that

$$x' = \frac{1}{\sqrt{1-v^2}}\left[\frac{v_1}{v^2}(v_1x+v_2y+v_3z)-v_1t\right]+x-\frac{v_1}{v^2}(v_1x+v_2y+v_3z),$$
$$y' = \frac{1}{\sqrt{1-v^2}}\left[\frac{v_2}{v^2}(v_1x+v_2y+v_3z)-v_2t\right]+y-\frac{v_2}{v^2}(v_1x+v_2y+v_3z),$$
$$z' = \frac{1}{\sqrt{1-v^2}}\left[\frac{v_3}{v^2}(v_1x+v_2y+v_3z)-v_3t\right]+z-\frac{v_3}{v^2}(v_1x+v_2y+v_3z).$$

Rewriting these relations in compact, vector form, we finally arrive at the following expressions of the Lorentz group in the general, three-dimensional case:

$$t' = \frac{t-\mathbf{v}\cdot\mathbf{r}}{\sqrt{1-v^2}},$$
$$\mathbf{r}' = \mathbf{r}+\frac{\mathbf{v}}{\sqrt{1-v^2}}\left(\frac{\mathbf{v}\cdot\mathbf{r}}{v^2}-t\right)-\frac{1}{v^2}\mathbf{v}(\mathbf{v}\cdot\mathbf{r}).$$

2.8. Differential and Integral Invariants of the Lorentz Group

Let us proceed to calculating *invariants of the Lorentz group*. The invariants play an extremely important role. These are quantities that depend on time, the coordinates, the velocities, and the accelerations and their numerical values do not depend on the reference frame in which the invariants are calculated. Hence, it is invariants rather than the choice of the reference frame which characterize the physics of phenomena. If a theory is to be constructed which is invariant under the Lorentz transformations, then this theory must be expressed in terms of their invariants.

The aforesaid is illustrated by the equations of relativistic mechanics which are studied in the current chapter.

The differential invariants G up to the second order inclusive can be found from the equation

$$\overset{(2)}{U} = 0, \quad \text{or} \quad x\frac{\partial G}{\partial t}+t\frac{\partial G}{\partial x}+(1-\dot{x}^2)\frac{\partial G}{\partial \dot{x}}-3\dot{x}\ddot{x}\frac{\partial G}{\partial \ddot{x}} = 0,$$

which is equivalent to the following system of ordinary differential equations:

$$\frac{dt}{x} = \frac{dx}{t} = \frac{d\dot{x}}{1-\dot{x}^2} = \frac{d\ddot{x}}{-3\dot{x}\ddot{x}}.$$

This system has the first integrals

$$G_1 = \sqrt{t^2-x^2}, \quad G_2 = (t+x)\sqrt{\frac{1-\dot{x}}{1+\dot{x}}}, \quad G_3 = \frac{\ddot{x}}{(1-\dot{x}^2)^{3/2}},$$

which are the desired differential invariants up to the second order inclusive. The zeroth order invariant G_1 is referred to as the interval.

To determine the integral invariants J, we proceed from the equation (see Section 1.10)

$$\overset{(2)}{U}J+\dot{x}J = 0, \quad \text{or} \quad x\frac{\partial J}{\partial t}+t\frac{\partial J}{\partial x}+(1-\dot{x}^2)\frac{\partial J}{\partial \dot{x}}-3\dot{x}\ddot{x}\frac{\partial J}{\partial \ddot{x}}+\dot{x}J = 0,$$

which is equivalent to the system

$$\frac{dt}{x} = \frac{dx}{t} = \frac{d\dot{x}}{1-\dot{x}^2} = \frac{d\ddot{x}}{-3\dot{x}\ddot{x}} = \frac{dJ}{-\dot{x}J}.$$

Apart from the above first integrals G_1, G_2, and G_3, this system has one more first integral,

$$G_4 = \frac{J}{\sqrt{1-\dot{x}^2}}.$$

The desired integral invariant J is determined implicitly by the equation

$$N(G_1, G_2, G_3, G_4) = 0 \quad \Longrightarrow \quad J = \sqrt{1-\dot{x}^2}\, M(G_1, G_2, G_3),$$

where N and M are arbitrary functions of their arguments.

Thus, the integral invariants of the Lorentz group have the general form

$$\int_a^b \sqrt{1-\dot{x}^2}\, M(G_1, G_2, G_3)\, dt.$$

In particular, the invariant with $M \equiv 1$,

$$\int_0^t \sqrt{1-\dot{x}^2}\, dt,$$

is referred to as the *intrinsic time* of a particle moving at the velocity $\dot{x}(t)$. Explain the reason for the introduction of this name.

Assume that, in a special case, the point particle moves at a constant velocity $\dot{x} = v$. Let us take the position of the particle to be the origin of the fixed inertial frame (t', x'). The new variables are related to the old ones by the Lorentz transformation,

$$t' = \frac{t - vx}{\sqrt{1-v^2}}, \qquad x' = \frac{x - vt}{\sqrt{1-v^2}}.$$

These transformations allows one, in particular, to find out how time passes in the clock placed at the moving point. For this point we have $x' \equiv 0$, and hence $x = vt$ and $t' = \sqrt{1-v^2}\, t$, which coincides with the value of the integral invariant in the case of constant velocity.

This reasoning justifies the meaning of intrinsic time ascribed to the invariant $\int_0^t \sqrt{1-\dot{x}}\, dt$ but does not prove anything.

Consider, for instance, the invariant

$$\int_0^t \sqrt{1-\frac{\dot{x}^2}{c^2}} \left[1 + \frac{\ddot{x}}{c^2(1-\dot{x}^2/c^2)^{3/2}}\right] dt,$$

in which we return to an arbitrary scale of the light speed c for the purpose of the further reasoning.

This invariant has the property that for $\ddot{x} \equiv 0$ it coincides with the intrinsic time at constant velocity and tends to the "Newtonian" absolute time as $c \to \infty$.

The reference frame associated with a nonuniformly moving point particle is noninertial, and prior to speaking about time at a point in this frame (in particular, at the origin), one should specify the law of transition from an inertial to a noninertial frame. To construct such laws, one has to introduce new postulates. For this reason, the identification of the expression $\int_0^t \sqrt{1-\dot{x}^2}\, dt$ with the intrinsic time of a particle moving at the velocity $\dot{x}(t)$ should be regarded as a new postulate independent of those introduced so far.

2.9. Relativistic Equations of Motion of a Particle

Now that the invariants have been found, we can proceed to deriving the generalized Newton equations. According to Axiom B of Section 2.5, these equations must have the same form in any inertial reference frame. The general form of second order equations invariant under a prescribed group (see Section 1.11) is the following:

$$G_3 = \Phi(G_1, G_2, F, m),$$

where G_1, G_2, and G_3 are the Lorentz group invariants found in the previous section. By virtue of the homogeneity of space (admissibility of a translation group), the function Φ must be independent of G_1 and G_2, since the functions $G_1 = \sqrt{t^2 - x^2}$ and $G_2 = (t + x)\sqrt{(1 - \dot{x})/(1 + \dot{x})}$ are not invariant under translations. Therefore an equation of dynamics can only have the form $G_3 = \Phi(F, m)$. Note, that by virtue of Axiom B, F and m are the parameters of the limit equation as $c \to \infty$, i.e., force and mass in Newton's equation, respectively. With the expression of G_3 found above, we have (with arbitrary scale of the light speed)

$$\frac{\ddot{x}}{[1 - (\dot{x}/c)^2]^{3/2}} = \Phi(F, m).$$

From the condition

$$\lim_{c\to\infty}\left\{\frac{\ddot{x}}{[1 - (\dot{x}/c)^2]^{3/2}} - \Phi(F, m)\right\} = \ddot{x} - \frac{F}{m}$$

it follows that $\Phi(F, m) = F/m$.

It should be noted that the relativistic generalization of Newton's equation admits, as in the classical case, the Lagrangian description. Indeed, rewrite the above equation of relativistic dynamics of a point particle as

$$\frac{d}{dt}\left(\frac{m\dot{x}}{\sqrt{1 - \dot{x}^2}}\right) = F \qquad (c = 1).$$

Assuming, just as in the classical case, that this equation identifies the rate of change of the momentum with the force applied,

$$\frac{dp}{dt} = F,$$

we obtain the momentum in the form

$$p = \frac{m\dot{x}}{\sqrt{1 - \dot{x}^2}}.$$

The momentum is, in turn, the derivative of the Lagrangian function with respect to the velocity,

$$\frac{\partial L}{\partial \dot{x}} = \frac{m\dot{x}}{\sqrt{1 - \dot{x}^2}}.$$

Whence,

$$L = -m\sqrt{1 - \dot{x}^2}.$$

The Hamiltonian action,

$$S = -m\int \sqrt{1 - \dot{x}^2}\, dt,$$

must be an integral invariant of the Lorentz group, which is indeed the case, since it turns out to be proportional to the intrinsic time.

The Lagrangian description of the relativistic dynamics of a point particle is convenient by that it permits one, for example, to perform the transition to curvilinear coordinates or, in particular, to the spatial case.

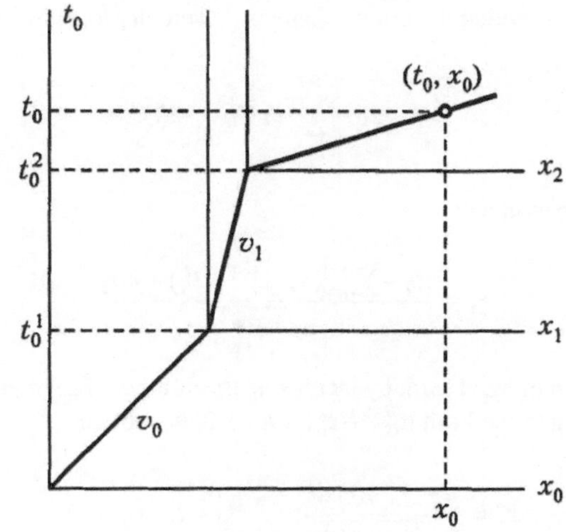

Figure 2.3

The three-dimensional generalization of the Lagrangian function which takes into account its invariance under rotations is given by

$$L = -m\sqrt{1 - \dot{x}^2 - \dot{y}^2 - \dot{z}^2}.$$

Hence, the three-dimensional equations of relativistic mechanics have the from

$$\frac{d}{dt}\left(\frac{m\dot{x}}{\sqrt{1-u^2}}\right) = F_x, \quad \frac{d}{dt}\left(\frac{m\dot{y}}{\sqrt{1-u^2}}\right) = F_y, \quad \frac{d}{dt}\left(\frac{m\dot{z}}{\sqrt{1-u^2}}\right) = F_z,$$

where $u^2 = \dot{x}^2 + \dot{y}^2 + \dot{z}^2$.

When rewriting Lagrange's equations in the new variables, $(t, x, y, z) \to (t', x', y', z')$, the equations must also be reprojected onto the new axes. Therefore the law of transformation of the left-hand sides of the above equations forces the same law upon the transformation of the right-hand sides. For the three-dimensional case, in passing from one inertial reference frame to another, the forces on the right-hand sides alter. For the one-dimensional case this is not so—the force F remains the same in any reference frames and is the conventional Newtonian force. In the one-dimensional case the equation of the dynamics of a particle is invariant under the Lorentz group, whereas in the three-dimensional case the equations are covariant.

2.10. Noninertial Reference Frames

To arrive at the law of transition from an inertial reference frame to a noninertial frame, one may proceed as follows.

Consider a fixed reference frame and denote it by (t_0, x_0). Let a point particle having the coordinate $x_0(t_0)$ move at a piecewise constant velocity relative to this reference frame (see Fig. 2.3).

Introduce a fixed reference frame, (t_i, x_i), at the beginning of each new segment of motion (Fig. 2.3). We will study the motion of the particle at a velocity v_i relative to the original reference frame in the frame (t_i, x_i).

Denote the axes of the moving frame associated with the particle by (τ_i, ξ_i). Obviously,

$$\tau_i = \frac{t_i - v_i x_i}{\sqrt{1 - v_i^2}}, \quad \xi_i = \frac{x_i - v_i t_i}{\sqrt{1 - v_i^2}}.$$

Denote the instant t_0 at which the points changes its velocity to v_i by t_0^i ($t_0^0 = 0$). Then,

$$x_i = x_0 - \sum_{k=0}^{i-1} v_k(t_0^{k+1} - t_0^k),$$

and the expression of ξ_i becomes

$$\xi_i = \frac{x_0 - \sum_{k=0}^{i-1} v_k(t_0^{k+1} - t_0^k) - v_i t_i}{\sqrt{1-v_i^2}}.$$

The numerator contains an integral sum; by increasing the number of segments of piecewise uniform motion and by proceeding to the limit $|t_0^{k+1} - t_0^k| \leq h \to 0$, we obtain

$$x^* = \frac{x - \int_0^t v(t)\,dt}{\sqrt{1-v^2(t)}} \qquad (t_0 \equiv t, \;\; x_0 \equiv x).$$

Similar considerations, under the additional assumption that in the transition from one segment of uniform motion to another the time at all point of space changes continuously, lead to the relation

$$t^* = \int_0^t \sqrt{1-v^2(t)}\,dt - v_0 \frac{x - \int_0^t v(t)\,dt}{\sqrt{1-v^2(t)}}.$$

The above relations for t^* and x^* determine the desired law of transition from an inertial reference frame to a noninertial one, $(t, x) \to (t^*, x^*)$.

If $v = v_0 = \text{const}$, these transformations reduce to the Lorentz transformations. The time passing at the origin of the moving noninertial frame coincides with the intrinsic time introduced above,

$$t^* = \int_0^t \sqrt{1-v^2(t)}\,dt.$$

Chapter 3

Application of Group Methods to Problems of Mechanics

3.1. Perturbation Theory for Configuration Manifolds of Resonant Systems

3.1.1. Statement of the Problem

We consider quasilinear oscillatory systems of the form

$$A\ddot{q} + Bq = \varepsilon Q(t, q, \dot{q}),$$

where A and B are $n \times n$ symmetric positive-definite matrices, q is an n-vector, and the right-hand side is a small perturbation, realized with the aid of the small factor ε. Without loss of generality, A is assumed to be the identity matrix and B a diagonal matrix.

If there is no perturbation ($\varepsilon = 0$), then any trajectory of the system in the configuration space is dense everywhere in a multidimensional parallelepiped, provided that the natural frequencies are incommensurable (nonresonant case). If a resonance takes place, then there exists a subspace in which any trajectory is a closed curve. Such curves are referred to as Lissajous figures. They are unstable with respect to infinitesimal perturbations, specifically, they are deformed to either a point or figures of other shape.

For these unstable trajectories the following three problems are stated in what follows:

(1) describe the evolutions of the figure when perturbations appear;
(2) classify the external perturbations according to a feature of the evolution they generate; and
(3) construct a control that guarantees the stability of the figure.

As will be established later, the manifolds of evolutions of unstable configuration manifolds of resonant systems possess pronounced group properties. Taking this into account makes the analysis of the entire problem more efficient and comprehensive.

We will consider only principal resonances, i.e., those having the least order for the multiplicity prescribed. This corresponds to the presence of multiple frequencies of natural oscillations in the nonperturbed system (with $\varepsilon = 0$). Moreover, the unperturbed system has a proper subspace corresponding to a multiple root in which any trajectories are ellipses lying in some linear two-dimensional manifold. For multiplicity two, this manifold coincides with the proper subspace.

There exist only four basic types of infinitesimal evolutions of the ellipse under perturbations. These are: (i) the change in the principal semiaxes, (ii) the change in the orientation of the ellipse's axes with respect to the subspace basis (let us refer to this type of evolutions as the shape precession), (iii) the change in the velocity of motion of the point particle in the ellipse, and (iv) the transformation of the ellipse into a figure which cannot be obtained by evolutions of the first three types (let us refer to this type as the shape destruction).

The geometric and algebraic properties of these types of evolutions depend fundamentally on the dimension of the proper subspace where these evolutions occur. Therefore these properties must be studied for each value of the multiplicity of a natural frequency separately.

By way of example we consider below the simple cases of double ($k = 2$) and triple ($k = 3$) roots.

3.1.2. The Case of Double Natural Frequency

Here, we focus on the following subsystem of the system $A\ddot{q} + Bq = \varepsilon Q$:

$$\ddot{q}_i + q_i = \varepsilon Q_i(t, q, \dot{q}), \qquad i = 1, 2.$$

Perform in this subsystem the change of the phase variables $(q_1, q_2, \dot{q}_1, \dot{q}_2) \to (x_1, x_2, x_3, x_4)$ according to the formulas

$$q = (E\cos t,\ E\sin t)\,x, \quad \dot{q} = (-E\sin t,\ E\cos t)\,x,$$

where $E = \begin{pmatrix} 1 & 0 \\ 0 & 1 \end{pmatrix}$ is the identity matrix.

This change of variables is the method of variation of Lagrange's constants in which the new variables are constrained by the additional condition

$$(E\cos t,\ E\sin t)\,\dot{x} = 0.$$

Taking into account the equation

$$(-E\sin t,\ E\cos t)\,\dot{x} = \varepsilon Q$$

obtained from the original system, we arrive at a system from which $\dot{x}$ can be expressed explicitly in the form

$$\dot{x} = \varepsilon \begin{pmatrix} E\cos t & -E\sin t \\ E\sin t & E\cos t \end{pmatrix} \begin{pmatrix} 0 \\ Q \end{pmatrix} \equiv \varepsilon \begin{pmatrix} -E\sin t \\ E\cos t \end{pmatrix} \begin{pmatrix} Q_1 \\ Q_2 \end{pmatrix}.$$

On the right-hand side, the system obtained contains the small factor ε. In accordance with the theory of perturbations, we average the system with respect to the explicit time to obtain

$$\dot{x} = \varepsilon X(x) \equiv \varepsilon \frac{1}{2\pi} \int_0^{2\pi} \begin{pmatrix} -E\sin t \\ E\cos t \end{pmatrix} \begin{pmatrix} Q_1 \\ Q_2 \end{pmatrix} dt.$$

Note that the subsystem $\ddot{q}_i + q_i = \varepsilon Q_i$ ($i = 1, 2$) is coupled with the other equations of the system $A\ddot{q} + Bq = \varepsilon Q$. However, if the right-hand sides $Q_i(t, q, \dot{q})$ are linear in all variables q and $\dot{q}$, then the averaged equations depend only on x.

If $\varepsilon = 0$, then $x \equiv \text{const}$ is a solution of the system $\dot{x} = \varepsilon X(x)$ and the trajectory $q = (E\cos t,\ E\sin t)\,x$ is an ellipse in the configuration space of the original system. Thus, each point of the phase space x determines a particular elliptic trajectory in the q-space.

The most important characteristics of the motion of a point particle in an elliptic trajectory are the oscillation energy, the angular momentum, and the angle of inclination of the major semiaxis to the q_1-axis.

Twice the value of the oscillation energy coincides with the square of the norm of x: $q_1^2 + q_2^2 + \dot{q}_1^2 + \dot{q}_2^2 = x_1^2 + x_2^2 + x_3^2 + x_4^2$. The angular momentum relative to the origin, $K = q_1\dot{q}_2 - q_2\dot{q}_1$, is proportional to the area of the ellipse and, for this reason, is alternatively called the quadrature. In terms of x, the angular momentum is expressed as $K = x_1x_4 - x_2x_3$.

To establish how the ellipse orientation is expressed in terms of x, we subject the vector q to the orthogonal transformation that brings the major semiaxis into coincidence with the q_1-axis:

$$\begin{pmatrix} \cos\theta & \sin\theta \\ -\sin\theta & \cos\theta \end{pmatrix} \begin{pmatrix} x_1\cos t + x_3\sin t \\ x_2\cos t + x_4\sin t \end{pmatrix} = \begin{pmatrix} a\cos(t+\tau) \\ b\sin(t+\tau) \end{pmatrix}.$$

Here, the unknowns are the angle θ, the major and minor semiaxes of the ellipse, a and b, and the time phase shift τ. Matching the coefficient of $\cos t$ and $\sin t$, we obtain four equation for these unknowns. The quantities a and b can easily be eliminated from these equations. After that, τ must be eliminated from two resulting equations. The remaining equation yields

$$\cos 2\theta = \frac{K_1}{\sqrt{K_1^2 + K_2^2}}, \quad \sin 2\theta = \frac{K_2}{\sqrt{K_1^2 + K_2^2}}.$$

Along with x^2 and K, the following two quadratic forms arise:

$$K_1 = \tfrac{1}{2}(x_1^2 - x_2^2 + x_3^2 - x_4^2), \quad K_2 = x_1 x_2 + x_3 x_4.$$

Among all elliptic trajectories, there are degenerate trajectories of two types. These are either rectilinear segments or circles.

In the case of a rectilinear segment, the quadrature vanishes,

$$K = x_1 x_4 - x_2 x_3 = 0.$$

In the case of a circle, the variables x_i are related by the constraints $x_1 = x_4$ and $x_2 = -x_3$ or $x_1 = -x_4$ and $x_2 = x_3$, which transform the change of variables $q = (E \cos t,\ E \sin t)\,x$ into a rotation group $(x_1, x_2) \to (q_1, q_2)$.

The two constraints can be combined into a single condition,

$$L = (x_1 \pm x_4)^2 + (x_2 \mp x_3)^2 \equiv x^2 \pm 2K = 0.$$

3.1.3. The Manifold of Degenerate Forms. Local Evolution Basis

In the phase space x the equation $K \equiv x_1 x_4 - x_2 x_3 = 0$ determines a three-dimensional conical surface to each point of which there correspond rectilinear oscillations in the configuration space q. The equation $L = x^2 \pm 2K = 0$ consists of two components: $L_+ = x^2 + 2K = 0$ and $L_- = x^2 - 2K = 0$. To each point of the manifold L there correspond motions in circles in the q-space. The other points of the x-space which belong to neither the cone $K = 0$ nor the manifold L define elliptic trajectories in the q-space.

Any trajectory of the system $\dot{x} = X(x)$ determines the evolution of an elliptic trajectory in the q-space. This evolution can be resolved into four elementary evolutions: (i) the change in the oscillation amplitude $x^2 = x_1^2 + x_2^2 + x_3^2 + x_4^2$, (ii) the change in the quadrature $K = x_1 x_4 - x_2 x_3$, (iii) the change in the orientation of the major semiaxis of the ellipse, which is determined by the angle θ with respect to the q_1-axis, and (iv) the change in the velocity of motion of the particle in an elliptic trajectory. The amplitude squared is proportional to the oscillation energy, the quadrature is proportional to the area of the ellipse described by the particle and equal to the angular momentum of the particle about the center, the change of the angle θ is called the precession of the elliptic shape, and the velocity of motion of the particle in the ellipse determines the oscillation frequency.

At each point of the phase space x, to each elementary evolution there corresponds a certain direction of the change of x.

To construct the direction vector determining the change in the amplitude, we apply a similarity group to the vector q:

$$q' = (1+\mu)q = \begin{pmatrix} (1+\mu)x_1 \cos t + (1+\mu)x_3 \sin t \\ (1+\mu)x_2 \cos t + (1+\mu)x_4 \sin t \end{pmatrix} = \begin{pmatrix} x_1' \cos t + x_3' \sin t \\ x_2' \cos t + x_4' \sin t \end{pmatrix}.$$

Whence,

$$x_1' = (1+\mu)x_1, \quad x_2' = (1+\mu)x_2, \quad x_3' = (1+\mu)x_3, \quad x_4' = (1+\mu)x_4.$$

This means that a similarity group in the q-space induces a similarity group in the phase space. Hence, the desired direction is determined by the vector

$$e_1 = \left.\frac{dx'}{d\mu}\right|_{\mu=0} = (x_1, x_2, x_3, x_4)^{\mathrm{T}}.$$

The direction of the change of the quadrature is determined by the normal to the manifold $K = x_1x_4 - x_2x_3$:

$$e_2 = \frac{dK}{dx} = (x_4, -x_3, -x_2, x_1)^{\mathrm{T}}.$$

If $x \in K = 0$, then this direction is the direction in which the destruction of the rectilinear shape of oscillations is the fastest.

To establish the direction determining the precession, we apply an orthogonal group $q \to q'$: $q' = \mathcal{A}q$, where $\mathcal{A}$ is a rotation matrix of the form

$$\mathcal{A} = \begin{pmatrix} \cos\alpha & \sin\alpha \\ -\sin\alpha & \cos\alpha \end{pmatrix}.$$

An orthogonal group acting in the q-space induces an orthogonal group $x \to x'$ in the phase space x; the latter group can be calculated as follows:

$$\mathcal{A}q = \mathcal{A}(E\cos t, E\sin t)\,x = \begin{pmatrix} \cos\alpha & \sin\alpha \\ -\sin\alpha & \cos\alpha \end{pmatrix} \begin{pmatrix} x_1\cos t + x_3\sin t \\ x_2\cos t + x_4\sin t \end{pmatrix} = \begin{pmatrix} x_1'\cos t + x_3'\sin t \\ x_2'\cos t + x_4'\sin t \end{pmatrix}.$$

Whence we obtain the relationship between x' and x:

$$\begin{aligned} x_1' &= x_1\cos\alpha + x_2\sin\alpha, \\ x_2' &= -x_1\sin\alpha + x_2\cos\alpha, \\ x_3' &= x_3\cos\alpha + x_4\sin\alpha, \\ x_4' &= -x_3\sin\alpha + x_4\cos\alpha. \end{aligned}$$

This orthogonal group determines in the x-space the vector field

$$e_3 = \left.\frac{dx'}{d\alpha}\right|_{\alpha=0} = (x_2, -x_1, x_4, -x_3)^{\mathrm{T}}.$$

To construct the direction determining the change of the frequency, one should apply the group of translations in time, $t \to t' = t + \tau$, to $q = (E\cos t, E\sin t)\,x$. In doing so, we obtain

$$\begin{aligned} x_1' &= x_1\cos\tau + x_3\sin\tau, \\ x_2' &= x_2\cos\tau + x_3\sin\tau, \\ x_3' &= -x_1\sin\tau + x_4\cos\tau, \\ x_4' &= -x_2\sin\tau + x_4\cos\tau. \end{aligned}$$

Hence, a group of translations in time induces in the phase space x an orthogonal group determining the vector field

$$e_4 = \left.\frac{dx'}{d\tau}\right|_{\tau=0} = (x_3, x_4, -x_1, x_2)^{\mathrm{T}}.$$

Thus, the basis of local infinitesimal evolutions is as follows:

$$e_1 = \begin{pmatrix} x_1 \\ x_2 \\ x_3 \\ x_4 \end{pmatrix}, \quad e_2 = \begin{pmatrix} x_4 \\ -x_3 \\ -x_2 \\ x_1 \end{pmatrix}, \quad e_3 = \begin{pmatrix} x_2 \\ -x_1 \\ x_4 \\ -x_3 \end{pmatrix}, \quad e_4 = \begin{pmatrix} x_3 \\ x_4 \\ -x_1 \\ -x_2 \end{pmatrix}.$$

Recall that e_1, e_2, e_3, and e_4 are the directions of change of the amplitude, quadrature, precession, and frequency, respectively.

3.1.4. Algebra of Local Evolutions

We will point out the most important properties of the basis just constructed. Consider the matrix composed of the basis vectors:

$$A = (e_1, e_2, e_3, e_4).$$

PROPERTY 1. The determinant of the matrix A is expressed as

$$\Delta \equiv \det A = (x_1^2 - x_2^2 + x_3^2 - x_4^2)^2 + 4(x_1x_2 + x_3x_4)^2 = 4(K_1^2 + K_2^2).$$

PROPERTY 2. The matrix of inner products of the basis vectors (Gramian matrix) has the form

$$\mathcal{G} = A^{\mathrm{T}}A = \begin{pmatrix} (e_1 \cdot e_1) & \cdots & (e_1 \cdot e_4) \\ \vdots & \ddots & \vdots \\ (e_4 \cdot e_1) & \cdots & (e_4 \cdot e_4) \end{pmatrix} = \begin{pmatrix} x^2 & 2K & 0 & 0 \\ 2K & x^2 & 0 & 0 \\ 0 & 0 & x^2 & -2K \\ 0 & 0 & -2K & x^2 \end{pmatrix}.$$

The determinant of the Gramian matrix is equal to

$$\det \mathcal{G} = \Delta^2 = (x^4 - 4K^2)^2 = (x^2 - 2K)^2(x^2 + 2K)^2 = L_-^2L_+^2.$$

As follows from the form of the Gramian matrix, the evolution basis is orthogonal on the cone $K = 0$ and degenerate on $L = 0$, i.e.,

$$\det \mathcal{G}\big|_{K=0} = x^8, \quad \det \mathcal{G}\big|_{L=0} = 0.$$

PROPERTY 3. The evolution basis in question generates a four-parameter commutative (abelian) Lie group whose representation as a group of automorphisms in $\mathbb{R}^4$ coincides with the maximal commutative subgroup of the GL(4, $\mathbb{R}$) group.

Indeed, the vector field $e_i(x)$ generates four one-parameter groups with

$$U_1 = x_1\frac{\partial}{\partial x_1} + x_2\frac{\partial}{\partial x_2} + x_3\frac{\partial}{\partial x_3} + x_4\frac{\partial}{\partial x_4}, \quad U_2 = x_4\frac{\partial}{\partial x_1} - x_3\frac{\partial}{\partial x_2} - x_2\frac{\partial}{\partial x_3} + x_1\frac{\partial}{\partial x_4},$$

$$U_3 = x_2\frac{\partial}{\partial x_1} - x_1\frac{\partial}{\partial x_2} + x_4\frac{\partial}{\partial x_3} - x_3\frac{\partial}{\partial x_4}, \quad U_4 = x_3\frac{\partial}{\partial x_1} + x_4\frac{\partial}{\partial x_2} - x_1\frac{\partial}{\partial x_3} - x_2\frac{\partial}{\partial x_4}.$$

Their commutators are zero, $[U_i, U_k] = 0$, for any i and k.

PROPERTY 4 (GLOBAL EVOLUTIONS OF THE CONE $K = 0$ ALONG THE VECTOR FIELDS e_i). Calculations show that

$$U_1K = 2K, \quad U_2K = x^2, \quad U_3K = 0, \quad U_4K = 0.$$

This means that K is an invariant of two groups of rotations, U_3 and U_4, and an invariant manifold of the similarity group U_1.

For the group U_2, which determines the evolution of the destruction of the rectilinear oscillation shape, we have

$$K' = e^{\pm U_2\tau}K = \left(1 \pm \tau U_2 + \tfrac{1}{2!}\tau^2U_2^2 \pm \cdots\right) = K\cosh 2\tau \pm \tfrac{1}{2}x^2\sinh 2\tau,$$

since $U_2x^2 = 4K$. As $\tau \to \infty$, the manifold

$$K\sinh 2\tau \pm \tfrac{1}{2}x^2\sinh 2\tau = 0$$

tends to the manifold $2K \pm x^2 = 0$, that is, to the manifold of circular trajectories $L = 0$.

PROPERTY 5. The basis $\{e_i\}$ is nonholonomic. Out of the four vector fields, only two fields are potential,

$$e_1 = \frac{dS}{dx}, \quad e_2\frac{dK}{dx},$$

where the equation $S = \frac{1}{2}(x^2 - 1) = \frac{1}{2}(x_1^2 + x_2^2 + x_3^2 + x_4^2 - 1)$ defines a sphere of radius 1, $S = 0$.

3.1.5. Classification of Perturbations

It is reasonable to begin the analysis of the perturbations Q_i acting in the system $\ddot{q}_i + q_i = \varepsilon Q_i$ $(i = 1, 2)$ with those linear in the coordinates and velocities:

$$\begin{pmatrix} Q_1 \\ Q_2 \end{pmatrix} = P \begin{pmatrix} q_1 \\ q_2 \end{pmatrix} + R \begin{pmatrix} \dot{q}_1 \\ \dot{q}_2 \end{pmatrix}.$$

The arbitrary matrix P of positional forces and the arbitrary matrix R of velocity forces can be uniquely decomposed into a symmetric and a skew-symmetric component. In turn, the symmetric components of these matrices can be uniquely decomposed into a scalar matrix and a traceless matrix. Thus, we have

$$P = C + H + N, \quad R = D + G + \Gamma,$$

where

$$C = \begin{pmatrix} c & 0 \\ 0 & c \end{pmatrix}, \quad H = h \begin{pmatrix} \cos 2\alpha & \sin 2\alpha \\ \sin 2\alpha & -\cos 2\alpha \end{pmatrix}, \quad N = \begin{pmatrix} 0 & n \\ -n & 0 \end{pmatrix},$$
$$D = \begin{pmatrix} d & 0 \\ 0 & d \end{pmatrix}, \quad G = g \begin{pmatrix} \cos 2\beta & \sin 2\beta \\ \sin 2\beta & -\cos 2\beta \end{pmatrix} \quad \Gamma = \begin{pmatrix} 0 & \gamma \\ -\gamma & 0 \end{pmatrix}.$$

The above six types of forces are termed as follows:

- Cq potential forces of the spherical type;
- Hq potential forces of the hyperbolic type;
- Nq several terms are used to name these forces: circular forces, pseudogyroscopic forces, proper nonconservative forces, and radial correction forces;
- $D\dot{q}$ dissipative forces of the spherical type, provided that $\dot{q}^{\mathrm{T}} D\dot{q} \le 0$;
- $G\dot{q}$ velocity forces of the hyperbolic type;
- $\Gamma\dot{q}$ gyroscopic forces.

The symmetric matrices H and G with zero trace are called deviators.

The coefficients h and g determine the norms of the deviators of hyperbolic forces, and the angles α and β identify the orientation of the principal stiffness and damping axes relative to the q_1- and q_2-axes, respectively.

Let us transform the listed forces into the respective right-hand sides of the system $\dot{x} = \varepsilon X(x)$ by using the formula

$$X(x) = \frac{1}{2\pi} \int_0^{2\pi} \begin{pmatrix} -E \sin t \\ E \cos t \end{pmatrix} \begin{pmatrix} Q_1 \\ Q_2 \end{pmatrix} dt, \qquad E = \begin{pmatrix} 1 & 0 \\ 0 & 1 \end{pmatrix}.$$

Taking into account the change of variables $q = (E\cos t, E\sin t)x$, $\dot{q} = (-E\sin t, E\cos t)x$, we obtain

$$Cq\colon\ X(x) = \tfrac{1}{2}c(-x_3, -x_4, x_1, x_2)^{\mathrm{T}},$$
$$Hq\colon\ X(x) = \tfrac{1}{2}h\big[(-x_3, x_4, x_1, -x_2)^{\mathrm{T}} \cos 2\alpha + (-x_4, -x_3, x_2, x_1)^{\mathrm{T}} \sin 2\alpha\big],$$
$$Nq\colon\ X(x) = \tfrac{1}{2}n(-x_4, x_3, x_2, -x_1)^{\mathrm{T}},$$
$$D\dot{q}\colon\ X(x) = \tfrac{1}{2}d(x_1, x_2, x_3, x_4)^{\mathrm{T}},$$
$$G\dot{q}\colon\ X(x) = \tfrac{1}{2}g\big[(x_1, -x_2, x_3, -x_4)^{\mathrm{T}} \cos 2\beta + (x_2, x_1, x_4, x_3)^{\mathrm{T}} \sin 2\beta\big],$$
$$\Gamma\dot{q}\colon\ X(x) = \tfrac{1}{2}\gamma(x_2, -x_1, x_4, -x_3)^{\mathrm{T}}.$$

Now, to find out what oscillation shape evolution is brought about by each of these forces, it is necessary to project them onto the basis vectors.

If we are interested in the local evolution of a rectilinear oscillation shape, then by virtue of the orthogonality of the basis $\{e_i\}$ on the cone $K = 0$, it suffices to calculate only the inner products of the above vectors $X(x)$ and the vectors e_i.

Outside the cone $K = 0$, the basis $\{e_i\}$ is not orthogonal, and hence we have to use the formula $X(x) = \frac{1}{\|x\|} AX_e(x)$ to project the forces on the basis $\{e_i\}$. We have $X_e = \|x\| A^{-1} X$.

The matrix A here is that specified above; its columns are the vectors e_i: $A = (e_1, e_2, e_3, e_4)$. To calculate the inverse of A, we use the relation $A^{\mathrm{T}}A = \mathcal{G}$. It follows that $A^{-1} = \mathcal{G}^{-1}A^{\mathrm{T}}$. The inverse of the Gramian matrix can be easily calculated:

$$\mathcal{G}^{-1} = \frac{1}{\Delta}\begin{pmatrix} x^2 & -2K & 0 & 0 \\ -2K & x^2 & 0 & 0 \\ 0 & 0 & x^2 & 2K \\ 0 & 0 & 2K & x^2 \end{pmatrix} \qquad (\Delta = x^4 - 4K^2).$$

The results of this projecting are presented in the table below, with X_e divided by $\|x\|$:

	Cq	Hq	Nq	$D\dot{q}$	$G\dot{q}$	$\Gamma\dot{q}$
e_1	0	$\frac{hK}{\sqrt{\Delta}}\sin[2(\theta-\alpha)]$	0	$\frac{\alpha}{2}$	$\frac{gx^2}{2\sqrt{\Delta}}\cos[2(\theta-\beta)]$	0
e_2	0	$-\frac{hx^2}{2\sqrt{\Delta}}\sin[2(\theta-\alpha)]$	$-\frac{n}{2}$	0	$-\frac{gK}{\sqrt{\Delta}}\cos[2(\theta-\beta)]$	0
e_3	0	$-\frac{hK}{\sqrt{\Delta}}\cos[2(\theta-\alpha)]$	0	0	$\frac{gx^2}{2\sqrt{\Delta}}\sin[2(\theta-\beta)]$	$\frac{\gamma}{2}$
e_4	$-\frac{c}{2}$	$-\frac{hx^2}{2\sqrt{\Delta}}\cos[2(\theta-\alpha)]$	0	0	$\frac{gK}{2\sqrt{\Delta}}\sin[2(\theta-\beta)]$	0

As follows from the table, four of the six types of forces lead to pure oscillation shape evolutions. Specifically, the dissipative forces $D\dot{q}$ give rise to the change in the amplitude alone, the circular forces lead only to the quadrature accumulation, the gyroscopic forces bring about only the oscillation shape precession, and the potential forces of the spherical type produce only changes in the frequency.

If one deals with a problem of control of the oscillation shape, then it is the forces of these four types which should be taken to control the corresponding evolution shapes.

In general, the hyperbolic forces Hq and $G\dot{q}$ bring about shape evolutions of all types. However, for a rectilinear shape ($K = 0$) the positional hyperbolic forces lead to the destruction of this shape and a change in the frequency, whereas the velocity hyperbolic forces cause the shape precession and a change in the amplitude.

The intensity of the evolution change in the case of the hyperbolic forces depends on the orientation of the shape relative to the axes of the corresponding deviator. For example, if a rectilinear oscillation shape is aligned with one of the principal stiffness axes ($\theta = \alpha$), then such a perturbation results only in a change of the frequency.

3.1.6. The Problem of Stabilization of the Oscillation Shape

Consider the problem of stabilization of a rectilinear oscillation shape. For instance, such a problem can arise for a Foucault pendulum. In this case, the rectilinear oscillation shape is the reference manifold relative to which the rotation of the pendulum base in an inertial space is measured. It is required that the oscillation shape is not destroyed, i.e., does not change in amplitude. The precession must be determined by only the rotation of the base.

By the introduction of amplitude and quadrature feedbacks, these requirements can be satisfied. The law of formation of these feedbacks must satisfy the following conditions:

(1) asymptotic stability of the rectilinear shape (reference manifold),
(2) invariance under the rotation group in the (q_1, q_2) plane,
(3) absence of interference of different control channels,
(4) invariance under the phase flow of the unperturbed system, $\ddot{q} + q = 0$, and
(5) invariance under the translation in time.

The first two requirements are dictated by the purpose of using the rectilinear oscillation shape. The remaining three requirements determine the quality of the control. If, for instance, the amplitude or quadrature control causes a precession of the shape (interference of channels), then this leads to an error in measuring the base rotation.

The invariance under the phase flow of the unperturbed system is convenient in that the control depends on the slow rather than fast variables.

Finally, the invariance with respect to time means that the device properties do not change with time.

In order that the control satisfy the condition of invariance under the rotation group, the control must only depend on the invariants of the group. The rotation group has the following three independent invariants of the first order:

$$I_0 = q_1^2 + q_2^2, \quad I_1 = \dot{q}_1^2 + \dot{q}_2^2, \quad I_3 = q_1\dot{q}_2 - q_2\dot{q}_1.$$

The invariance under the phase flow of the unperturbed system requires that these invariants be simultaneously first integrals of the system $\ddot{q} + q = 0$. The unperturbed system has also three independent first integrals, specifically,

$$E = \tfrac{1}{2}(I_0 + I_1), \quad K = I_2, \quad L = q_1q_2 + \dot{q}_1\dot{q}_2.$$

Thus, the first two integrals, the energy E and the quadrature K, satisfy the condition of invariance under the rotation group.

In accordance with the table presented above, one can use only the forces $D\dot{q}$ to control the amplitude, provided that there is no channel interference. Denote by E_0 the energy corresponding to the oscillation amplitude being stabilized.

To implement the simple amplitude feedback, it suffices to set the coefficient d occurring in the scalar matrix D equal to $\varepsilon(E_0 - E)$.

To control the quadrature, without bringing about any other shape evolutions, it is necessary to utilize the circular forces. In the simplest case, the coefficient n can be set equal to μK.

Thus, we obtain the following equations of the controlled Foucault pendulum:

$$\ddot{q}_1 + q_1 = \varepsilon(E - E_0)\dot{q}_1 + \mu K q_2,$$
$$\ddot{q}_2 + q_2 = \varepsilon(E - E_0)\dot{q}_2 - \mu K q_1,$$

where

$$E = \tfrac{1}{2}(q_1^2 + q_2^2 + \dot{q}_1^2 + \dot{q}_2^2), \quad K = q_1\dot{q}_2 - q_2\dot{q}_1.$$

This system of nonlinear equations is a two-dimensional analogue of the Van der Pol equation. To the above equations in terms of q there correspond the following equations in terms of x:

$$\dot{x} = -\varepsilon S e_1 - \mu K e_2.$$

Without loss of generality, we set $E_0 = \frac{1}{2}$.

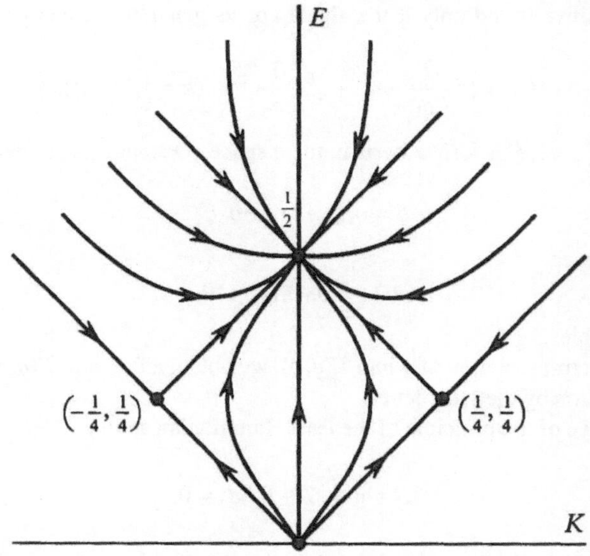

Figure 3.1

This fourth order system can easily be reduced to a second order system for S and K. Indeed, we have

$$\dot{S} = \frac{dS}{dx} \cdot \dot{x} = -\varepsilon S(e_1 \cdot e_1) - \mu K(e_1 \cdot e_2) \qquad \left(\frac{dS}{dx} = e_1\right).$$

Similarly, by differentiating K, we obtain

$$\dot{K} = \frac{dK}{dx} \cdot \dot{x} = -\varepsilon S(e_2 \cdot e_1) - \mu K(e_2 \cdot e_2) \qquad \left(\frac{dK}{dx} = e_2\right).$$

Previously, we found that $(e_1 \cdot e_2) = 2K$ (see the Gramian matrix). In addition, $(e_1 \cdot e_1) = (e_2 \cdot e_2) = x^2 = 2S + 1$. As a result, we obtain

$$\dot{S} = -\varepsilon S(2S + 1) - 2\mu K^2,$$
$$\dot{K} = -2\varepsilon SK - \mu K(2S + 1).$$

The rectilinear oscillations of fixed amplitude in the original system are asymptotically stable if the point $S = K = 0$ of the system obtained is asymptotically stable. For this to be the case, it suffices that $\varepsilon > 0$ and $\mu > 0$. Figure 3.1 shows the phase portrait of this system for $\varepsilon = \mu$ in terms of K and E.

3.2. Poincaré's Equation on Lie Algebras

Consider a mechanical system with kinetic energy T and potential energy U. Let $q_1, \ldots, q_n$ be the local (generalized) coordinates of this system and let $\dot{q}_1, \ldots, \dot{q}_n$ be the generalized velocities.

Assume that a Lie group with respect to the generalized coordinates acts in the domain of definition of the system. Assume also that this Lie group possesses the transitivity property (a group is said be transitive if for any two points of the space there exists a transformation from the group which transfers one point to the other).

A group is transitive if and only if the algebra of its generators contains n linearly unrelated generators,

$$U_k = \xi_1^k \frac{\partial}{\partial q_1} + \cdots + \xi_n^k \frac{\partial}{\partial q_n} \qquad (k = 1, \ldots, n).$$

The vectors $\xi^k = \{\xi_1^k, \ldots, \xi_n^k\}$ form a basis in the q space. Decompose the generalized velocity in this basis to obtain

$$\dot{q} = \eta_1 \xi^1 + \cdots + \eta_n \xi^n.$$

Let us take the variables

$$\{q_1, \ldots, q_n, \eta_1, \ldots, \eta_n\}$$

to be the phase variables.

Substituting the representation of $\dot{q}$ into $T(q, \dot{q})$, we obtain a function $T(q, \eta)$ (for simplicity we denote the new function by the same letter).

We take advantage of the principle of the least Hamiltonian action:

$$\delta S = \delta \int (T - U)\, dt = 0.$$

Calculate the variations of the kinetic and potential energies; we have

$$\delta T = \sum_{i=1}^{n} \frac{\partial T}{\partial \eta_i} \delta \eta_i + \sum_{i=1}^{n} \frac{\partial T}{\partial q_i} \delta q_i, \qquad \delta U = \sum_{i=1}^{n} \frac{\partial U}{\partial q_i} \delta q_i.$$

Projecting the variation δq_i onto the basis $\xi^1, \ldots, \xi^n$, we find that

$$\delta q_i = \delta\omega_1\, \xi_i^1 + \cdots + \delta\omega_n\, \xi_i^n, \qquad \delta q = \delta\omega_1\, \xi^1 + \cdots + \delta\omega_n\, \xi^n.$$

The variation $\delta\eta$ must also be expressed in terms of $\delta\omega$. To this end, we differentiate the last relation to obtain

$$\delta\dot{q} = \delta\dot{\omega}_1\, \xi^1 + \cdots + \delta\dot{\omega}_n\, \xi^n + \delta\omega_1 \left(\frac{\partial \xi^1}{\partial q_1} \dot{q}_1 + \cdots + \frac{\partial \xi^1}{\partial q_n} \dot{q}_n \right) + \cdots + \delta\omega_n \left(\frac{\partial \xi^n}{\partial q_1} \dot{q}_1 + \cdots + \frac{\partial \xi^n}{\partial q_n} \dot{q}_n \right).$$

Replacing $\dot{q}$ in this relation by η yields

$$\delta\eta = \delta\dot{\omega}_1\, \xi^1 + \cdots + \delta\dot{\omega}_n\, \xi^n + \delta\omega_1 \sum_{k=1}^{n} \frac{\partial \xi^1}{\partial q_k} \sum_{i=1}^{n} \eta_i \xi_k^i + \cdots + \delta\omega_n \sum_{k=1}^{n} \frac{\partial \xi^n}{\partial q_k} \sum_{i=1}^{n} \eta_i \xi_k^i.$$

On the other hand, by calculating the variation of $\dot{q} = \sum_{k=1}^{n} \eta_i \xi^k$, we arrive at the relation

$$\delta\dot{q} = \delta\eta_1\, \xi^1 + \cdots + \delta\eta_n\, \xi^n + \eta_1 \left(\frac{\partial \xi^1}{\partial q_1} \delta q_1 + \cdots + \frac{\partial \xi^1}{\partial q_n} \delta q_n \right) + \cdots + \eta_n \left(\frac{\partial \xi^n}{\partial q_1} \delta q_1 + \cdots + \frac{\partial \xi^n}{\partial q_n} \delta q_n \right).$$

Thus, we have obtained two representations of $\delta\dot{q}$. Equating them with each other, we find that

$$(\delta\eta_1 - \delta\dot{\omega}_1)\xi^1 + \cdots + (\delta\eta_n - \delta\dot{\omega}_n)\xi^n + \sum_{s=1}^{n} \eta_s \sum_{k=1}^{n} \frac{\partial \xi^s}{\partial q_k} \sum_{i=1}^{n} \delta\omega_i\, \xi_k^i - \sum_{s=1}^{n} \delta\omega_s \sum_{k=1}^{n} \frac{\partial \xi^s}{\partial q_k} \sum_{i=1}^{n} \eta_i\, \xi_k^i = 0.$$

Renaming the indices $i \leftrightarrow s$ in the first term with sums yields

$$(\delta\eta_1 - \delta\dot{\omega}_1)\xi^1 + \cdots + (\delta\eta_n - \delta\dot{\omega}_n)\xi^n + \sum_{i=1}^{n} \eta_i \sum_{k=1}^{n} \frac{\partial \xi^i}{\partial q_k} \sum_{s=1}^{n} \delta\omega_s\, \xi_k^s - \sum_{s=1}^{n} \delta\omega_s \sum_{k=1}^{n} \frac{\partial \xi^s}{\partial q_k} \sum_{i=1}^{n} \eta_i\, \xi_k^i = 0.$$

Or, equivalently,

$$(\delta\eta_1 - \delta\dot{\omega}_1)\xi^1 + \cdots + (\delta\eta_n - \delta\dot{\omega}_n)\xi^n + \sum_{i,s=1}^{n} \eta_i\,\delta\omega_s \sum_{k=1}^{n}\left(\frac{\partial\xi^i}{\partial q_k}\xi_k^s - \frac{\partial\xi^s}{\partial q_k}\xi_k^i\right) = 0.$$

Note that

$$\sum_{k=1}^{n}\left(\frac{\partial\xi^i}{\partial q_k}\xi_k^s - \frac{\partial\xi^s}{\partial q_k}\xi_k^i\right)$$

is the vector with the components of the commutator $[U_s, U_i]$.

Since the generators of a group form an algebra, we have

$$[U_s, U_i] = \sum_{l=1}^{n} C_{si}^l U_l,$$

where the C_{si}^l are structural constants of the group.

Hence,

$$(\delta\eta_1 - \delta\dot{\omega}_1)\xi^1 + \cdots + (\delta\eta_n - \delta\dot{\omega}_n)\xi^n + \sum_{i,s=1}^{n} \eta_i\,\delta\omega_s\,C_{si}^1\xi^1 + \cdots + \sum_{i,s=1}^{n} \eta_i\,\delta\omega_s\,C_{si}^n\xi^n = 0.$$

It follows that

$$\delta\eta_k = \delta\dot{\omega}_k - \sum_{i,s=1}^{n} \eta_i\,\delta\omega_s\,C_{si}^k.$$

Thus, the variation of the Hamiltonian action acquires the form

$$\int\left[\sum_{k=1}^{n}\frac{\partial T}{\partial\eta_k}\delta\dot{\omega}_k - \sum_{k,i,s=1}^{n}\eta_i\frac{\partial T}{\partial\eta_k}\,\delta\omega_s\,C_{si}^k + \sum_{k,s=1}^{n}\left(\frac{\partial T}{\partial q_k}\,\delta\omega_s\,\xi_k^s - \frac{\partial U}{\partial q_k}\,\delta\omega_s\,\xi_k^s\right)\right]dt.$$

Integrate the first term by parts to obtain

$$\int\frac{\partial T}{\partial\eta_k}\,\delta\dot{\omega}_k\,dt = \frac{\partial T}{\partial\eta_k}\,\delta\omega_k\Big|_{t_0}^{t} - \int\frac{d}{dt}\frac{\partial T}{\partial\eta_k}\,\delta\omega_k\,dt.$$

Since the variation $\delta\omega_k$ is independent, it follows that

$$\frac{d}{dt}\frac{\partial T}{\partial\eta_k} + \sum_{i,s=1}^{n}\frac{\partial T}{\partial\eta_i}\eta_s C_{ks}^i = \Omega_k,$$

where

$$\Omega_k = \sum_{s=1}^{n}\left(\frac{\partial T}{\partial q_s}\xi_s^k - \frac{\partial U}{\partial q_s}\xi_s^k\right)$$

is the generalized force.

The equations obtained are referred to as Poincaré's equations. Obviously, Lagrange's equations are a special case of Poincaré's equations.

In the dynamics of a rigid body the role of the variables η_i can be played by the components of the angular velocity of the body in projections onto the body-associated axes. In this case, Poincaré's equations coincide with Euler's equations.

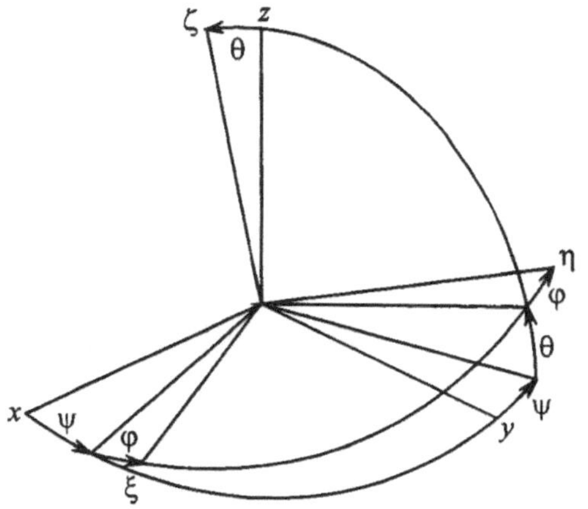

Figure 3.2

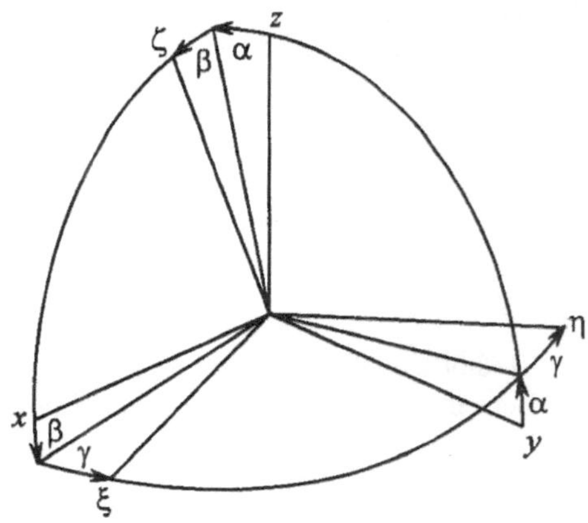

Figure 3.3

3.3. Kinematics of a Rigid Body

3.3.1. Ways of Specifying the Orientation of a Rigid Body

A collection of point particles which consists of more than one particle is called a *rigid body* (sometimes, absolutely rigid body) if the distance between any two particles of this collection does not change. The *rotation* of a rigid body is its motion such that at least one particle of the body is fixed all the time in a chosen reference frame.

Angles of finite rotation. With any rigid body a coordinate trihedral can be associated rigidly fixed in it, so that all particles of the body do not move relative to the trihedral. It is convenient to place the origin of the coordinate trihedral at a fixed point. The orientation of a rigid body is identified as the orientation of one trihedral (rigidly attached to the body) relative to another trihedral taken to be fixed.

Historically, the first way of specifying the orientation of a rigid body was the use of Eulerian angles (see Fig. 3.2). The Eulerian angles ψ, θ, and φ determining the orientation of the moving trihedral $\xi\eta\zeta$ relative to the fixed trihedral xyz are the angles of plane rotations through which the fixed trihedral must be mentally turned to superpose it with the moving trihedral. The first rotation is performed in the xy-plane through the angle ψ, the second in the yz-plane through the angle θ, and the third in the xy-plane again through the angle φ. Note that, after the first mental rotation, the yz-plane changes its orientation relative to the fixed space. The same is true for the xy-plane after the second rotation.

The angle ψ is referred to as the precession angle, θ the nutation angle, and φ the angle of proper rotation. It is important to bear in mind the following. The three numbers ψ, θ, and φ determining the orientation of a rigid body are not observable quantities unlike the Cartesian coordinates of a particle. The above rotations assigning a meaning to the angles ψ, θ, and φ form an imaginary structure. A different structure can be used to specify the orientation of a rigid body. For example, Fig. 3.3 illustrates one of the sequences of rotations widely used nowadays—the Krylov–Bulgakov angles α, β, and γ. Here the first rotation is performed about the first axis through the angle α, the second about the second axis through β, and the third about the third axis through γ, thus allowing one to superpose mentally the fixed frame with the moving frame. Symbolically, this sequence of rotations may be represented as 1–2–3. One can imagine other sequences of rotations; for example, the symbol 3–1–2 indicates that the first rotation is performed about the third axis, the second about the first axis, and the last about the second axis.

For the Eulerian angles the sequence of rotations is represented as 3–1–3. The totality of possible sequences is called the *system of angles of finite rotation*. All sequences of rotations can divided into the following four classes:

I	1–2–3,	3–1–2,	2–3–1;
II	1–3–2,	2–1–3,	3–2–1;
III	1–2–1,	2–3–2,	3–1–3;
IV	1–3–1,	3–2–3,	2–1–2.

In classes I and II the rotations are performed about all three axes and in classes III and IV one of the axes is left out. For convenience of describing the properties of the above classes, we change the designation of the sequence of rotations in classes III and IV so that (i) the first digit indicates the axis about which two rotations are made, (ii) the second digit refers to the axis about which one rotation is performed, and (iii) the third digit is the number of the axis about which no rotation is made. For example, in the case of the Eulerian angles, 3–1–3, the new designation of this sequence of rotations is 3–1–2. Thus, the last two classes are now represented as

III	1–2–3,	2–3–1,	3–1–2;
IV	1–3–2,	3–2–1,	2–1–3.

In this notation, each class includes all sequences of rotations which can be obtained one from another by an even substitution. Recall that a substitution is even (odd) if it consists of an even (odd) number of exchanges of adjacent elements. The transition from class I to class II, as well as from class III to class IV, is performed by an odd substitution.

All sequences of rotations belonging to the same class are equivalent to each other in the sense that, for the same numerical values of the angles α, β, and γ, the body orientation defined by one sequence and that defined by another sequence differ by the rotation about the bisector of a coordinate angle through an angle of $\pm 120°$.

Thus, among the twelve possible sequences of rotations about coordinate axes, only four are essentially different. The sequences of classes I and II are referred to as *angles of finite rotation of the first kind*. The sequences of classes III and IV are called *angles of finite rotation of the second kind*.

The theorem below proves that to identify the orientation of a rigid body three angles suffice.

THEOREM. *Any orientation of a rigid body can be achieved by three sequential plane rotations from any initial orientation in any of the above ways.*

Proof. For any of the four classes the proof is performed in the same manner. To be specific, we consider the first class (the Krylov–Bulgakov angles 1–2–3). Look at Fig. 3.4. Let the trihedral $\xi\eta\zeta$ is oriented in an arbitrary fashion relative to the trihedral xyz. In the generic case, the yz-plane intersects with the $\xi\eta$-plane in a straight line l. If the two planes coincide, then this is a special case where the trihedral xyz can be superposed with the trihedral $\xi\eta\zeta$ by using two plane rotations or even one. In this special case, one or even two angles are zero. In the generic case, the line of intersection, l, exists; then by a plane rotation of the trihedral xyz about the x-axis the y-axis can superposed with the straight line l.

Since there is no direction defined on the line l, the minimum angle of this rotation lies in the interval $0 \leq \alpha \leq \pi$. In the new position of the trihedral xyz, the y-axis is perpendicular to the ζ-axis (since the line l is perpendicular to ζ). Hence, the z-axis can be superposed with the ζ-axis by the second rotation of the trihedral xyz about the l-axis (or the y-axis). Since we also must provide the coincidence of the directions of these axes, the minimum angle of this rotation must lie in the range $0 \leq \beta \leq 2\pi$. After the second rotation, the xy- and $\xi\eta$-planes coincide. For the trihedrals to be completely superposed, it suffices to rotate the trihedral xyz about the z-axis through an angle γ ($0 \leq \gamma \leq 2\pi$). This proves the theorem.

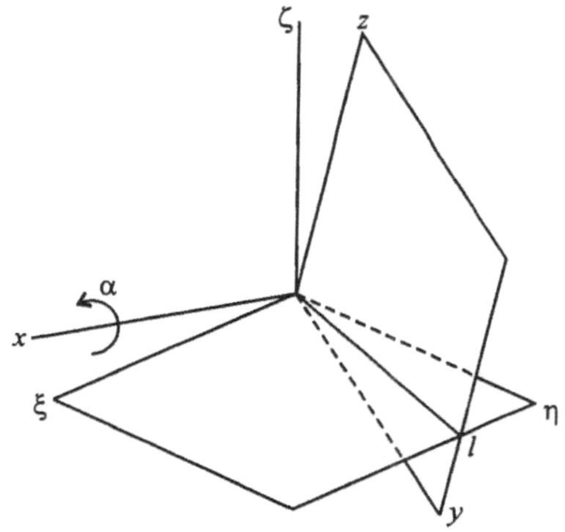

Figure 3.4

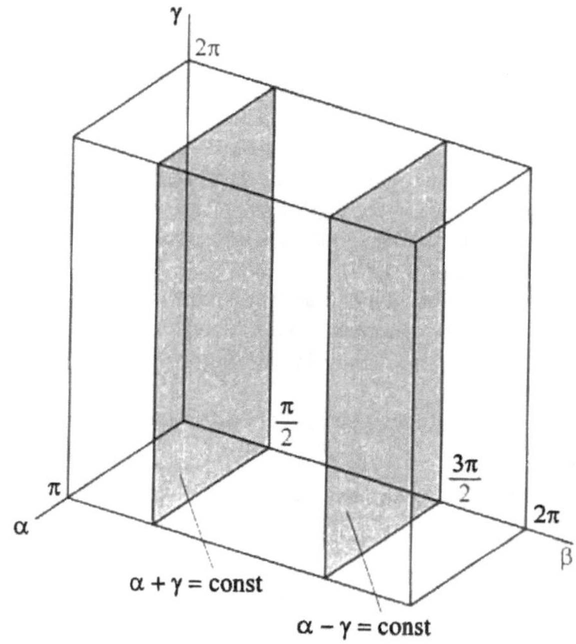

Figure 3.5

Figure 3.5 depicts a parallelepiped whose opposite faces are identified with each other and to each point of which there corresponds a unique orientation of a rigid body. By virtue of the above theorem, *all* orientations of the body turn out to be exhausted. However, there is no one-to-one correspondent between the orientations of the body and the points of the parallelepiped.

If $\beta = \frac{1}{2}\pi$ or $\frac{3}{2}\pi$, then the z-axis takes up, after the second rotation, the position of the x-axis and the third rotation is added to the first, thus resulting in the single rotation through the angle $\alpha + \gamma$. In other words, to the set of points $\alpha + \gamma = \text{const}$, $\beta = \frac{1}{2}\pi$ there corresponds a unique position of the rigid body. If $\beta = \frac{3}{2}\pi$, then the set $\alpha - \gamma = \text{const}$ possesses this property. This fact is common for the angles of finite rotation of the first kind. For the angles of the second kind, the planes in the space $\alpha\beta\gamma$ where the representation of the positions of a rigid body by three angles is not unique are the planes $\beta = 0$ and $\beta = \pi$.

This inconvenience of the finite rotation angles sends one in search of other ways of specifying the positions of a rigid body which are free of the above degenerations.

A mathematical object which is in one-to-one correspondence with the set of all positions of a rigid body with a fixed point is the set of all 3×3 orthogonal matrices. This set forms a three-parameter Lie group, possible ways of whose local parametrization have just been considered.

Orthogonal matrices. Mathematically, the rotation of a rigid body can be specified as a linear map of the three-dimensional Euclidean space into itself, with the distance between any two points being preserved under this map. Such maps are called *orthogonal*. Let a vector $\mathbf{R}$ be given in a trihedral xyz. It is convenient here to represent this vector as a column matrix,

$$\mathbf{R} = \begin{pmatrix} x \\ y \\ z \end{pmatrix}.$$

Let this vector be mapped by a rotation transformation into a vector $\mathbf{R}'$,

$$\mathbf{R}' = \begin{pmatrix} x' \\ y' \\ z' \end{pmatrix}.$$

The transformation itself is defined in the trihedral xyz by a matrix A,

$$A = \begin{pmatrix} a_{11} & a_{12} & a_{13} \\ a_{21} & a_{22} & a_{23} \\ a_{31} & a_{32} & a_{33} \end{pmatrix},$$

so that

$$\mathbf{R}' = A\mathbf{R}.$$

It follows from the definition of a rotation that the inner squares of $\mathbf{R}$ and $\mathbf{R}'$ are equal to each other,

$$\mathbf{R}^{\mathrm{T}}\mathbf{R} = (\mathbf{R}')^{\mathrm{T}}\mathbf{R}'.$$

Whence,

$$\mathbf{R}^{\mathrm{T}}\mathbf{R} = (A\mathbf{R})^{\mathrm{T}}A\mathbf{R} = \mathbf{R}^{\mathrm{T}}A^{\mathrm{T}}A\mathbf{R}.$$

It is evident that $A^{\mathrm{T}}A = E$, i.e., the product of the transpose of a rotation transformation matrix by the matrix itself is the identity matrix. This condition implies that the columns of A form an orthonormal system (i.e., the sum of the squares of the column entries is equal to unity, and the inner products of different columns by each other are zero). Such matrices are said to be *orthogonal*.

Example. Consider the transformation of rotation through an angle φ about the x-axis. Since the first coordinate of the vector $\mathbf{R}$ does not change under this transformation, we assume that $\mathbf{R}$ lies in the yz-plane entirely (see Fig. 3.6).

To establish the relationship between the coordinates of the transformed vector, y' and z', and those of the original vector, y and z, we rotate the triangle $Oy\mathbf{R}$ through the angle φ so as to superpose $\mathbf{R}$ with $\mathbf{R}'$ (see Fig. 3.7). It immediately follows from the figure that $y' = y\cos\varphi - z\sin\varphi$ and $z' = y\sin\varphi + z\cos\varphi$. If the vector $\mathbf{R}$ does not lie in the plane of rotation entirely, then the relation $x' = x$ must be added to the above two. Hence, the matrix of rotation about the x-axis has the from

$$A = \begin{pmatrix} 1 & 0 & 0 \\ 0 & \cos\varphi & -\sin\varphi \\ 0 & \sin\varphi & \cos\varphi \end{pmatrix}.$$

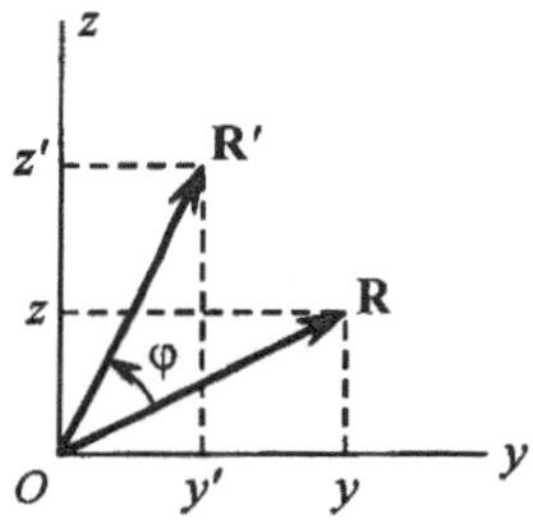

Figure 3.6

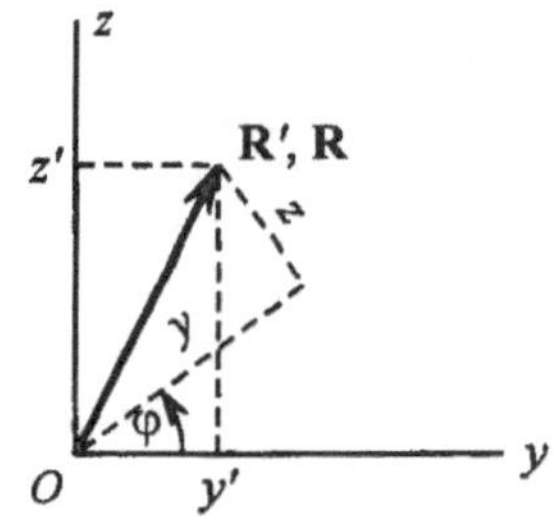

Figure 3.7

Similarly, one can establish that the matrices of rotation about the y- and z-axes have the form

$$\begin{pmatrix} \cos\varphi & 0 & \sin\varphi \\ 0 & 1 & 0 \\ -\sin\varphi & 0 & \cos\varphi \end{pmatrix} \quad \text{and} \quad \begin{pmatrix} \cos\varphi & -\sin\varphi & 0 \\ \sin\varphi & \cos\varphi & 0 \\ 0 & 0 & 1 \end{pmatrix},$$

respectively. Here, it is taken into account that positive in the zx-plane is the rotation from the x-axis to the z-axis.

Point out the most important properties of orthogonal matrices.

PROPERTY 1. From the relation $\det(A^{\mathrm{T}}A) = 1$ it follows that $\det A = \pm 1$.

PROPERTY 2. As is known from algebra, if $\det A \neq 0$, then there exists a unique matrix, denoted by A^{-1}, such that $A^{-1}A = AA^{-1} = E$. This means that $A^{\mathrm{T}} = A^{-1}$ and $AA^{\mathrm{T}} = E$ for orthogonal matrices, i.e., not only the columns but also rows of A form an orthonormal system.

PROPERTY 3. The product of two orthogonal matrices is an orthogonal matrix.

Indeed, let A and B be orthogonal matrices. Consider $C = AB$. We have $C^{\mathrm{T}} = (AB)^{\mathrm{T}} = B^{\mathrm{T}}A^{\mathrm{T}}$. This implies that $C^{\mathrm{T}}C = B^{\mathrm{T}}A^{\mathrm{T}}AB = E$.

PROPERTY 4. Since the operation of matrix multiplication is associative, it follows from Properties 2 and 3 that the set of all orthogonal matrices forms a group. In Section 1.2, the group O(n) of orthogonal transformations acting in the n-dimensional Euclidean space was introduced, as well as its subgroup SO(n) with positive determinant. The O(n) group is isomorphic to the group of $n \times n$ orthogonal matrices; for this reason, these groups are not distinguished in applications. For mechanics, of most significance are the just introduced group O(3) of 3×3 orthogonal matrices and its special subgroup SO(3).

PROPERTY 5. There is a one-to-one correspondence between all elements of the SO(3) group (all orthogonal matrices A with $\det A = 1$) and all rotations of a rigid body.

Indeed, consider the transform of the first unit vector of the trihedral xyz,

$$\mathbf{i}' = A\mathbf{i} = \begin{pmatrix} a_{11} & a_{12} & a_{13} \\ a_{21} & a_{22} & a_{23} \\ a_{31} & a_{32} & a_{33} \end{pmatrix} \begin{pmatrix} 1 \\ 0 \\ 0 \end{pmatrix} = \begin{pmatrix} a_{11} \\ a_{21} \\ a_{31} \end{pmatrix}.$$

That is, the transform of $\mathbf{i}$ is the first column of A. Analogously, the transforms of the unit vectors $\mathbf{j}$ and $\mathbf{k}$ are the second and third columns of A, respectively. What this means is that if an orthogonal matrix is prescribed, then the position of the new trihedral is determined uniquely. Conversely, given the position of the new trihedral, by taking its unit vectors to be the columns of the matrix A, one can uniquely determine the matrix associated with the new trihedral. In this case, the number of the column corresponds to the number of the unit vector, which implies that $\det A = 1$.

Thus, the SO(3) group is the main mathematical model for the set of rotations of a rigid body. The SO(3) group defines the configuration manifold of a rigid body with one fixed point.

PROPERTY 6. The entries of an orthogonal matrix are equal to their cofactors.

Indeed, applying the rule of calculation of the inverse of a matrix, we have $A^{-1} = \{A_{ij}\}^{\mathrm{T}}/\det A$. Considering only matrices with $\det A = 1$ and taking into account the fact that $A^{\mathrm{T}} = A^{-1}$, we obtain $a_{ij} = A_{ij}$.

PROPERTY 7. Establish how an orthogonal transformation acts on a complex vector:

$$A(\mathbf{P} + i\mathbf{Q}) = A\mathbf{P} + i\mathbf{Q},$$

where **P** and **Q** are real vectors.

If the norm of a complex vector is defined as

$$|\mathbf{P} + i\mathbf{Q}| = \sqrt{(\mathbf{P} + i\mathbf{Q})^*(\mathbf{P} + i\mathbf{Q})} = \sqrt{P^{\mathrm{T}}P + Q^{\mathrm{T}}Q},$$

then the property of the orthogonal transformations to preserve the norm of a real vector extends also to the complex vectors. (The asterisk denotes the Hermitian conjugate of the vector, which can be obtained by substituting $-i$ for i in the original vector and calculating the transpose of the result.)

PROPERTY 8 (EIGENVECTORS AND EIGENVALUES OF ORTHOGONAL TRANSFORMATIONS). The eigenvectors satisfy the condition $A\mathbf{R} = \lambda\mathbf{R}$, where λ is a number called an eigenvalue. Since the orthogonal transformations are length-preserving, we have $|\lambda| = 1$. The eigenvalues are determined by the equation $\det(A - \lambda E) = 0$, with

$$\det(A - \lambda E) = -\lambda^3 + (a_{11} + a_{22} + a_{33})\lambda^2 - (A_{11} + A_{22} + A_{33})\lambda + \det A,$$

where A_{11}, A_{22}, and A_{33} are the cofactors of a_{11}, a_{22}, and a_{33}.

Taking into account Property 6, we find that

$$\det(A - \lambda E) = -\lambda^3 + \lambda^2 \operatorname{tr} A - \lambda \operatorname{tr} A + 1 = 0.$$

It is evident that $\lambda_1 = 1$ is a root of this equation. For the other two roots we obtain

$$\lambda_{2,3} = \tfrac{1}{2}(\operatorname{tr} A - 1) \pm \sqrt{\tfrac{1}{4}(\operatorname{tr} A - 1)^2 - 1}.$$

Since the modulus of each eigenvalue must be equal to unity, we can use the following trigonometric representation of the complex numbers λ_2 and λ_3:

$$\lambda_{2,3} = \cos\varphi \pm i \sin\varphi.$$

Hence,

$$\cos\varphi = \tfrac{1}{2}(\operatorname{tr} A - 1).$$

To determine three eigenvectors $\mathbf{R}_k$ $(k = 1, 2, 3)$ corresponding the three eigenvalues λ_k, one must solve three homogeneous systems of equations $A\mathbf{R}_k = \lambda_k \mathbf{R}_k$. The first eigenvalue is real; hence, the first eigenvector is also real:

$$\mathbf{R}_1 \equiv \mathbf{R} = \begin{pmatrix} x \\ y \\ z \end{pmatrix}.$$

Thus, we arrive at the following system for x, y, and z:

$$\begin{aligned}
(a_{11} - 1)x + a_{12}y + a_{13}z &= 0, \\
a_{21}x + (a_{22} - 1)y + a_{23}z &= 0, \\
a_{31}x + a_{32}y + (a_{33} - 1)z &= 0, \\
x^2 + y^2 + z^2 &= 1.
\end{aligned}$$

The last nonlinear equation added is connected with the desire to obtain an eigenvector of unit norm. It is not difficult to verify that the system has two solutions that differ in sign:

$$x = \pm\frac{a_{32} - a_{23}}{2\sin\varphi}, \quad y = \pm\frac{a_{13} - a_{31}}{2\sin\varphi}, \quad z = \pm\frac{a_{21} - a_{12}}{2\sin\varphi}.$$

To choose the sign we impose the following condition: a rotation about a coordinate axis through a positive angle must be made counterclockwise when viewed from the side where the axis points to. For the matrix of rotation about the x-axis from the last example, we have

$$a_{32} = \sin\varphi, \quad a_{23} = -\sin\varphi, \quad a_{13} = a_{31} = a_{21} = a_{12} = 0.$$

Hence, $x = \pm 1$, $y = 0$, and $z = 0$. Thus, in the solution obtained, the plus sign must be selected.

The eigenvectors for the complex conjugate eigenvalues $\lambda_{2,3} = \cos\varphi \pm i\sin\varphi$ are also complex conjugate, $\mathbf{R}_{2,3} = \mathbf{P} \pm i\mathbf{Q}$. Let us demonstrate that the three real vectors $\mathbf{P}$, $\mathbf{Q}$, and $\mathbf{R}$ form a right-handed orthogonal system of vectors.

Calculate first the product $(A\mathbf{R}_1)^{\mathrm{T}}\mathbf{R}_2$. On the one hand,

$$(A\mathbf{R}_1)^{\mathrm{T}}\mathbf{R}_2 = \mathbf{R}_1^{\mathrm{T}}\mathbf{R}_2,$$

since $A\mathbf{R}_1 = \mathbf{R}_1$. On the other hand,

$$(A\mathbf{R}_1)^{\mathrm{T}}\mathbf{R}_2 = \mathbf{R}_1^{\mathrm{T}}A^{\mathrm{T}}\mathbf{R}_2 = \lambda_3\mathbf{R}_1^{\mathrm{T}}\mathbf{R}_2,$$

since $A^{\mathrm{T}}\mathbf{R}_2 = \lambda_2^{-1}\mathbf{R}_2$ and $\lambda_2\lambda_3 = 1$.

On comparing both results, one can see that if $\lambda_3 \neq 1$, then $\mathbf{R}_1^{\mathrm{T}}\mathbf{R}_2 = 0$; whence it follows that $\mathbf{R}^{\mathrm{T}}\mathbf{P} = 0$ and $\mathbf{R}^{\mathrm{T}}\mathbf{Q} = 0$.

Consider now the product $(A\mathbf{R}_2)^{\mathrm{T}}\mathbf{R}_2$. On the one hand,

$$(A\mathbf{R}_2)^{\mathrm{T}} = \lambda_2\mathbf{R}_2^{\mathrm{T}}\mathbf{R}_2.$$

On the other hand,

$$(A\mathbf{R}_2)^{\mathrm{T}} = \mathbf{R}_2^{\mathrm{T}}A^{\mathrm{T}}\mathbf{R}_2 = \lambda_3\mathbf{R}_2^{\mathrm{T}}\mathbf{R}_2.$$

Comparing the results, one can see that if $\lambda_2 \neq \lambda_3$, then $\mathbf{R}_2^{\mathrm{T}}\mathbf{R}_2 = 0$, or $\mathbf{P}^2 - \mathbf{Q}^2 + 2i\mathbf{P}^{\mathrm{T}}\mathbf{Q} = 0$; whence, $\mathbf{P}^2 = \mathbf{Q}^2$ and $\mathbf{P}^{\mathrm{T}}\mathbf{Q} = 0$.

The case $\lambda_1 = \lambda_2 = \lambda_3 = 1$ corresponds to the identity transformation $A = E$, for which any vector is an eigenvector, and hence three mutually orthogonal vectors can always be found.

It remains to show that the system $\mathbf{R}, \mathbf{P}, \mathbf{Q}$ is right-handed. To this end, we will find eigenvectors for complex eigenvalues in the special case of rotation about the x-axis, which was considered in the last example. The system of equations for $\mathbf{R}_2$ has the form $A\mathbf{R}_2 = \lambda_2\mathbf{R}_2$, or

$$\begin{aligned}(1 - \cos\varphi - i\sin\varphi)x &= 0,\\ -i\sin\varphi\, y - \sin\varphi\, z &= 0,\\ \sin\varphi\, y - i\sin\varphi\, z &= 0.\end{aligned}$$

It follows that $x = 0$, $y = 1$, and $z = -i$. Thus, the eigenvector $\mathbf{R}_2 = \mathbf{P} - i\mathbf{Q}$ has the form

$$\mathbf{R}_2 = \begin{pmatrix}0\\1\\-i\end{pmatrix}, \quad \mathbf{P} = \begin{pmatrix}0\\1\\0\end{pmatrix}, \quad \mathbf{Q} = \begin{pmatrix}0\\0\\1\end{pmatrix}.$$

The system

$$\mathbf{R} = \begin{pmatrix}1\\0\\0\end{pmatrix}, \quad \mathbf{P} = \begin{pmatrix}0\\1\\0\end{pmatrix}, \quad \mathbf{Q} = \begin{pmatrix}0\\0\\1\end{pmatrix}$$

is right-handed. Since the general case differs from that just considered by the transformation with positive determinant, the proven holds true in the general case as well.

PROPERTY 9. Any orthogonal transformation of space is equivalent to a rotation of space about an eigenvector **R** through an angle φ.

Indeed, rewrite the equation $A\mathbf{R}_k = \lambda_k \mathbf{R}_k$ for each of the eigenvectors; we have

$$\begin{aligned} A\mathbf{R} &= \mathbf{R}, \\ A\mathbf{P} &= \mathbf{P}\cos\varphi + \mathbf{Q}\sin\varphi, \\ A\mathbf{Q} &= -\mathbf{P}\sin\varphi + \mathbf{Q}\cos\varphi. \end{aligned}$$

If the axes of the original trihedral, **i**, **j**, and **k**, are selected to point along the vectors **R**, **P**, and **Q**, respectively, then

$$\mathbf{R} = \begin{pmatrix} 1 \\ 0 \\ 0 \end{pmatrix}, \quad \mathbf{P} = \begin{pmatrix} 0 \\ 1 \\ 0 \end{pmatrix}, \quad \mathbf{Q} = \begin{pmatrix} 0 \\ 0 \\ 1 \end{pmatrix}.$$

It follows from the relations obtained that

$$(A\mathbf{R}, A\mathbf{P}, A\mathbf{Q}) = A = \begin{pmatrix} 1 & 0 & 0 \\ 0 & \cos\varphi & -\sin\varphi \\ 0 & \sin\varphi & \cos\varphi \end{pmatrix}.$$

This means that, in the basis **R**, **P**, **Q**, the orthogonal matrix is the matrix of plane rotation about the vector **R** through the angle φ.

COROLLARY (EULER'S THEOREM). *Any movement of a rigid body with one fixed point can be replaced by a plane rotation about some axis through some angle.*

In the presentation of Property 9, the direction of the axis (Eulerian axis) was found and the angle (Eulerian angle) was determined.

Quaternions. Consider a four-dimensional vector space over the field of real numbers. Any element Λ of this space has the form

$$\Lambda = \lambda_0 i_0 + \lambda_1 i_1 + \lambda_2 i_2 + \lambda_3 i_3$$

in some basis i_0, i_1, i_2, i_3. The real numbers λ_0, λ_1, λ_2, and λ_3 are called the coordinates of the vector Λ in the specified basis.

A vector space turns into an *algebra* if a multiplication of two vectors which assigns a third vector to them is defined in this space. The multiplication law must be associative,

$$(\Lambda \circ \mathrm{M}) \circ \mathrm{N} = \Lambda \circ (\mathrm{M} \circ \mathrm{N}),$$

for any three vectors. It must also be distributive,

$$(\Lambda + \mathrm{M}) \circ (\mathrm{N} + \mathrm{R}) = \Lambda \circ \mathrm{N} + \mathrm{M} \circ \mathrm{N} + \Lambda \circ \mathrm{R} + \mathrm{M} \circ \mathrm{R},$$

for any four vectors. Finally, this law must satisfy the condition

$$(\lambda\Lambda) \circ (\mu\mathrm{M}) = \lambda\mu\Lambda \circ \mathrm{M},$$

where λ and μ are arbitrary real numbers, and Λ and M are arbitrary vectors. Note that this definition of an algebra is suited for a vector space of arbitrary dimension. In this case, the dimension of the vectors space is called the dimension of the algebra. Note also that no commutativity of the multiplication operation is presumed.

By virtue of these requirements, to determine the product of any two vectors, it is sufficient to know the multiplication table for the basis vectors. Four-dimensional vectors are said to be *quaternions* if their multiplication satisfies the listed requirements and is determined by the following table of multiplication of unit vectors:

$$i_0 \circ i_k = i_k \circ i_0 = i_k, \qquad k = 0, 1, 2, 3,$$
$$i_k \circ i_k = -i_0, \qquad k = 1, 2, 3,$$
$$i_1 \circ i_2 = i_3, \quad i_2 \circ i_3 = i_1, \quad i_3 \circ i_1 = i_2,$$
$$i_2 \circ i_1 = -i_3, \quad i_3 \circ i_2 = -i_1, \quad i_1 \circ i_3 = -i_2.$$

The introduced abstract algebra of quaternions allows for various interpretations. Address the algebra of complex numbers to find out the difference between the abstract definition of an algebra and its interpretation. The algebra of complex numbers is based on the two-dimensional vector space, $\Lambda = \lambda_0 i_0 + \lambda_1 i_1$, with the following table of multiplication of unit vectors:

$$i_0 \circ i_0 = i_0, \quad i_0 \circ i_1 = i_1 \circ i_0 = i_1, \quad i_1 \circ i_1 = -i_0.$$

Below we give three interpretations of the thus defined algebra of complex numbers.

Geometric interpretation. Each vector Λ is considered to mean a directed segment of a straight line in the geometric plane. The product of two such vectors is the vector whose length is equal to the product of the lengths of the factors and the angle of inclination to the abscissa axis is the sum of the corresponding angles of the factors.

Matrix interpretation. If i_0 is taken to mean the matrix $\left(\begin{smallmatrix}1 & 0\\ 0 & 1\end{smallmatrix}\right)$ and i_1 the matrix $\left(\begin{smallmatrix}0 & -1\\ 1 & 0\end{smallmatrix}\right)$, then the vector space of the matrices of the form

$$\Lambda = \lambda_0 \begin{pmatrix}1 & 0\\ 0 & 1\end{pmatrix} + \lambda_1 \begin{pmatrix}0 & -1\\ 1 & 0\end{pmatrix} = \begin{pmatrix}\lambda_0 & -\lambda_1\\ \lambda_1 & \lambda_0\end{pmatrix}$$

with the customary rule of matrix multiplication obeys the above rules of multiplication in abstract form.

Numerical interpretation. If by i_0 the real number 1 is meant and by i_1 the imaginary unit $\sqrt{-1}$ is understood, then the vector space of the numbers of the form

$$\Lambda = \lambda_0 + \lambda_1\sqrt{-1}$$

obeys the same rules of multiplication as well.

Now return to quaternions. The most important interpretation of quaternions is the *geometric-numerical* interpretation. According to it, the unit vector i_0 is identified with the real unit and the unit vectors i_1, i_2, and i_3 form a basis in the three-dimensional Euclidean space. Thus, any quaternion can be represented as

$$\Lambda = \lambda_0 + \lambda_1\mathbf{i}_1 + \lambda_2\mathbf{i}_2 + \lambda_3\mathbf{i}_3 = \lambda_0 + \boldsymbol{\lambda}, \qquad \lambda_0 \in \mathbb{R}^1, \quad \boldsymbol{\lambda} \in \mathbb{E}^3.$$

If now the rules of multiplication of the three-dimensional unit vectors are defined as $\mathbf{i}_k \circ \mathbf{i}_k = -1$ and $\mathbf{i}_k \circ \mathbf{i}_l = \mathbf{i}_k \times \mathbf{i}_l$ $(k \neq l)$, where $\mathbf{i}_k \times \mathbf{i}_l$ is the customary vector product, then the product of any two quaternions satisfies the above rules of multiplication in abstract form. In what follows, we will often use the traditional notation for the unit vectors: $\mathbf{i}_1 = \mathbf{i}$, $\mathbf{i}_2 = \mathbf{j}$, and $\mathbf{i}_3 = \mathbf{k}$. This interpretation allows us to write out the following formula for the product of any two quaternions:

$$\Lambda \circ \mathrm{M} = \lambda_0\mu_0 - \boldsymbol{\lambda}\cdot\boldsymbol{\mu} + \lambda_0\boldsymbol{\mu} + \mu_0\boldsymbol{\lambda} + \boldsymbol{\lambda}\times\boldsymbol{\mu}, \qquad \Lambda = \lambda_0 + \boldsymbol{\lambda}, \quad \mathrm{M} = \mu_0 + \boldsymbol{\mu}.$$

The *matrix interpretation of quaternions* (Pauli's spin matrices) will be discussed later in this section.

The table of multiplication of basis vectors needed for the vector space to turn into an algebra can be chosen arbitrarily. The particular table presented above possesses the unique property that it and only it permits one to introduce the operation of division in the above algebra of quaternions, i.e., to define the *inverse* of the operation of multiplication introduced.

Note that there are only three cases in a finite-dimensional space where an algebra with division can be introduced. These cases include: (i) the algebra of real numbers, (ii) the algebra of complex numbers, and (iii) the algebra of quaternions. The first two algebras are commutative (Frobenius' theorem).

Introduce the notion of the conjugate quaternion,

$$\Lambda = \lambda_0 + \boldsymbol{\lambda}, \quad \bar{\Lambda} = \lambda_0 - \boldsymbol{\lambda},$$

and the norm of a quaternion,

$$\|\Lambda\| = \Lambda \circ \bar{\Lambda} = \bar{\Lambda} \circ \Lambda = \lambda_0^2 + \lambda_1^2 + \lambda_2^2 + \lambda_3^2.$$

The basic properties of the operation of multiplication of quaternions are given below.

PROPERTY 1. The multiplication introduced is noncommutative,

$$\Lambda \circ \mathrm{M} \neq \mathrm{M} \circ \Lambda,$$

since $\mathbf{i}_k \circ \mathbf{i}_l \neq \mathbf{i}_l \circ \mathbf{i}_k$ $(k \neq l)$.

PROPERTY 2. The conjugate of the product of two or more quaternions is equal to the product of the conjugates of the quaternions in the reverse order: $\overline{\Lambda \circ \mathrm{M}} = \bar{\mathrm{M}} \circ \bar{\Lambda}$.

Indeed,

$$\overline{(\lambda_0 + \boldsymbol{\lambda}) \circ (\mu_0 + \boldsymbol{\mu})} = \lambda_0\mu_0 - \boldsymbol{\lambda} \cdot \boldsymbol{\mu} - \lambda_0\boldsymbol{\mu} - \mu_0\boldsymbol{\lambda} - \boldsymbol{\lambda} \times \boldsymbol{\mu},$$

$$(\mu_0 - \boldsymbol{\mu}) \circ (\lambda_0 - \boldsymbol{\lambda}) = \lambda_0\mu_0 - \boldsymbol{\lambda} \cdot \boldsymbol{\mu} - \lambda_0\boldsymbol{\mu} - \mu_0\boldsymbol{\lambda} + \boldsymbol{\mu} \times \boldsymbol{\lambda}.$$

PROPERTY 3. The norm of the product is equal to the product of the norms:

$$\|\Lambda \circ \mathrm{M}\| = (\Lambda \circ \mathrm{M}) \circ \overline{(\Lambda \circ \mathrm{M})} = \Lambda \circ \mathrm{M} \circ \bar{\mathrm{M}} \circ \bar{\Lambda} = \|\Lambda\| \, \|\mathrm{M}\|.$$

PROPERTY 4. The operation of multiplication of quaternions is invariant under the orthogonal transformations of the vector components of the quaternions. By this the following is meant. Let two quaternions $\Lambda = \lambda_0 + \boldsymbol{\lambda}$ and $\mathrm{M} = \mu_0 + \boldsymbol{\mu}$ be multiplied to give $N = n_0 + \mathbf{n}$. If, prior to multiplication, the vector components of Λ and M are subjected to an orthogonal transformation, $\boldsymbol{\lambda}' = A\boldsymbol{\lambda}$ and $\boldsymbol{\mu}' = A\boldsymbol{\mu}$, and after that the resulting quaternions $\Lambda' = \lambda_0 + \boldsymbol{\lambda}'$ and $\mathrm{M}' = \mu_0 + \boldsymbol{\mu}'$ are multiplied, then the product is equal to $N' = n_0 + A\mathbf{n}$. Thus, the operations of multiplication of quaternions and orthogonal transformation of their vector components are permutable.

This property follows from the fact that the scalar product $\boldsymbol{\lambda} \cdot \boldsymbol{\mu}$ in the formula for the product of quaternions does not change under the orthogonal transformations; the vector product $\boldsymbol{\lambda} \times \boldsymbol{\mu}$ is known to possess the invariance property.

PROPERTY 5. For any nonzero quaternion ($\|\Lambda\| \neq 0$) there exists its inverse. One can readily see that the quaternion

$$\Lambda^{-1} = \bar{\Lambda}/\|\Lambda\|$$

possesses this property.

In the numerical interpretation, the quaternions are also referred to as *hypercomplex numbers*. The basic theorem of algebra[3] does not hold for hypercomplex numbers.

[3] In brief, the basic theorem of algebra states that any polynomial of order n has exactly n complex roots.

Example. Solve the equation $x^2 + x + 1 = 0$ over the fields of real, complex, and hypercomplex numbers.

(a) In real numbers, there is no solution.

(b) In complex numbers, there are two solutions:

$$x = -\tfrac{1}{2} \pm i\tfrac{\sqrt{3}}{2}.$$

(c) Find all solutions in hypercomplex numbers $x = x_0 + \mathbf{x}$. We have

$$x^2 = (x_0 + \mathbf{x}) \circ (x_0 + \mathbf{x}) = x_0^2 - \mathbf{x}^2 + 2x_0\mathbf{x}.$$

Substituting this expression into the equation yields

$$x_0^2 - \mathbf{x}^2 + 2x_0\mathbf{x} + x_0 + \mathbf{x} + 1 = 0.$$

A quaternion is zero if and only if its both scalar and vector components are zero:

$$x_0^2 - \mathbf{x}^2 + x_0 + 1 = 0, \quad (2x_0 + 1)\mathbf{x} = \mathbf{0}.$$

If $\mathbf{x} = \mathbf{0}$, then the equation for x_0 has no solutions; if $\mathbf{x} \neq 0$, then $x_0 = -\frac{1}{2}$ and from the first equation it follows that $\mathbf{x}^2 = \frac{3}{4}$.

Thus, any quaternion of the form $x = -\frac{1}{2} + \mathbf{x}$, where $\|\mathbf{x}\| = \frac{\sqrt{3}}{2}$, is a solution of the original equation.

The number of solutions is infinite. The basic theorem of algebra, according to which the number of solutions in complex numbers must be equal to the order of the equation, does not hold for hypercomplex numbers (quaternions).

The use of quaternions for specifying the orientation of a rigid body with one fixed point is based on the properties of the *adjoint map* introduced below:

$$R \to R': \quad R' = \operatorname{Ad} R \equiv \Lambda \circ R \circ \overline{\Lambda},$$

where $\operatorname{Ad} R$ stands for the adjoint of R. This map of the algebra of quaternions into itself (R is an arbitrary quaternion) is determined by a fixed quaternion Λ having the unit norm, $\|\Lambda\| = 1$.

The adjoint map possesses the following properties.

PROPERTY 1. The map $R' = \operatorname{Ad} R$ does not change the scalar component of the quaternion,

$$\Lambda \circ R \circ \overline{\Lambda} = \Lambda \circ (r_0 + \mathbf{r}) \circ \overline{\Lambda} = \Lambda \circ r_0 \circ \overline{\Lambda} + \Lambda \circ \mathbf{r} \circ \overline{\Lambda} = r_0 + \Lambda \circ \mathbf{r} \circ \overline{\Lambda}.$$

It remains to show that $\Lambda \circ \mathbf{r} \circ \overline{\Lambda}$ has zero scalar component. Indeed, $\overline{\Lambda \circ \mathbf{r} \circ \overline{\Lambda}} = \Lambda \circ \overline{\mathbf{r}} \circ \overline{\Lambda} = -\Lambda \circ \mathbf{r} \circ \overline{\Lambda}$. The conjugate of a quaternion is equal to the negative of the quaternion only if its scalar component is zero.

PROPERTY 2. The effect of the adjoint map on the vector component of a quaternion is equivalent to that of a rotation transformation: $\mathbf{r}' = A\mathbf{r}$ $(A^{\mathrm{T}}A = E)$.

Indeed, consider the norm of R':

$$\|R'\| = \|\Lambda \circ R \circ \overline{\Lambda}\| = \|\Lambda\|\,\|R\|\,\|\overline{\Lambda}\| = \|R\|.$$

Hence, $r_0'^2 + r_1'^2 + r_2'^2 + r_3'^2 = r_0^2 + r_1^2 + r_2^2 + r_3^2$. Since $r_0' = r_0$ by virtue of Property 1, we have $\|\mathbf{r}'\| = \|\mathbf{r}\|$.

Thus, along with orthogonal matrices, unit quaternions with $\|\Lambda\| = 1$ can also be used to define orthogonal transformations of vectors in the three-dimensional vector space. We write such a quaternion in the form

$$\Lambda = \lambda_0 + \lambda\mathbf{e},$$

where λ is the modulus of the vector $\boldsymbol{\lambda}$ and $\mathbf{e}$ is the unit vector of its direction, $\mathbf{e} = \boldsymbol{\lambda}/\lambda$.

It follows from the condition $\|\Lambda\| = 1$ that $\lambda_0^2 + \lambda^2 = 1$. Two scalar quantities related by the equation of a circle of unit radius can always be represented in the from $\lambda_0 = \cos\frac{\varphi}{2}$, $\lambda = \sin\frac{\varphi}{2}$. Then the expression $\Lambda = \cos\frac{\varphi}{2} + \mathbf{e}\sin\frac{\varphi}{2}$ with the condition $|\mathbf{e}| = 1$ determine the general form of a quaternion with unit norm.

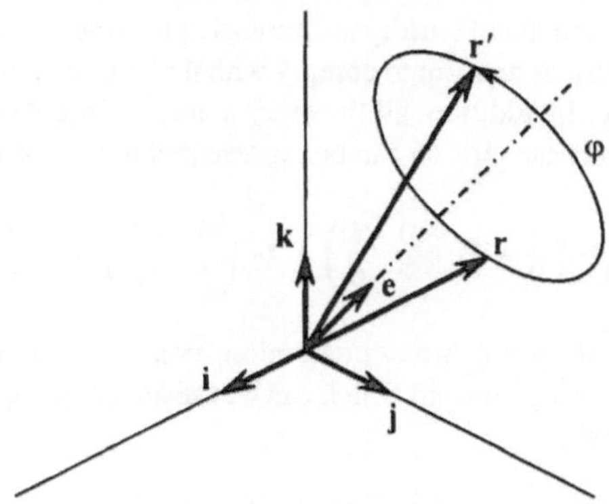

Figure 3.8

THEOREM. *The rotation defined by the quaternion* Λ *in* $\mathbb{E}^3$ *is the rotation about the vector* **e** *through the angle* φ *(see Fig. 3.8).*

Proof. Calculate directly the matrix A of the rotation for the prescribed quaternion $\Lambda = \cos\frac{\varphi}{2} + \mathbf{e}\sin\frac{\varphi}{2}$. In accordance with Property 5 of orthogonal matrices, the columns of A are the transforms of the original basis vectors. Let us calculate these transforms. Let the original basis be selected so that $\mathbf{i} = \mathbf{e}$, $\mathbf{j} \perp \mathbf{e}$, and $\mathbf{k} \perp \mathbf{e}$. (Freedom in choosing a basis in $\mathbb{E}^3$ is guaranteed by Property 4 of the operation of multiplication of quaternions.) Then,

$$\begin{aligned}
\mathbf{i}' &= \left(\cos\tfrac{\varphi}{2} + \mathbf{i}\sin\tfrac{\varphi}{2}\right) \circ \mathbf{i} \circ \left(\cos\tfrac{\varphi}{2} - \mathbf{i}\sin\tfrac{\varphi}{2}\right) = \mathbf{i},\\
\mathbf{j}' &= \left(\cos\tfrac{\varphi}{2} + \mathbf{i}\sin\tfrac{\varphi}{2}\right) \circ \mathbf{j} \circ \left(\cos\tfrac{\varphi}{2} - \mathbf{i}\sin\tfrac{\varphi}{2}\right) = \mathbf{j}\cos\varphi + \mathbf{k}\sin\varphi,\\
\mathbf{k}' &= \left(\cos\tfrac{\varphi}{2} + \mathbf{i}\sin\tfrac{\varphi}{2}\right) \circ \mathbf{k} \circ \left(\cos\tfrac{\varphi}{2} - \mathbf{i}\sin\tfrac{\varphi}{2}\right) = -\mathbf{j}\sin\varphi + \mathbf{k}\cos\varphi.
\end{aligned}$$

Hence,

$$A = \begin{pmatrix} 0 & 0 & 1 \\ 0 & \cos\varphi & -\sin\varphi \\ 0 & \sin\varphi & \cos\varphi \end{pmatrix},$$

which proves the theorem.

Thus, in the expression of the unit quaternion $\cos\frac{\varphi}{2} + \mathbf{e}\sin\frac{\varphi}{2}$ defining the adjoint map, **e** is a unit vector of the axis of Eulerian rotation in $\mathbb{E}^3$, and φ is the angle of this rotation. Componentwise, the quaternion can be represented as

$$\Lambda = \left\{\cos\tfrac{\varphi}{2},\ x\sin\tfrac{\varphi}{2},\ y\sin\tfrac{\varphi}{2},\ z\sin\tfrac{\varphi}{2}\right\},$$

where x, y, and z are the components of **e**. The elements of the quaternion,

$$\lambda_0 = \cos\tfrac{\varphi}{2}, \quad \lambda_1 = x\sin\tfrac{\varphi}{2}, \quad \lambda_2 = y\sin\tfrac{\varphi}{2}, \quad \lambda_3 = z\sin\tfrac{\varphi}{2},$$

are referred to as the *Rodrigues–Hamilton parameters* of finite rotation.

Pauli's spin matrices. Pauli's spin matrices are associated with a matrix interpretation of quaternions in which the rules determining the products of the basis vectors are most natural. Identify the basis vectors with the following matrices:

$$i_0 \equiv \begin{pmatrix} 1 & 0 \\ 0 & 1 \end{pmatrix}, \quad i_1 \equiv \begin{pmatrix} 0 & i \\ i & 0 \end{pmatrix}, \quad i_2 \equiv \begin{pmatrix} 0 & -1 \\ 1 & 0 \end{pmatrix}, \quad i_3 \equiv \begin{pmatrix} i & 0 \\ 0 & -i \end{pmatrix},$$

where $i = \sqrt{-1}$. It can be verified that if here by the multiplication the customary matrix multiplication is understood, then these matrices turn out to comply with the table of multiplication of basis vectors which was introduced above. In addition, all the other axioms of algebra of quaternions turn out to be satisfied. It follows that any quaternion can be represented in the form

$$\Lambda = \lambda_0 \begin{pmatrix} 1 & 0 \\ 0 & 1 \end{pmatrix} + \lambda_1 \begin{pmatrix} 0 & i \\ i & 0 \end{pmatrix} + \lambda_2 \begin{pmatrix} 0 & -1 \\ 1 & 0 \end{pmatrix} + \lambda_3 \begin{pmatrix} i & 0 \\ 0 & -i \end{pmatrix}.$$

Specific interpretations of the algebra of quaternions possess a number of properties which are beyond the basic axioms of the algebra and which can be used to advantage. In our case, the matrices can be added together to obtain

$$\Lambda = \begin{pmatrix} \lambda_0 + i\lambda_3 & -\lambda_2 + i\lambda_1 \\ \lambda_2 + i\lambda_1 & \lambda_0 - i\lambda_3 \end{pmatrix}.$$

After that the multiplication of quaternions can be carried out as the customary multiplication of matrices. The conjugate of a quaternion is expressed as

$$\overline{\Lambda} = \begin{pmatrix} \lambda_0 - i\lambda_3 & \lambda_2 - i\lambda_1 \\ -\lambda_2 - i\lambda_1 & \lambda_0 + i\lambda_3 \end{pmatrix}.$$

It is apparent that the conjugate of a quaternion, in the sense defined previously, considered now in matrix form is the Hermit conjugate matrix (complex conjugate and transposed matrix), i.e.,

$$\overline{\Lambda} \equiv \Lambda^*.$$

The norm of the quaternion is equal in this case to the determinant of the corresponding matrix:

$$\|\Lambda\| = \det \Lambda.$$

Thus, the quaternions with unit norm, which serve to identify the orientation of a rigid body, are described with 2×2 complex matrices satisfying the conditions $\Lambda\Lambda^* = E$ and $\det \Lambda = 1$.

The set of such quaternions forms a group which is called a *special unitary group* and denoted by SU(2). The entries of unitary matrices, which are complex combinations of the quaternion components, are referred to as the *Cayley–Klein parameters*,

$$a = \lambda_0 + i\lambda_3, \quad b = \lambda_2 + i\lambda_1.$$

In this notation, the quaternion is represented by the matrix

$$\Lambda = \begin{pmatrix} a & -\bar{b} \\ b & \bar{a} \end{pmatrix}.$$

Some ideas of other properties of the Cayley–Klein parameters are given by linear-fractional transformations.

Linear-fractional transformations. Consider the rotation of a rigid body as a transformation of a unit sphere into itself (see Fig. 3.9). The rotation is made about the axis defined by a unit vector **e** through an angle φ and, hence, is determined by the quaternion $\Lambda = \cos\frac{\varphi}{2} + \mathbf{e}\sin\frac{\varphi}{2}$.

Under stereographic projecting of the sphere onto the xy-plane, to the rotation there corresponds some transformation of the plane. Let us find out what this transformation is. We will demonstrate that any circle on the sphere transforms under stereographic projecting into a circle on the plane (the straight lines on the plane are regarded as circles passing through the infinite point). To this end, we

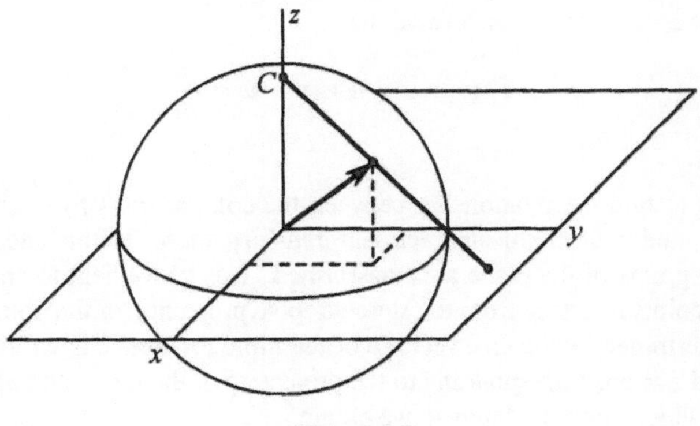

Figure 3.9

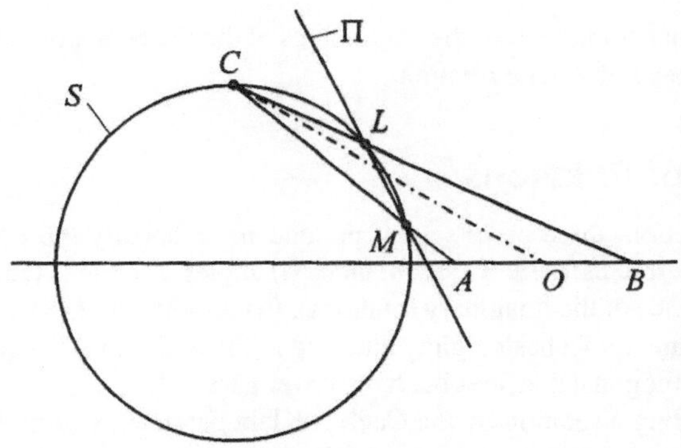

Figure 3.10

consider an arbitrary circle on the xy-plane and construct a conic surface based on this circle with vertex at the pole C of the sphere. This cone intersects the sphere in a curve which is the prototype of the circle on the plane under stereographic projecting. If the circle on the xy-plane passes through the infinite point, i.e., is a straight line, then the cone turns into a plane and the intersection of a plane and a sphere is always a circle (or a point).

Consider the section of the sphere by the plane passing through the cone axis and perpendicular to the xy-plane. This section is shown in Fig. 3.10. The points L and M lying in this section belong to the cone and sphere simultaneously. Draw a plane Π through these points perpendicularly to the plane of the section. Since the angle ABC is equal to half the angle subtended by the arc CL and the angle CML is based on this arc, the angles ABC and CML are equal to each other. Hence, the plane Π has the same orientation relative to the cone as the xy-plane (i.e., by the rotation about the cone axis through the angle π, the plane Π can be transferred to a position in which it is parallel to the xy-plane). Consequently, the section of the cone by the plane Π is also a circle. The plane Π intersects with the sphere in a circle; since the points L and M are common to these circles, the two circles coincide.

Under the rotation specified by the quaternion Λ, any circle on the sphere transforms into another circle. According to the preceding, the transforms of these circles are also circles on the plane. But this means that to the transformation of rotation of the sphere there corresponds a transformation of the plane which takes any circle to a circle. It is known from the theory of functions of a complex variable that the linear-fractional transformations

$$z = x + iy \to w = u + iv: \quad w = \frac{p + qz}{m + nz}$$

possess this property.

It is of interest to find the relationship between the components of the quaternion Λ and the parameters p, q, m, and n of the linear-fractional transformation. To this end, it suffices to know how three arbitrary points of the plane are transformed. It is convenient to choose such points as follows: two fixed points resulting from the stereographic projecting of the points of intersection of the straight line determined by the unit vector **e** of the finite rotation axis with a sphere and one of the points $z = 0$ and $z = \infty$, corresponding to the projecting of the lower and upper pole. Omitting the manipulations, which are of no interest, we obtain

$$w = \frac{az - \overline{b}}{bz + \overline{a}}.$$

Thus, the Cayley–Klein parameters are the coefficients of the linear-fractional transformation which is the stereographic image of a finite rotation.

3.3.2. Addition of Rotations

In the previous subsection, three basic sets of parameters to specify the position of a rigid body with one fixed point were presented. These include: (i) angles of finite rotation (with the necessary indication of the sequence of the imaginary rotations), (ii) orthogonal matrices, which map the basis of a fixed reference frame into a basis rigidly attached to the body, and (iii) quaternions, which play the same role as the orthogonal matrices but have fewer parameters.

In essence, specifying a rotation by the Cayley–Klein parameters or by linear-fractional transformations is equivalent to specifying it by quaternions; there are only slight distinctions in the rules of manipulating these parameters.

The most important problem of the kinematics of a rigid body is the problem of the addition of two or more rotations, i.e., the problem of calculating the parameters of the resulting rotation, provided that the parameters of its constituent rotations are known. For example, let the position of the trihedral $x'y'z'$ be determined by the Eulerian angles ψ_1, θ_1, and φ_1 relative to the fixed trihedral xyz and let the position of the rigid body, to which a frame $\xi\eta\zeta$ is rigidly attached, be determined by the Eulerian angles ψ_2, θ_2, and φ_2 relative to the trihedral $x'y'z'$. Find the angles ψ_3, θ_3, and φ_3 determining the position of the body relative to the fixed reference frame xyz. For the angles of finite rotation in general and Eulerian angles in particular, the solution of this problem is quite cumbersome. However, in terms of orthogonal matrices or quaternions, the operation of addition of rotations is the basic group operation of the corresponding group (the SO(3) group or the group of unit quaternions SU(2)).

The group of orthogonal transformations SO(3). If a transformation of rotation $\mathbf{R} \to \mathbf{R}'$ with orthogonal matrix A is performed,

$$\mathbf{R}' = A\mathbf{R},$$

followed by a transformation $\mathbf{R}' \to \mathbf{R}''$ with matrix B,

$$\mathbf{R}'' = B\mathbf{R}',$$

then the resulting transformation is obviously expressed as

$$\mathbf{R}'' = B\mathbf{R}' = BA\mathbf{R}.$$

The matrix $C = BA$ equal to the product of the matrices of the constituent rotations in the *reverse order* is just the matrix determining the total rotation.

So far we have viewed an orthogonal transformation as a generator that maps space into itself. The reference frame in which the matrix of this transformation has been written out has not changed. This is the so-called *active point of view* on the transformation. However, one can view the transforms $\mathbf{i}'$, $\mathbf{j}'$, and $\mathbf{k}'$ of the basis vectors $\mathbf{i}$, $\mathbf{j}$, and $\mathbf{k}$ as the basis of a new reference frame $x'y'z'$ and the orthogonal transformation as a transformation to the new reference frame. The vectors $\mathbf{R}$ themselves are considered to be unchanged (the *passive point of view*). To stress the fact that the vector $\mathbf{R}$ remains the same (change only its coordinates with respect to the new basis), we will use the designation $\mathbf{R}^{(\prime)}$. Find out how the coordinates of $\mathbf{R}$ are expressed in the new basis. The transforms of the original basis vectors have the form

$$\mathbf{i}' = A\mathbf{i}, \quad \mathbf{j}' = A\mathbf{j}, \quad \mathbf{k}' = A\mathbf{k}.$$

Substituting the inverse transforms

$$\mathbf{i} = A^{\mathrm{T}}\mathbf{i}', \quad \mathbf{j} = A^{\mathrm{T}}\mathbf{j}', \quad \mathbf{k} = A^{\mathrm{T}}\mathbf{k}'$$

into the decomposition

$$\mathbf{R} = x\mathbf{i} + y\mathbf{j} + z\mathbf{k},$$

we obtain

$$\mathbf{R} = xA^{\mathrm{T}}\mathbf{i}' + yA^{\mathrm{T}}\mathbf{j}' + zA^{\mathrm{T}}\mathbf{k}' = A^{\mathrm{T}}(x\mathbf{i}' + y\mathbf{j}' + z\mathbf{k}').$$

In the new basis its unit vectors are expressed as

$$\mathbf{i}' = \begin{pmatrix} 1 \\ 0 \\ 0 \end{pmatrix}, \quad \mathbf{j}' = \begin{pmatrix} 0 \\ 1 \\ 0 \end{pmatrix}, \quad \mathbf{k}' = \begin{pmatrix} 0 \\ 0 \\ 1 \end{pmatrix}.$$

Hence, the last formula for $\mathbf{R}$ can be rewritten in the form

$$\mathbf{R}^{(\prime)} = \begin{pmatrix} x' \\ y' \\ z' \end{pmatrix} = A^{\mathrm{T}} \begin{pmatrix} x \\ y \\ z \end{pmatrix}, \qquad \mathbf{R}^{(\prime)} = A^{\mathrm{T}}\mathbf{R}.$$

Now let us pass from the basis $\mathbf{i}', \mathbf{j}', \mathbf{k}'$ to a new basis $\mathbf{i}''\mathbf{j}''\mathbf{k}''$ by means of an orthogonal transformation with matrix B such that $\mathbf{i}'' = B\mathbf{i}'$, $\mathbf{j}'' = B\mathbf{j}'$, $\mathbf{k}'' = B\mathbf{k}'$. Then, in accordance with the preceding, we have $\mathbf{R}^{(\prime\prime)} = B^{\mathrm{T}}\mathbf{R}^{(\prime)} = B^{\mathrm{T}}A^{\mathrm{T}}\mathbf{R} = C^{\mathrm{T}}\mathbf{R}$. Thus, the matrix of the resulting transformation of the coordinates is

$$C = AB.$$

This is just the formula for the adding of rotations from the standpoint of the transformation of coordinates.

Let us point out the principal differences of the two points of view on the adding of rotations.

Active point of view. The matrices of consecutive rotations are multiplied in the *reverse order*. All matrices are represented in the common basis $\mathbf{i}, \mathbf{j}, \mathbf{k}$.

Passive point of view. The matrices of consecutive rotations are multiplied in the *direct order*. Each matrix is treated in the basis rotated by it (the matrix A is treated in $\mathbf{i}, \mathbf{j}, \mathbf{k}$, the matrix B in $\mathbf{i}', \mathbf{j}', \mathbf{k}'$, etc.).

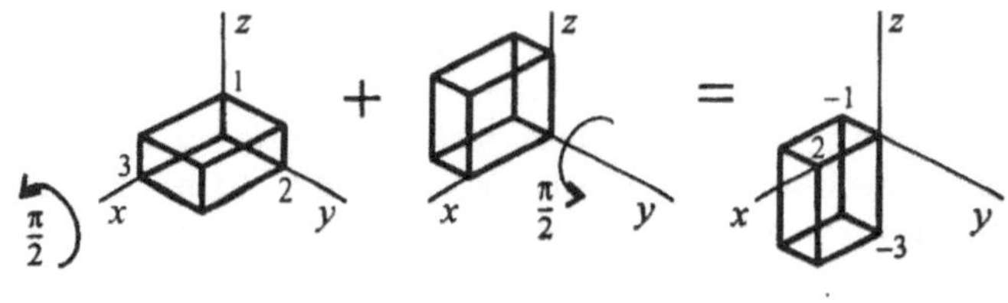

Figure 3.11

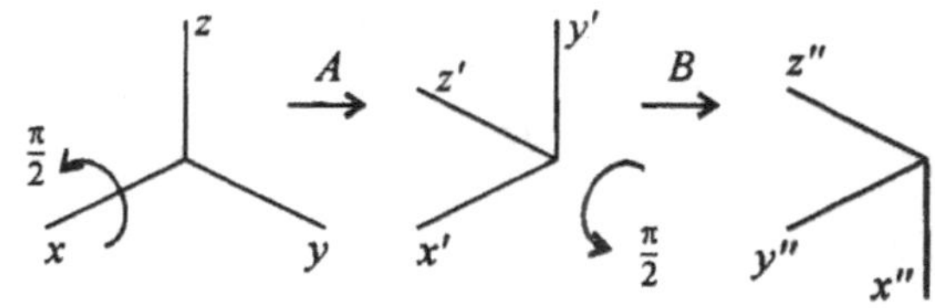

Figure 3.12

Example. A body was first turned about the x-axis through the angle $\frac{\pi}{2}$ and then about the y-axis through the angle $\frac{\pi}{2}$ (see Fig. 3.11). Find the matrix of the resulting rotation.

Judging from the statement of the problem, the active point of view on the transformation is implied. The matrix of the first rotation has the form

$$A = \begin{pmatrix} 1 & 0 & 0 \\ 0 & \cos\varphi & -\sin\varphi \\ 0 & \sin\varphi & \cos\varphi \end{pmatrix}, \qquad \varphi = \frac{\pi}{2}.$$

The matrix of the second rotation is expressed as

$$B = \begin{pmatrix} \cos\varphi & 0 & \sin\varphi \\ 0 & 1 & 0 \\ -\sin\varphi & 0 & \cos\varphi \end{pmatrix}, \qquad \varphi = \frac{\pi}{2}.$$

For the matrix of the resulting rotation we have

$$C = BA = \begin{pmatrix} 0 & 0 & 1 \\ 0 & 1 & 0 \\ -1 & 0 & 0 \end{pmatrix} \begin{pmatrix} 1 & 0 & 0 \\ 0 & 0 & -1 \\ 0 & 1 & 0 \end{pmatrix} = \begin{pmatrix} 0 & 1 & 0 \\ 0 & 0 & -1 \\ -1 & 0 & 0 \end{pmatrix}.$$

In particular, the vector determining a diagonal of the parallelepiped depicted in Fig. 3.11 is transferred to the new position

$$\mathbf{R}' = C\mathbf{R} = \begin{pmatrix} 0 & 1 & 0 \\ 0 & 0 & -1 \\ -1 & 0 & 0 \end{pmatrix} \begin{pmatrix} 3 \\ 2 \\ 1 \end{pmatrix} = \begin{pmatrix} 2 \\ -1 \\ -3 \end{pmatrix}.$$

This example can also be treated from the passive point of view (see Fig. 3.12).

The matrix of the first rotation is the same as that in the previous case. The matrix of the second rotation must be rewritten in the new frame $x'y'z'$; we have

$$B = \begin{pmatrix} \cos\varphi & -\sin\varphi & 0 \\ \sin\varphi & \cos\varphi & 0 \\ 0 & 0 & 1 \end{pmatrix}, \qquad \varphi = -\frac{\pi}{2}.$$

The matrix of the resulting rotation is obtained by multiplying the matrices of the constituent rotations in the direct order:

$$C = AB = \begin{pmatrix} 1 & 0 & 0 \\ 0 & 0 & -1 \\ 0 & 1 & 0 \end{pmatrix} \begin{pmatrix} 0 & 1 & 0 \\ -1 & 0 & 0 \\ 0 & 0 & 1 \end{pmatrix} = \begin{pmatrix} 0 & 1 & 0 \\ 0 & 0 & -1 \\ -1 & 0 & 0 \end{pmatrix}.$$

In the example just considered, both ways of calculating the matrix C turn out to be equivalent in complexity. However, one or the other approach may prove preferable in particular problems.

Example. Express a matrix of finite rotation in terms of the Eulerian angles.

Recall that the rotations are made here in the following order:

$$xyz \xrightarrow{\psi} x'y'z' \xrightarrow{\theta} x''y''z'' \xrightarrow{\varphi} \xi\eta\zeta.$$

The angle ψ is the angle of rotation about the z-axis, θ the angle of rotation of the trihedral $x'y'z'$ about the x'-axis, and φ the angle of rotation of the trihedral $x''y''z''$ about the z''-axis (see Fig. 3.2).

In solving this problem, one can use either the active or passive representation of the orthogonal transformation. In this example, the passive representation turns out to be much more convenient, since the matrices of the constituent rotations appear much simpler in the intermediate axes rather than the original frame xyz.

The matrix of the first rotation:

$$\Psi = \begin{pmatrix} \cos\psi & -\sin\psi & 0 \\ \sin\psi & \cos\psi & 0 \\ 0 & 0 & 1 \end{pmatrix}.$$

The matrix of the second rotation:

$$\Theta = \begin{pmatrix} 1 & 0 & 0 \\ 0 & \cos\theta & -\sin\theta \\ 0 & \sin\theta & \cos\theta \end{pmatrix}.$$

The matrix of the last rotation:

$$\Phi = \begin{pmatrix} \cos\varphi & -\sin\varphi & 0 \\ \sin\varphi & \cos\varphi & 0 \\ 0 & 0 & 1 \end{pmatrix}.$$

The resulting rotation matrix is determined by multiplying by the constituent rotation matrices in the direct order:

$$A = \Psi\Theta\Phi = \begin{pmatrix} A_{11} & A_{12} & A_{13} \\ A_{21} & A_{22} & A_{23} \\ A_{31} & A_{32} & A_{33} \end{pmatrix},$$

where

$$A_{11} = \cos\varphi\cos\psi - \sin\varphi\cos\theta\sin\psi, \quad A_{12} = -\sin\varphi\cos\psi - \cos\varphi\cos\theta\sin\psi, \quad A_{13} = \sin\theta\sin\psi,$$
$$A_{21} = \cos\varphi\sin\psi + \sin\varphi\cos\theta\cos\psi, \quad A_{22} = -\sin\varphi\sin\psi + \cos\varphi\cos\theta\cos\psi, \quad A_{23} = -\sin\theta\cos\psi,$$
$$A_{31} = \sin\varphi\sin\theta, \quad A_{32} = \cos\varphi\sin\theta, \quad A_{33} = \cos\theta.$$

Quaternion addition of rotations. Just as in the case of matrices, to the addition of rotations there corresponds the multiplication of unit quaternions (the group operation of the SU(2) group). However, the active and passive points of view on the transformations have considerable distinctions.

Active point of view. Let the first rotation be specified by a quaternion Λ,

$$\mathbf{R}' = \Lambda \circ \mathbf{R} \circ \overline{\Lambda},$$

and let the second rotation be defined by a quaternion M,

$$\mathbf{R}'' = \mathrm{M} \circ \mathbf{R}' \circ \overline{\mathrm{M}}.$$

Consequently, the resulting rotation

$$\mathbf{R}'' = \mathrm{M} \circ \Lambda \circ \mathbf{R} \circ \overline{\Lambda} \circ \overline{\mathrm{M}} = \mathrm{M} \circ \Lambda \circ \mathbf{R} \circ \overline{(\mathrm{M} \circ \Lambda)}$$

is specified by the quaternion $N = \mathrm{M} \circ \Lambda$.

Just as in the case of orthogonal matrices, to the addition of rotations in the case of the active representation there corresponds the product of the quaternions of the constituent rotations in the *reverse order*. In this case, all quaternions are specified in the common basis $\mathbf{i}_k$ $(k = 1, 2, 3)$.

Passive point of view. Let us find out first how the representation of a vector changes under the transition to the new basis $\mathbf{i}'_k = \Lambda \circ \mathbf{i}_k \circ \overline{\Lambda}$.

The decomposition of a vector $\mathbf{R}$ in the old basis has the form

$$\mathbf{R} = x\mathbf{i}_1 + y\mathbf{i}_2 + z\mathbf{i}_3.$$

After the transition to the new basis, we obtain

$$\mathbf{R}^{(\prime)} = x\overline{\Lambda} \circ \mathbf{i}_1 \circ \Lambda + y\overline{\Lambda} \circ \mathbf{i}_2 \circ \Lambda + z\overline{\Lambda} \circ \mathbf{i}_3 \circ \Lambda = \overline{\Lambda} \circ (x\mathbf{i}'_1 + y\mathbf{i}'_2 + z\mathbf{i}'_3) \circ \Lambda = x'\mathbf{i}'_1 + y'\mathbf{i}'_2 + z'\mathbf{i}'_3.$$

Since the direction cosines of the finite rotation axis are the same in both the initial and rotated axes, we have $\Lambda = \lambda_0 + \lambda_1\mathbf{i}'_1 + \lambda_2\mathbf{i}'_2 + \lambda_3\mathbf{i}'_3 = \lambda_0 + \lambda_1\mathbf{i}_1 + \lambda_2\mathbf{i}_2 + \lambda_3\mathbf{i}_3$. Then one can consider the quaternions in the above expression of $\mathbf{R}^{(\prime)}$ to be specified in the basis $\mathbf{i}'_k$.

Let a quaternion M define another rotation of the basis, $\mathbf{i}''_k = \mathrm{M} \circ \mathbf{i}'_k \circ \overline{\mathrm{M}}$. Then, by analogy with the preceding, we obtain

$$\mathbf{R}^{(\prime\prime)} = \overline{\mathrm{M}} \circ (x'\mathbf{i}''_1 + y'\mathbf{i}''_2 + z'\mathbf{i}''_3) \circ \mathrm{M} = \overline{\mathrm{M}} \circ \overline{\Lambda} \circ (x\mathbf{i}''_1 + y\mathbf{i}''_2 + z\mathbf{i}''_3) \circ \Lambda \circ \mathrm{M}.$$

Recall that the designation $\mathbf{R}^{(\prime\prime)}$ indicates that the vector $\mathbf{R}$ remains the same, what is changed are only its coordinates in the new basis $\mathbf{i}''_1, \mathbf{i}''_2, \mathbf{i}''_3$. Thus, the quaternion of the resulting rotation is expressed as

$$N = \Lambda \circ \mathrm{M},$$

i.e., it is equal to the product of the constituent rotation quaternions in the direct order. In accordance with the preceding, the components of the quaternion are specified in the rotated bases. Although the quaternions to be multiplied are specified in different bases, they are formally represented in the last basis when being multiplied. The result turns out to be written in the original basis $\mathbf{i}_k$, since

$$N = n_0 + n_1\mathbf{i}''_1 + n_2\mathbf{i}''_2 + n_3\mathbf{i}''_3 = n_0 + n_1\mathbf{i}_1 + n_2\mathbf{i}_2 + n_3\mathbf{i}_3.$$

Example. Consider the penultimate example again (see Fig. 3.11). It is now required to find the quaternion of the resulting rotation, which will give information about the angle of the equivalent Eulerian rotation and the axis about which this rotation is to be made. Let us first follow the active point of view. The first rotation is specified by the quaternion $\Lambda = \cos\frac{\pi}{4} + \mathbf{i}_1 \sin\frac{\pi}{4}$ and the second by the quaternion $\mathrm{M} = \cos\frac{\pi}{4} + \mathbf{i}_2 \sin\frac{\pi}{4}$. The resulting rotation is given by the quaternion

$$N = \mathrm{M} \circ \Lambda = \left(\frac{\sqrt{2}}{2} + \mathbf{i}_2\frac{\sqrt{2}}{2}\right) \circ \left(\frac{\sqrt{2}}{2} + \mathbf{i}_1\frac{\sqrt{2}}{2}\right) = \frac{1}{2}(1 + \mathbf{i}_1 + \mathbf{i}_2 - \mathbf{i}_3),$$

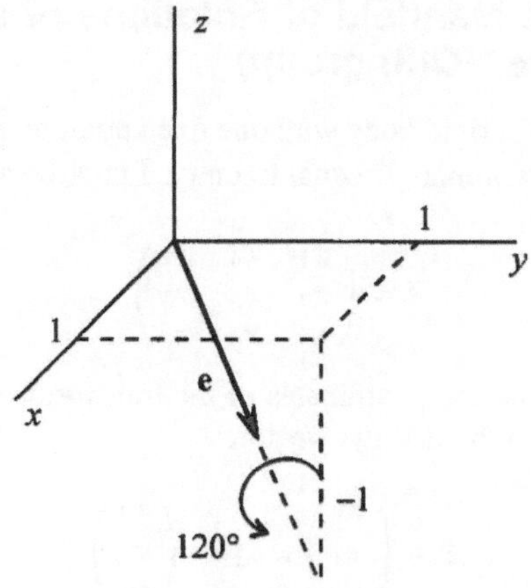

Figure 3.13

that is, the equivalent Eulerian rotation is performed about the vector $\mathbf{e} = \frac{1}{\sqrt{3}}(\mathbf{i}_1 + \mathbf{i}_2 - \mathbf{i}_3)$ through the angle 120° (see Fig. 3.13).

Carry out similar calculations from the passive point of view. The quaternion of the first rotation remains the same: $\Lambda = \cos\frac{\pi}{4} + \mathbf{i}_1 \sin\frac{\pi}{4}$. The quaternion of the second rotation, which is specified in the turned axes, is now represented differently: $\mathrm{M} = \cos\frac{\pi}{4} - \mathbf{i}_3 \sin\frac{\pi}{4}$. The resulting rotation is given by the quaternion

$$N = \Lambda \circ \mathrm{M} = \left(\frac{\sqrt{2}}{2} + \mathbf{i}_1'' \frac{\sqrt{2}}{2}\right) \circ \left(\frac{\sqrt{2}}{2} - \mathbf{i}_3'' \frac{\sqrt{2}}{2}\right) = \frac{1}{2}(1 + \mathbf{i}_1'' + \mathbf{i}_2'' - \mathbf{i}_3'') = \frac{1}{2}(1 + \mathbf{i}_1 + \mathbf{i}_2 - \mathbf{i}_3).$$

The composition of rotations can also be performed in terms of Cayley–Klein parameters. To the sum of rotations there corresponds the product of appropriate matrices. The rules of multiplication are analogous to those used to calculate the quaternion N.

Example. Find the relationship between the angles of finite rotation of the first kind (see Fig. 3.3) and the parameters of the equivalent Eulerian rotation. In this case, the quaternions of the consecutive rotations are the simplest in the passive representation. In accordance with the preceding, the quaternions must formally be written in the original basis and multiplied in the direct order:

$$A = \cos\frac{\alpha}{2} + \mathbf{i}_1 \sin\frac{\alpha}{2}, \quad B = \cos\frac{\beta}{2} + \mathbf{i}_2 \sin\frac{\beta}{2}, \quad C = \cos\frac{\gamma}{2} + \mathbf{i}_3 \sin\frac{\gamma}{2}.$$

The resulting rotation is determined by the quaternion

$$\Lambda = A \circ B \circ C.$$

Let us write out the result coordinatewise:

$$\lambda_0 \equiv \cos\frac{\varphi}{2} = \cos\frac{\alpha}{2}\cos\frac{\beta}{2}\cos\frac{\gamma}{2} - \sin\frac{\alpha}{2}\sin\frac{\beta}{2}\sin\frac{\gamma}{2},$$

$$\lambda_1 \equiv x\sin\frac{\varphi}{2} = \sin\frac{\alpha}{2}\cos\frac{\beta}{2}\cos\frac{\gamma}{2} + \cos\frac{\alpha}{2}\sin\frac{\beta}{2}\sin\frac{\gamma}{2},$$

$$\lambda_2 \equiv y\sin\frac{\varphi}{2} = \cos\frac{\alpha}{2}\sin\frac{\beta}{2}\cos\frac{\gamma}{2} - \sin\frac{\alpha}{2}\cos\frac{\beta}{2}\sin\frac{\gamma}{2},$$

$$\lambda_3 \equiv z\sin\frac{\varphi}{2} = \cos\frac{\alpha}{2}\cos\frac{\beta}{2}\sin\frac{\gamma}{2} + \sin\frac{\alpha}{2}\sin\frac{\beta}{2}\cos\frac{\gamma}{2}.$$

The Eulerian rotation angle φ can be found from the first relation. After that the projections x, y, and z of the unit vector determining the Eulerian rotation axis can be found.

3.3.3. Topology of the Manifold of Rotations of a Rigid Body (Topology of the SO(3) group)

Consider the set of positions of a rigid body with one fixed point or, what is the same, the set of all orthogonal matrices whose determinant is equal to unity. Let A be an orthogonal matrix with unit determinant,

$$A = \begin{pmatrix} x_1 & x_4 & x_7 \\ x_2 & x_5 & x_8 \\ x_3 & x_6 & x_9 \end{pmatrix}.$$

Recall that the columns of A are the coordinates of the transforms of the basis vectors under the transformation with matrix A. If the first two vectors

$$\mathbf{i}' = \begin{pmatrix} x_1 \\ x_2 \\ x_3 \end{pmatrix}, \quad \mathbf{j}' = \begin{pmatrix} x_4 \\ x_5 \\ x_6 \end{pmatrix}$$

are known, then the third vector is uniquely determined by the orthogonality condition and the condition $\det A = 1$. Therefore the set of all orthogonal matrices, which forms the SO(3) group, can be represented as the intersection of the following three hypersurfaces in a six-dimensional space:

$$\begin{aligned} &x_1^2 + x_2^2 + x_3^2 = 1, \\ &x_4^2 + x_5^2 + x_6^2 = 1, \\ &x_1 x_4 + x_2 x_5 + x_3 x_6 = 0. \end{aligned}$$

This is a geometric interpretation of the SO(3) group immersed into a six-dimensional space.

The most important topological characteristic of a manifold is the property of being simply connected. A manifold is said to be *simply connected* if any closed contour on it can be continuously deformed to a point. For example, this condition is satisfied for a sphere in the three-dimensional space and is not satisfied for a torus. A closed contour on the SO(3) group corresponds to a continuous rotation of a rigid body which starts from the initial position of the body and ends at the initial position again.

THEOREM. *The SO(3) group is not simply connected. All closed curves on it can be divided into two classes; the first class involves the curves that can be contracted into a point by a continuous deformation, and the second class comprises the curves deformable to a curve corresponding to the rotation of the body through the angle 2π about a fixed axis.*

Proof. It is convenient to represent the configuration manifold of the rigid body as follows. By Euler's theorem, any rotation can be determined by the axis of finite rotation, $\mathbf{e}$, and the angle of the right-handed rotation about it, φ. Let us form a vector $\varphi\mathbf{e}$, $0 \leq \varphi \leq \pi$, in the three-dimensional space. There is no one-to-one correspondence between the set of positions of the body and the point of the ball introduced, since the rotation about $\mathbf{e}$ through the angle π results in the same position of the body as the rotation about $-\mathbf{e}$ through the angle π. If, however, we identify the diametrically opposite points of the ball surface (i.e., the points of the sphere of radius π) with each other, then we obtain a set which is in one-to-one correspondence with the rotations. All closed trajectories in the ball can be classified into two types: internal trajectories (see Fig. 3.14a) and trajectories that reach the surface and then issue from the diametrically opposite identical point (Fig. 3.14b).

The trajectories of the first type are continuously deformable to the origin. The trajectories of the second type cannot be deformed to the origin, since they are discontinuous (the rotation itself is continuous). These can be deformed to the diameter of the ball which corresponds to the rotation of the body though the angle 2π about the diameter. Figure 3.14b shows a trajectory with one emergence on the surface. The general case includes two subcases: (i) an even and (ii) an odd

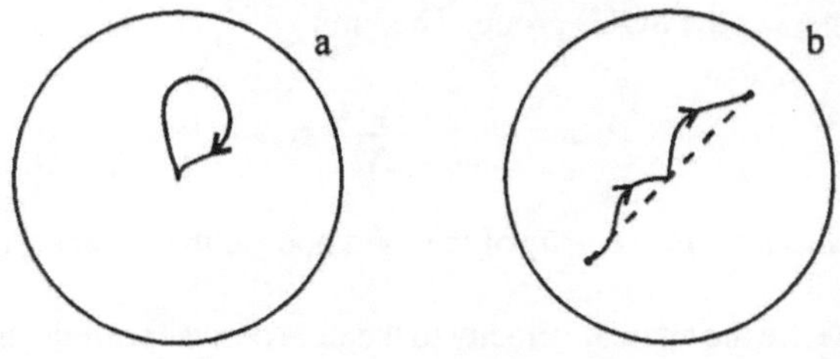

Figure 3.14

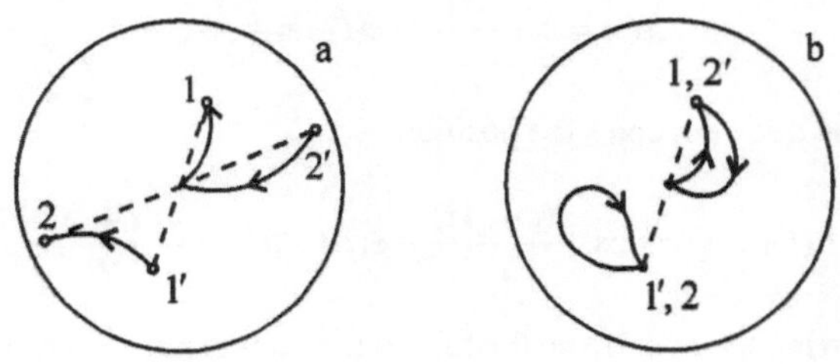

Figure 3.15

number of emergences. It is not difficult to understand that in the first case the curve is continuously deformable to the origin and in the second case, to the diameter.

Consider an example of two emergences on the surface (see Fig. 3.15a). The curve first reaches the surface at point 1 and then issues from point 1′ to reach the surface for the second time at point 2. After that the curves issues from point 2′ and finally returns to the starting point. Obviously, point 2 can be continuously deformed to point 1′; in this case, point 2′ is continuously transferred to point 1 (Fig. 3.15b). After that the closed loop at 1′, 2 continuously deforms to point 1′, 2 and the other loop deforms to the origin. This proves the theorem.

The set of rotations represented by quaternions with unit norm is a manifold which is a three-dimensional sphere, $\lambda_0^2+\lambda_1^2+\lambda_2^2+\lambda_3^2=1$, in the four-dimensional space $\lambda_0, \lambda_1, \lambda_2, \lambda_3$. This manifold is simply connected but is not in one-to-one correspondence with the SO(3) group. The point is that two quaternions Λ and $-\Lambda$ specify the same position of the rigid body. The group of unit quaternions is referred to as a *simply connected covering* of the rotation group. Without going into details of this term, we note that a covering of a group is constructed as follows: each element of the covering group is an element of the original group which is completed by the indication of a trajectory that connects this element with one a priori fixed. All such trajectories connecting the two elements are declared equivalent if they are continuously deformable to each other. A covering manifold of a group is called a *development* of this group. For example, the straight line $\mathbb{R}^1$ is a development of a circle (SO(2) group).

3.3.4. Angular Velocity of a Rigid Body

Definition of the angular velocity. Let a rigid body continuously rotate about a fixed point. Define the notion of the instantaneous angular velocity at an instant of time t as follows. Take the position of the body at this instant to be the initial position. According to Euler's theorem, the position of the body at some close instant $t+\Delta t$ can be obtained by a rotation of the body from the initial position about some axis identified by a unit vector $\varepsilon(t)$ through an angle $\Delta\varphi(t)$. Hence, this position can be described by the vector of finite rotation, $\Delta\varphi(t)\,\varepsilon(t)$, introduced above in proving the theorem of non-simple-connectedness of the SO(3) group. The limit

$$\omega = \lim_{\Delta t \to 0} \frac{\Delta\varphi(t)}{\Delta t}\varepsilon(t)$$

is called the *instantaneous angular velocity* of the rigid body at the instant t, provided that this limit exists.

It is convenient to relate the angular velocity to the time derivative of the body position specified in one of the ways presented in the previous subsection. The above definition complies best of all with the identification of the current position of the body with the quaternion

$$\Lambda(t) = \cos\frac{\varphi(t)}{2} + \mathbf{e}(t)\sin\frac{\varphi(t)}{2}.$$

At the instant $t + \Delta t$ the body occupies the position

$$\Lambda(t+\Delta t) = \cos\frac{\varphi(t+\Delta t)}{2} + \mathbf{e}(t+\Delta t)\sin\frac{\varphi(t+\Delta t)}{2}.$$

The quaternion transferring the body from the first to the second position is given by

$$\Delta\Lambda = \cos\frac{\Delta\varphi}{2} + \varepsilon\sin\frac{\Delta\varphi}{2} \approx 1 + \varepsilon\frac{\Delta\varphi}{2}.$$

From the active point of view, in accordance with the law of addition of rotations, we have

$$\Lambda(t+\Delta t) = \Delta\Lambda \circ \Lambda(t) \approx \left(1 + \varepsilon\frac{\Delta\varphi}{2}\right) \circ \Lambda(t).$$

Whence,

$$\dot{\Lambda} = \lim_{\Delta t \to 0} \frac{\Lambda(t+\Delta t) - \Lambda(t)}{\Delta t} = \frac{1}{2}\omega \circ \Lambda(t).$$

Thus, the angular velocity is expressed in terms of the derivative of the quaternion as

$$\omega = 2\dot{\Lambda} \circ \overline{\Lambda}.$$

Here, both the vector ω and the quaternion Λ are defined in the fixed frame.

From the passive point of view, the law of addition of rotations yields

$$\Lambda(t+\Delta t) = \Lambda \circ \Delta\Lambda \approx \Lambda \circ \left(1 + \varepsilon\frac{\Delta\varphi}{2}\right).$$

Whence,

$$\omega = 2\overline{\Lambda} \circ \dot{\Lambda}.$$

Here, both ω and Λ are defined in the frame rigidly attached to the body. As was indicated previously, the quaternion Λ has the same form in both cases, but this is not the case for the instantaneous angular velocity ω.

Addition of angular velocities. The problem is stated as follows. A rigid body rotates arbitrarily relative to a fixed reference frame xyz. The body is represented by a frame $x'y'z'$ rigidly attached to it. The position of the trihedral $x'y'z'$ relative to the trihedral xyz is specified by a quaternion $\Lambda(t)$. The angular velocity of the frame $x'y'z'$ relative to the frame xyz is given by $\omega_1 = 2\dot{\Lambda} \circ \overline{\Lambda}$. In turn, relative to this body rotates another body represented by a trihedral $x''y''z''$. The position of $x''y''z''$

relative to $x'y'z'$ is specified by a quaternion M(t), and the angular velocity of the trihedral $x''y''z''$ with respect to $x'y'z'$ is given by $\omega_2 = 2\dot{\mathrm{M}} \circ \overline{\mathrm{M}}$ (in the frame $x'y'z'$).

The question is: What is the angular velocity of the trihedral $x''y''z''$ relative to the fixed trihedral xyz?

In accordance with the rule of addition of rotations (the passive point of view), the position of $x''y''z''$ relative to xyz is determined by

$$N = \Lambda \circ \mathrm{M}.$$

Consequently, the desired angular velocity is expressed as

$$\omega = 2\dot{N} \circ \overline{N} = 2(\dot{\Lambda} \circ \mathrm{M} + \Lambda \circ \dot{\mathrm{M}}) \circ \overline{\mathrm{M}} \circ \overline{\Lambda} = 2\dot{\Lambda} \circ \overline{\Lambda} + 2\Lambda \circ \dot{\mathrm{M}} \circ \overline{\mathrm{M}} \circ \overline{\Lambda}.$$

The first term is ω_1 in projections onto xyz and the second term is ω_2 in projections onto $x'y'z'$ ($\omega_2 = 2\dot{\mathrm{M}} \circ \overline{\mathrm{M}}$) which is projected onto xyz. Thus, if the angular velocities ω_1 and ω_2 are projected onto the same frame, then

$$\omega = \omega_1 + \omega_2.$$

Euler's formula. Euler's formula specifies how the linear velocities of the points of a rotating rigid body are distributed. Let $\mathbf{r}$ be the initial position of a point of the body in the fixed reference frame xyz and let $\mathbf{r}'(t)$ denote the position of this point in the course of the rotation. We have

$$\mathbf{r}'(t) = \Lambda(t) \circ \mathbf{r} \circ \overline{\Lambda}(t),$$

where $\Lambda(t)$ is the quaternion determining the rotation from $\mathbf{r}$ to $\mathbf{r}'(t)$. The velocity of the point is given by

$$\dot{\mathbf{r}}'(t) = \dot{\Lambda}(t) \circ \mathbf{r} \circ \overline{\Lambda}(t) + \Lambda(t) \circ \mathbf{r} \circ \dot{\overline{\Lambda}}(t).$$

Express this velocity in terms of the position $\mathbf{r}'$ of the point. To this end, we substitute the relation $\mathbf{r} = \overline{\Lambda} \circ \mathbf{r}' \circ \Lambda$ into this formula to obtain

$$\dot{\mathbf{r}}' = \dot{\Lambda} \circ \overline{\Lambda} \circ \mathbf{r}' + \mathbf{r}' \circ \Lambda \circ \dot{\overline{\Lambda}} = \tfrac{1}{2}(\omega \circ \mathbf{r}' + \mathbf{r}' \circ \overline{\omega}) = \omega \times \mathbf{r}'.$$

The velocity distribution law

$$\dot{\mathbf{r}}' = \omega \times \mathbf{r}'$$

is referred to as Euler's formula. By virtue of invariant representation of this formula, it is valid not only in the frame xyz but also in any other frame. Euler's formula can be rewritten in matrix form as

$$\dot{\mathbf{r}}' = \Omega \mathbf{r}'.$$

Here, the quantity Ω is referred to as the matrix of the angular velocity; in effect it determines the operator of the vector product. This matrix is skew-symmetric; the following two representations are widely used:

$$\Omega = \begin{pmatrix} 0 & -\omega_z & \omega_y \\ \omega_z & 0 & -\omega_x \\ -\omega_y & \omega_x & 0 \end{pmatrix}$$

in projections onto the fixed axes xyz and

$$\Omega = \begin{pmatrix} 0 & -r & q \\ r & 0 & -p \\ -q & p & 0 \end{pmatrix}$$

in projections onto the axes of a body-attached reference frame $\xi\eta\zeta$, where p, q, and r is the traditional (Eulerian) notation for these projections.

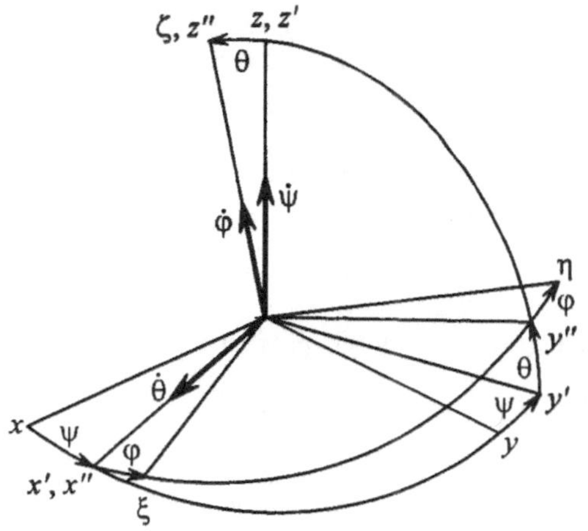

Figure 3.16

Euler's kinematic equations. These equations express the relationship between the projections of the angular velocity of the body onto the body-attached axes and the time derivatives of the Eulerian angles. This relationship can be established by representing an arbitrary rotation of the body as the composition of three plane rotations and by utilizing the above law of addition of angular velocities (see Fig. 3.16),

$$xyz \xrightarrow{\psi} x'y'z' \xrightarrow{\theta} x''y''z'' \xrightarrow{\varphi} \xi\eta\zeta.$$

The angular velocities of the constituent rotations are expressed as

$$\dot{\boldsymbol{\Psi}} = \begin{pmatrix} 0 \\ 0 \\ \dot{\psi} \end{pmatrix} \quad \text{in the frame } x'y'z',$$

$$\dot{\boldsymbol{\Theta}} = \begin{pmatrix} \dot{\theta} \\ 0 \\ 0 \end{pmatrix} \quad \text{in the frame } x''y''z'',$$

$$\dot{\boldsymbol{\Phi}} = \begin{pmatrix} 0 \\ 0 \\ \dot{\varphi} \end{pmatrix} \quad \text{in the frame } \xi\eta\zeta.$$

One should add these velocities together having them projected first onto the axes $\xi\eta\zeta$. In doing so, we obtain

$$\omega = \begin{pmatrix} \cos\varphi & \sin\varphi & 0 \\ -\sin\varphi & \cos\varphi & 0 \\ 0 & 0 & 1 \end{pmatrix} \begin{pmatrix} 1 & 0 & 0 \\ 0 & \cos\theta & \sin\theta \\ 0 & -\sin\theta & \cos\theta \end{pmatrix} \begin{pmatrix} 0 \\ 0 \\ \dot{\psi} \end{pmatrix} + \begin{pmatrix} \cos\varphi & \sin\varphi & 0 \\ -\sin\varphi & \cos\varphi & 0 \\ 0 & 0 & 1 \end{pmatrix} \begin{pmatrix} \dot{\theta} \\ 0 \\ 0 \end{pmatrix} + \begin{pmatrix} 0 \\ 0 \\ \dot{\varphi} \end{pmatrix}, \quad \omega = \begin{pmatrix} p \\ q \\ r \end{pmatrix}.$$

Whence,

$$p = \dot{\psi}\sin\varphi\sin\theta + \dot{\theta}\cos\varphi,$$
$$q = \dot{\psi}\cos\varphi\sin\theta - \dot{\theta}\sin\varphi,$$
$$r = \dot{\psi}\cos\theta + \dot{\varphi}.$$

If the angular velocity projections $p(t)$, $q(t)$, and $r(t)$ are specified and the body position in terms of the Eulerian angles is to be determined, then the following system of equations must be solved:

$$\dot{\psi} = \frac{p(\sin\varphi + q\cos\varphi)}{\sin\theta},$$
$$\dot{\theta} = p\cos\varphi - q\sin\varphi,$$
$$\dot{\varphi} = r - (p\sin\varphi + q\cos\varphi)\cot\theta.$$

This is a nonlinear system of ordinary differential equations with variable coefficients which has a singularity at $\theta = 0$.

Poisson's equations are more convenient for solving the problem determining the body position for given angular velocity.

Poisson's kinematic equations. Consider the rotation of a rigid body about a fixed point in terms of the basic variables determining the position of the body—the parameters $A(t)$ of the SO(3) group. The position of any point of the body is given by

$$\mathbf{r}' = A(t)\,\mathbf{r}$$

and the velocity is expressed as

$$\dot{\mathbf{r}}' = \dot{A}(t)\,\mathbf{r} = \dot{A}(t)\,A^{\mathrm{T}}(t)\,\mathbf{r}'.$$

Applying Euler's formula in matrix form, we obtain

$$\Omega\mathbf{r}' = \dot{A}(t)\,\mathbf{r} = \dot{A}(t)\,A^{\mathrm{T}}(t)\,\mathbf{r}'.$$

Whence,

$$\dot{A} = \Omega A.$$

This matrix formula represents Poisson's equations in projections onto the axes xyz. To rewrite these equations in projections onto the axes $\xi\eta\zeta$, we note that

$$\begin{pmatrix} 0 & -\omega_z & \omega_y \\ \omega_z & 0 & -\omega_x \\ -\omega_y & \omega_x & 0 \end{pmatrix} = A \begin{pmatrix} 0 & -r & q \\ r & 0 & -p \\ -q & p & 0 \end{pmatrix} A^{\mathrm{T}}.$$

Thus, in projections onto the axes $\xi\eta\zeta$,

$$\dot{A} = A\Omega.$$

Poisson's kinematic equations possess significant advantages compared with Euler's kinematic equations—the former are linear and have no singularities. The drawback is that the former equations have a higher dimensionality compared with the latter.

Linear kinematic equations of minimum dimensionality can be obtained in terms of quaternions. As was established previously, these equations have the form

$$2\dot{\Lambda} = \omega \circ \Lambda \quad \text{in the axes } xyz,$$
$$2\dot{\Lambda} = \Lambda \circ \omega \quad \text{in the axes } \xi\eta\zeta.$$

The main property of Poisson's equations is the following. To construct the general solution of Poisson's equations it suffices to know only one particular solution, as contrasted with the general case of linear systems, where the number of linearly independent particular solutions required is equal to the order of the system. We will demonstrate this.

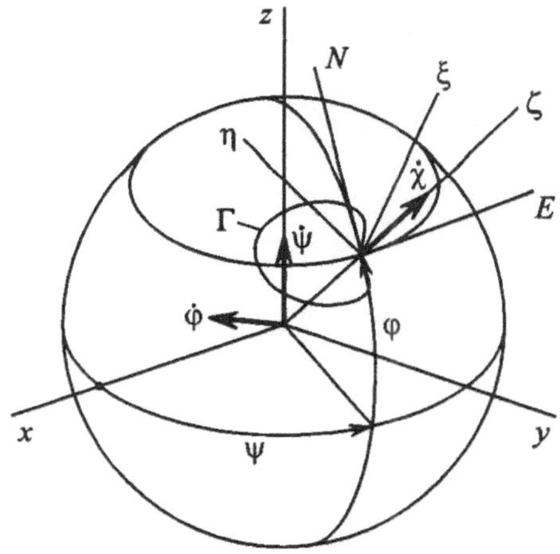

Figure 3.17

Suppose we know a particular solution $A_*(t)$. We will seek the general solution in the form

$$A(t) = CA_*(t),$$

where C is a constant orthogonal matrix. The general solution will be found if we succeed in determining C so as to satisfy any prescribed initial conditions $A(0) = A_0$. We have

$$A(0) = CA_*(0) = A_0 \quad \Longrightarrow \quad C = A_0 A_*^{\mathrm{T}}(0).$$

Hence, the general solution has the form

$$A = A_0 A_*^{\mathrm{T}}(0)\, A_*(t).$$

Thus, it turns out that all solutions are congruent to each other, since these can be obtained one from another by a rotation transformation.

Ishlinskii's theorem of solid angle.[4] As follows from the preceding, the kinematic equations for a rigid body—represented in one or another form (Euler's equations or Poisson's equations in terms of orthogonal matrices or quaternions)—have an appreciably more complicated structure than the kinematic equations for a point particle,

$$\dot{x} = v_x(t), \quad \dot{y} = v_y(t), \quad \dot{z} = v_z(t).$$

If the particle velocity has zero projection onto some axis (e.g., $v_x(t) \equiv 0$), then there is no change in the corresponding coordinate (in our case x).

This is not the case for a rigid body. If the angular velocity of the body has zero projection onto some axis, it cannot be stated that the body does not move about this axis.

Prior to formulating this fact exactly, we note that the manifold of positions of a rigid body with one fixed point (the SO(3) group) is not Euclidean, unlike a point particle, whose configuration manifold is a three-dimensional Euclidean space.

THEOREM (ISHLINSKII). *If some axis* $\mathbf{l}$, *rigidly fixed in the body, describes in the process of motion a closed conic surface, while the projection of the angular velocity of the body onto this axis is identically zero,* $\mathbf{l} \cdot \omega = 0$, *then after the axis returns to its initial position the body is found to be turned about this axis through the angle equal to the solid angle of the cone described by the axis* $\mathbf{l}$ *in the fixed space.*

Proof. To prove this theorem, it is convenient to use the "geographic" coordinates to specify the current position of the moving trihedral $\xi\eta\zeta$ (see Fig. 3.17). The orientation of the moving

[4] It was first proved by A. Yu. Ishlinskii in 1952.

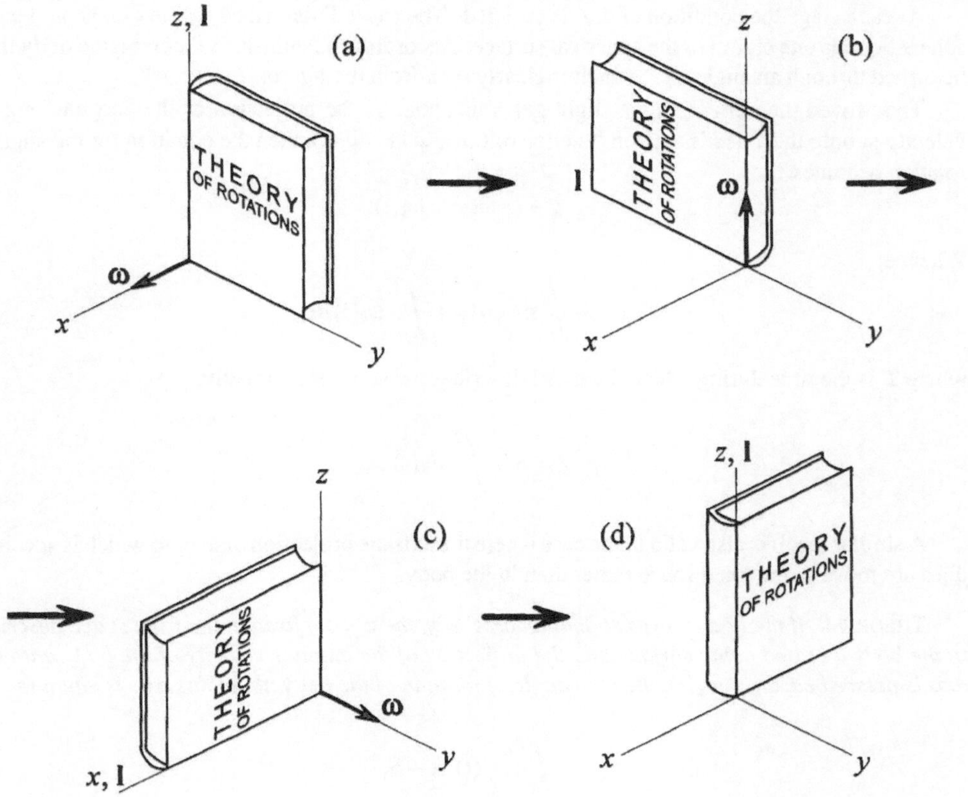

Figure 3.18

trihedral $\xi\eta\zeta$ relative to the fixed trihedral xyz is defined by three angles: ψ (longitude), φ (latitude), and χ (azimuth). To avoid overloading of the figure, we shifted the origin of the moving trihedral relative to that of the fixed trihedral. In the figure, the axis **l** in question coincides with the ζ-axis and describes a closed curve Γ on the surface of a unit sphere.

Express the condition $\mathbf{l} \cdot \boldsymbol{\omega} \equiv 0$ in terms of the angles ψ, φ, and χ explicitly. To this end, we use the rule of addition of angular velocities. The angular velocity $\dot{\varphi}$ is perpendicular to the meridional plane and, hence, has no projection onto the ζ-axis. The projection of $\dot{\psi}$ onto this axis is equal to $\dot{\psi} \sin\varphi$. The angular velocity $\dot{\chi}$ is parallel to the ζ-axis. Thus,

$$\mathbf{l} \cdot \boldsymbol{\omega} \equiv r = \dot{\chi} + \dot{\psi} \sin\varphi = 0.$$

Whence,

$$\chi = -\oint_{\Gamma} \sin\varphi \, d\psi.$$

Taking advantage of Green's theorem, we find that

$$\chi = \iint_S \cos\varphi \, d\varphi \, d\psi = \iint_S dS = S,$$

where S is the area of the domain on the sphere bounded by the curve Γ. By definition, this is just the solid angle of the cone with generatrix Γ. This proves the theorem.

Illustrate the theorem by the following example. Figure 18a shows a body in an initial position. The ζ-axis of the body-attached reference frame, which coincides initially with the z-axis, describes a closed conical surface resulting from three simple rotations shown in Figs. 18b–d.

At each stage the condition $\omega \perp \zeta$ is satisfied. The curve Γ described by the ζ-axis on the unit sphere bounds one eighth of the spherical surface. According to Ishlinskii's theorem, the body must be turned through an angle $\pi/2$, which is clearly seen from the figure.

The proved theorem admits a slight generalization. If the projection of the absolute angular velocity ω onto the axis $\mathbf{l}$ is a given function of time, $\omega \cdot \mathbf{l} = \omega_l(t)$, then the equation for the angle of rotation becomes

$$\dot{\chi} + \dot{\psi} \sin \varphi = \omega_l(t).$$

Whence,

$$\chi = -\oint_\Gamma \sin \varphi \, d\psi + \int_0^T \omega_l(t)\, dt,$$

where T is the time during which the axis $\mathbf{l}$ describes the contour Γ. Finally,

$$\chi = S + \int_0^T \omega_l(t)\, dt.$$

A similar result is also valid in the case where the axis the projection of ω onto which is specified does not move in the fixed space rather than in the body.

THEOREM. *If the body moves so that an axis (say, the z-axis) fixed in the fixed space describes in the body a closed conic surface and the projection of the angular velocity of the body onto this axis is prescribed,* $\omega_z = \omega_z(t)$, *then the angle of rotation of the body about this axis is equal to*

$$\int_0^T \omega_z(t)\, dt - S,$$

where S *is the solid angle of the cone described by the axis in the body and* T *is the time in which the axis describes this cone.*

Proof. The theorem can easily be proved by considering the rotating body to be fixed and the fixed space to be moving relative to the body. Then the angular velocity of the space relative to the body is equal to the negative of the angular velocity of the body relative to the space. For determining the orientation of the space relative to the body, Ishlinskii's theorem can be applied. We have

$$\chi = S - \int_0^T \omega_z(t)\, dt.$$

This proves the theorem, since the angle of rotation of the body relative to the space is equal to the negative of that of the space relative to the body.

3.4. Problems of Mechanics Admitting Similarity Groups

3.4.1. Suslov Problem

Consider the problem of a point particle sliding down an inclined rough plane (see Fig. 3.19). The equations of motion have the form

$$m\ddot{x} = mg \sin \alpha - \frac{fmg \cos \alpha \, \dot{x}}{\sqrt{\dot{x}^2 + \dot{y}^2}}, \quad m\ddot{y} = -\frac{fmg \cos \alpha \, \dot{y}}{\sqrt{\dot{x}^2 + \dot{y}^2}},$$

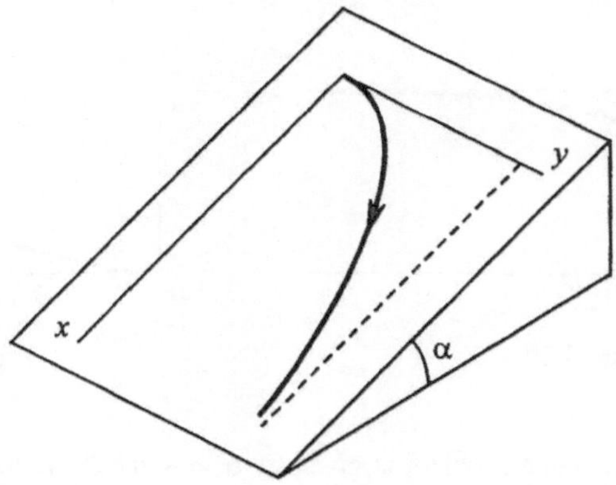

Figure 3.19

where m is the mass of the particle, α the angle of inclination of the plane to the horizon, and f the dry friction coefficient.

Introduce a scaled time t' so that $t = t'/\sqrt{g \sin \alpha}$ and denote $k = f \cot \alpha$. Then the above equations become

$$\ddot{x} = 1 - \frac{k\dot{x}}{\sqrt{\dot{x}^2 + \dot{y}^2}}, \quad \ddot{y} = -\frac{k\dot{y}}{\sqrt{\dot{x}^2 + \dot{y}^2}},$$

where the dots stand now for the differentiation with respect to t'.

For $k \neq 1$, the general solution of these equations can be found in Suslov (1946).[5] The case of $k = 1$ is critical—the friction force is equal to the projection of the gravitational force onto the plane or, equivalently, the angle of inclination of the plane to the horizon is equal to the so-called friction angle. If α exceeds the friction angle, then the particle cannot be at rest. If α is less than the friction angle, then the moving particle eventually comes to a state of rest for any initial conditions. The case $k = 1$ was not considered by Suslov. The expressions he obtained for the motion of the particle degenerate at $k = 1$. Here we consider this problem precisely for this case with the initial conditions $x(0) = y(0) = \dot{x}(0) = 0$ and $\dot{y}(0) = 1$. That is, the particle being at rest is subjected at the initial instant to an impulse directed horizontally. Introducing the notation $\dot{x} = u$, $\dot{y} = v$, and $\sqrt{\dot{x}^2 + \dot{y}^2} = w$, we arrive at the system

$$\dot{u} = 1 - \frac{u}{w}, \quad \dot{v} = -\frac{v}{w}; \qquad u(0) = 0, \quad v(0) = 1.$$

The group of symmetries of this system is obviously the similarity group with generator

$$U = u\frac{\partial}{\partial u} + v\frac{\partial}{\partial v}$$

and canonical coordinates

$$\alpha = \frac{u}{v}, \quad \beta = \ln v.$$

Switching to the new variables α and β, we reduce the system to the form

$$\dot{\alpha} = e^{-\beta}, \quad \dot{\beta} = -\frac{e^{-\beta}}{\sqrt{1 + \alpha^2}}; \qquad \alpha(0) = 0, \quad \beta(0) = 0.$$

Divide the first equation by the second to obtain

$$\frac{d\alpha}{d\beta} = -\sqrt{1 + \alpha^2}, \qquad \alpha(0) = 0.$$

[5] G. K. Suslov, *Theoretical Mechanics*, Gostekhizdat, Moscow, 1946 [in Russian].

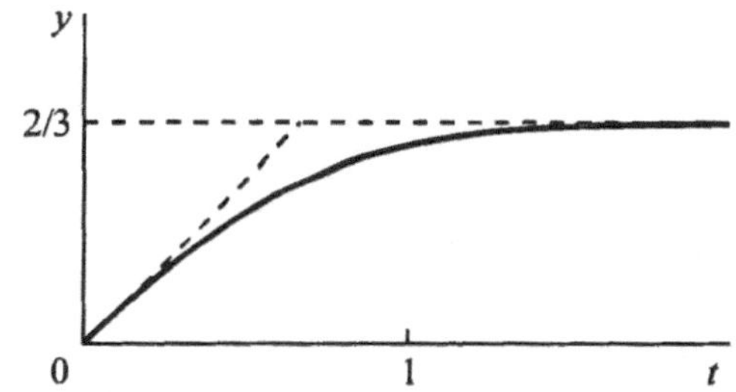

Figure 3.20

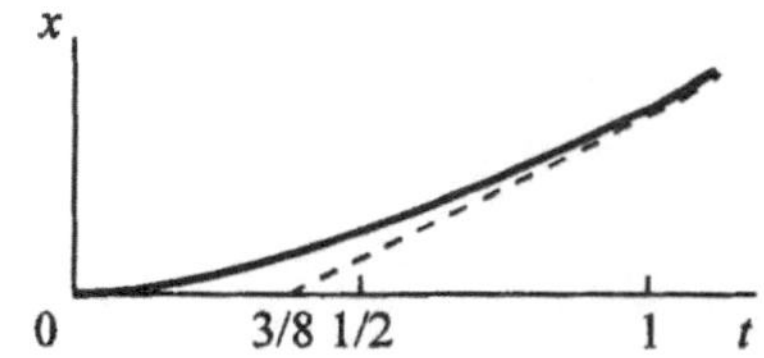

Figure 3.21

The solution of this initial value problem is given by $\alpha = -\sinh\beta$. Returning to the variables u and v, we find that

$$u = \tfrac{1}{2}(1 - v^2).$$

Substituting this expression into the equation $\dot{v} = -v/w$ yields

$$\dot{v} = -\frac{2v}{1+v^2}, \qquad v(0) = 1.$$

The solution of this equation has the form

$$\ln v + \frac{v^2}{2} = -2t + \frac{1}{2}.$$

Dividing the equation $\dot{y} = v$ by $\dot{v} = -2v/(1+v^2)$, we obtain

$$\frac{dy}{dv} = -\frac{1}{2}(1+v^2), \qquad y(1) = 0.$$

Whence,

$$y = -\frac{v}{2} - \frac{v^3}{6} + \frac{2}{3}.$$

Thus, the relations

$$t = \frac{1}{4} - \frac{v^2}{4} - \frac{1}{2}\ln v,$$
$$y = \frac{2}{3} - \frac{v}{2} - \frac{v^3}{6}$$

determine the solution $y(t)$ in parametric form. This solution is shown in Fig. 3.20.

Dividing $\dot{x} = u$ by $\dot{v} = -2v/(1+v^2)$ and taking into account the fact that $2u = 1 - v^2$, we arrive at the problem

$$\frac{dx}{dv} = -\frac{1-v^4}{4v}, \qquad x(1) = 0.$$

Whence follows the solution $x(t)$ in parametric form:

$$t = \frac{1}{4} - \frac{v^2}{4} - \frac{1}{2}\ln v,$$
$$x = -\frac{1}{16} + \frac{v^4}{16} - \frac{1}{4}\ln v.$$

This solution is shown in Fig. 3.21.

As $t \to \infty$, we have $v \to 0$ and $u \to \frac{1}{2}$. The coordinates $x(t)$ and $y(t)$ themselves are described by the asymptotic relations

$$x(t) \sim \frac{t}{2} - \frac{3}{16}, \quad y(t) \sim \frac{2}{3} - \frac{1}{2}e^{-2t}.$$

Thus, in the critical case considered, the motion of the point particle tends to a uniform motion with a speed of $1/2$ along the x-axis and tends to a limit value of $2/3$ in y.

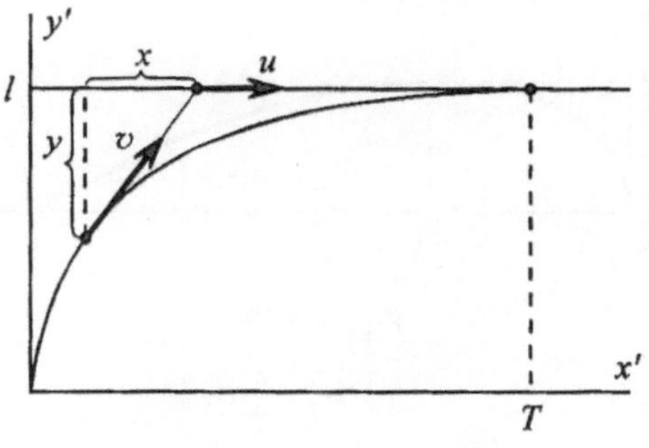

Figure 3.22

3.4.2. The Problem of the Follower Trajectory

Consider two points moving in a plane (Fig. 3.22). The followed point moves at a constant speed u in a straight line parallel to the horizontal axis and located at a distance l from it. The follower point moves at a constant speed $v > u$ in the direction of the followed point. At the initial instant the line connecting the two points is vertical and the follower point is at the horizontal axis so that the distance between the points is equal to l.

It is required to find the time in which the points meet.

The position of the followed point, $\{x, y\}$, will be considered relative to the reference frame with origin at the follower point. According to the law of addition of velocities, we have

$$v_x + \dot{x} = u, \quad v_y + \dot{y} = 0,$$

where v_x and v_y are the projections of the velocity of the follower point onto the coordinate axes; these projections are expressed as

$$v_x = \frac{vx}{\sqrt{x^2 + y^2}}, \quad v_y = \frac{vy}{\sqrt{x^2 + y^2}}.$$

Consequently, the equations of motion of the followed point in the reference frame associated with the follower point have the form

$$\dot{x} = u - \frac{vx}{\sqrt{x^2 + y^2}}, \quad \dot{y} = -\frac{vy}{\sqrt{x^2 + y^2}}.$$

These equations are identical with those in the Suslov problem considered above. All procedures applied in solving the Suslov problem can be applied to the problem in question.

By passing to canonical coordinates of the similarity group, $(x, y) \to (\alpha, \beta) : \alpha = x/y,\ \beta = \ln y$, we arrive at the problem

$$\frac{d\alpha}{d\beta} = -\frac{u}{v}\sqrt{1 + \alpha^2}, \qquad \beta(0) = \ln l, \quad \alpha(0) = 0.$$

Denoting $u/v = k < 1$, we write out the solution of this problem:

$$\alpha = -\sinh[k(\beta - \ln l)].$$

Then the equation for β resulting from the passage to the canonical coordinates,

$$\dot{\beta} = -\frac{v}{e^{\beta}\sqrt{1 + \alpha^2}},$$

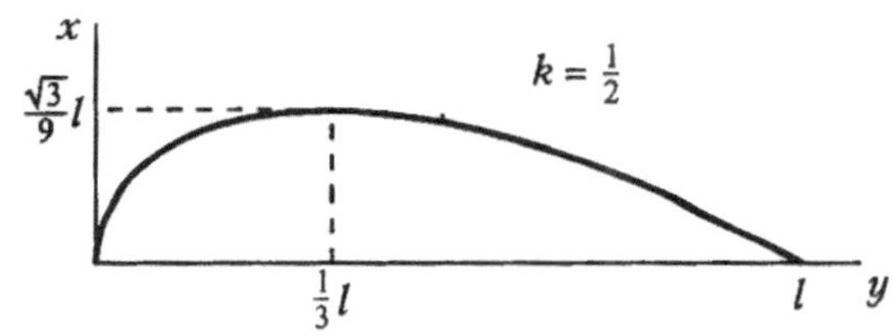

Figure 3.23

permits us to establish that

$$\int e^{\beta} \cosh[k(\beta - \ln l)]\, d\beta = C - vt,$$

or

$$(1-k)l^{-k}e^{\beta(1+k)} + (1+k)l^{k}e^{\beta(1-k)} = 2(1-k^2)(C - vt),$$

where C is the constant of integration.

Returning to the old variables, we obtain the relationship between y and t in implicit form:

$$(1-k)l^{-k}y^{1+k} + (1+k)l^{k}y^{1-k} = 2(1-k^2)(C - vt).$$

Since $y = l$ at $t = 0$, we find that

$$C = \frac{l}{1-k^2}.$$

At the meeting time $t = T$, we have $y = 0$; hence,

$$T = \frac{l}{v(1-k^2)} = \frac{lv}{v^2 - u^2}.$$

It is not difficult to determine the trajectory of motion of the followed point. To this end, we take l to be the scale of length, so that now $l = 1$. Then, returning to the original coordinates, from the equation $\alpha = -\sinh[k(\beta - \ln l)]$ we obtain

$$x = -\tfrac{1}{2}y(y^k - y^{-k}).$$

This is just the equation of the trajectory of the followed point in the reference frame associated with the follower point. Figure 3.23 depicts this trajectory for $k = 1/2$.

To find the trajectory of the follower point in the coordinates x', y', we substitute $x = ut - x'$ and $y = l - y'$ into the above equations to obtain

$$(1-k)(1-y')^{1+k} + (1+k)(1-y')^{1-k} = 2[1 - (1-k^2)vt],$$
$$ut - x' = -\tfrac{1}{2}(1-y')[(1-y')^k - (1-y')^{-k}].$$

Eliminating time from these equations, we arrive at the equation of the desired trajectory:

$$x' = \frac{2k + (1-k)(1-y')^{1+k} - (1+k)(1-y')^{1-k}}{1-k^2}.$$

The derivative has the form

$$\frac{dx'}{dy'} = (1-y')^{-k} - (1-y')^{k}.$$

It vanishes at $y' = 0$ and becomes infinite at $y' = 1$. Thus, the trajectory of the follower point has a vertical tangent at the beginning of the motion and a horizontal tangent at the meeting instant (see Fig. 3.22).

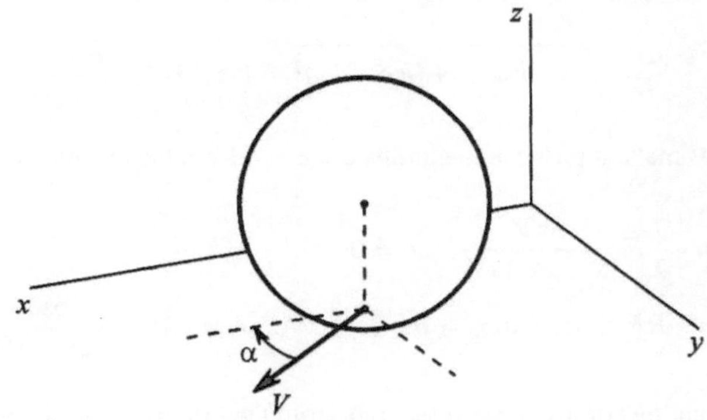

Figure 3.24

3.4.3. Rolling of a Homogeneous Ball Over a Rough Plane

Consider a homogeneous ball of radius R and mass m rolling over a rough plane. Introduce a fixed reference frame xyz such that the x- and y-axes lie in the plane of rolling (see Fig. 3.24). Then the equation of motion of the ball has the form

$$J\frac{d\omega}{dt} = \mathbf{M}, \quad m\ddot{\mathbf{r}} = \mathbf{F},$$

where $J = \frac{2}{5}mR^2$ is the moment of inertia of the ball about its center. The velocity of the point coinciding with the center of the contact spot is expressed as

$$V_x = \dot{x} - R\omega_y, \quad V_y = \dot{y} + R\omega_z, \qquad V = \sqrt{V_x^2 + V_y^2}.$$

It is assumed that the ball rolls by its own momentum, the xy-plane is horizontal, and the forces and torques on the right-hand sides of the equations of motion are determined by the force of dry friction at the point of contact.

In the case of combined slip (translational slip at a speed V and pure spinning at an angular velocity ω_z), the magnitude of the friction force and that of the friction torque are defined as[6]

$$F = F_0\frac{3\pi V}{8\varepsilon\omega_z + 3\pi V}, \qquad M = M_0\frac{16\varepsilon\omega_z}{16\varepsilon\omega_z + 15\pi V}.$$

These expressions are derived by applying Coulomb's dry friction hypothesis to the area element of the contact spot followed by integrating the force element and torque element of the area of the contact spot. Here, F_0 is the friction force in pure slip ($\omega_z \equiv 0$), M_0 the friction torque in pure spinning ($V \equiv 0$), and ε the radius of the contact spot.

Without loss of generality, the angular velocity of spinning ω_z (the projection of the absolute angular velocity ω of the ball onto the z-axis) in the expressions of F and M can be considered positive ($\omega_z > 0$).

Since $M_x = -RF_y$ and $M_y = RF_x$, where $F_x = F\cos\alpha$ and $F_y = F\sin\alpha$, the complete system of dynamic equations of a ball rolling by its own momentum takes the form

$$m\ddot{x} = -F\cos\alpha, \quad m\ddot{y} = -F\sin\alpha,$$
$$J\dot{\omega}_x = -RF\sin\alpha, \quad J\dot{\omega}_y = RF\cos\alpha, \quad J\dot{\omega}_z = -M_0\frac{16\varepsilon\omega_z}{16\varepsilon\omega_z + 15\pi V},$$

[6] V. Ph. Zhuravlev, *The model of dry friction in the problem of the rolling of rigid bodies*, J. Appl. Maths Mechs, Vol. 62, No. 5, pp. 705–710, 1998.

where

$$V = \sqrt{(\dot{x} - R\omega_y)^2 + (\dot{y} + R\omega_x)^2}, \quad \tan\alpha = \frac{\dot{y} + R\omega_x}{\dot{x} - R\omega_y}.$$

The last relations make it possible to eliminate $\dot{x}$ and $\dot{y}$ from the equations of motion. We have

$$m\dot{V} = -\frac{7}{2}F_0\frac{3\pi V}{8\varepsilon\omega_z + 3\pi V}, \quad \dot{\alpha} = 0,$$

$$J\dot{\omega}_x = -RF\sin\alpha, \quad J\dot{\omega}_y = RF\cos\alpha, \quad J\dot{\omega}_z = -M_0\frac{16\varepsilon\omega_z}{16\varepsilon\omega_z + 15\pi V}.$$

The first and last equations of this system separate from the others. If we succeed in solving them for V and ω_z, then we can determine ω_x and ω_y by quadrature.

Introduce the notation

$$v = 3\pi V, \quad u = 8\varepsilon\omega_z, \quad \tau = \frac{21\pi}{m}F_0 t; \quad k = \frac{40M_0\varepsilon}{21\pi F_0 R^2}.$$

Then the equations concerned become

$$\dot{v} = -\frac{v}{u+v}, \quad \dot{u} = -\frac{ku}{2u+5v}.$$

This system admits a similarity group and, hence, can be reduced to quadratures by the passage to canonical coordinates.

Performing the appropriate change of variables $(u, v) \to (q, p)$,

$$q = \frac{u}{v}, \quad p = \ln v,$$

we obtain

$$\dot{q} = \frac{5 - k + (2-k)q}{(1+q)(5+2q)}qe^{-p}, \quad \dot{p} = -\frac{1}{1+q}e^{-p}.$$

Whence,

$$\frac{dq}{dp} = -\frac{5 - k + (2-k)q}{5+2q}q.$$

This is a separable equation. Integrating it yields

$$q^{\frac{5}{5-k}}\left[5 - k + (2-k)q\right]^{\frac{2k}{(5-k)(2-k)}}e^p = \text{const}.$$

In terms of (u, v) this relation acquires the form

$$\left(\frac{u}{v}\right)^{\frac{5}{5-k}}\left[5 - k + (2-k)\frac{u}{v}\right]^{\frac{2k}{(5-k)(2-k)}}v = \text{const}.$$

This is a first integral of the system

$$\dot{v} = -\frac{v}{u+v}, \quad \dot{u} = -\frac{ku}{2u+5v},$$

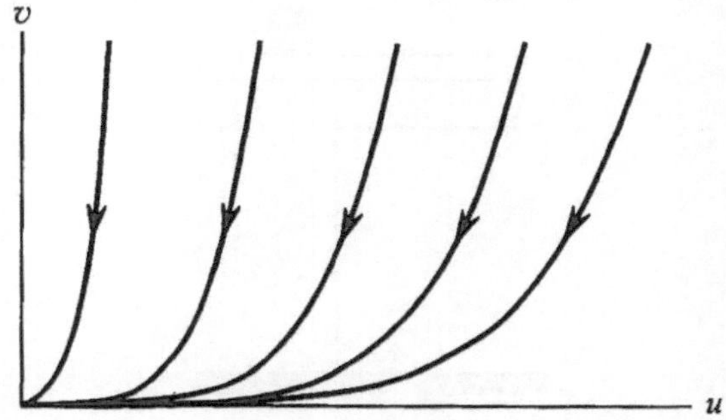

Figure 3.25

which determines a family of phase trajectories in the (u, v) plane (see Fig. 3.25). A phase point tends to the equilibrium $u = v = 0$ along a phase trajectory and reaches this equilibrium in a finite time.

To prove this, we consider the following analytic system of differential equations which has the same family of phase trajectories:

$$\frac{dv}{d\theta} = -v(2u + 5v), \qquad \frac{du}{d\theta} = -ku(u + v). \tag{$*$}$$

The time τ is related to the time θ by

$$\tau = \int_0^\theta (u + v)(2u + 5v)\, d\theta$$

and is a functional defined on the phase trajectories of system $(*)$, for which the time of reaching the equilibrium is infinite, $\theta \in [0, \infty)$.

System $(*)$ has an exact particular solution of the form

$$u = \frac{u(1)}{\theta}, \qquad v = \frac{v(1)}{\theta}.$$

For this solution, τ tends to a finite value as $\theta \to \infty$. The same holds true for any particular solution of system $(*)$, since it is not difficult to establish that

$$\sqrt{u^2 + v^2} \leq \frac{\sqrt{u_0^2 + v_0^2}}{1 + k\theta\sqrt{u_0^2 + v_0^2}}$$

along the trajectories of system $(*)$ for $k < 1$.

The above analysis allows us to formulate the following properties a ball moving by its own momentum over a horizontal rough plane in the case that the interaction of the ball with the plane obeys Coulomb's dry friction law.

PROPERTY 1. The relative slip velocity does not change in direction: $\alpha \overset{t}{\equiv} \text{const}$. Of course, this does not mean that the velocity of the ball center (or, what is the same, the velocity of the contact point) does not change its direction.

PROPERTY 2. The slip velocity V and the spinning angular velocity ω_z decrease monotonically with time and vanish simultaneously. This fact holds true for any initial conditions and any parameters of the problem.

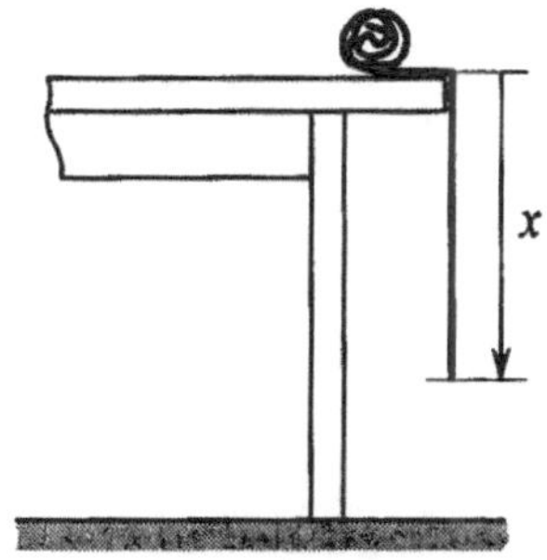

Figure 3.26

PROPERTY 3. The time in which V and ω_z vanish is finite for any initial conditions and parameters as well.

As was noted previously, the determination of $V(t)$ and $\omega_z(t)$ makes it possible to reduce the problem of finding the other unknowns, $\omega_x(t)$, $\omega_y(t)$, $x(t)$, and $y(t)$, to quadratures. In particular, to determine the trajectory of motion of the ball center, one should solve the kinematic equations

$$\dot{x} = V \sin\alpha + R\omega_y, \quad \dot{y} = V \cos\alpha - R\omega_x$$

by simple integration of the right-hand sides after $\omega_x(t)$, $\omega_y(t)$, and $V(t)$ have been found.

At the stage where the ball moves with slip, the trajectory of its center of mass is expressed by transcendental functions. After the completion of the this stage, the ball begins to perform a plane rolling, while describing a rectilinear trajectory.

3.5. Problems With Determinable Linear Groups of Symmetries

The examples considered above did not require symmetry groups to be calculated, since if a symmetry group of a system is a similarity group, then this fact is usually quite obvious.

However, even the existence of groups of nonuniform extensions is not so obvious and the determination of such groups requires the application of a calculation procedure. Groups of nonuniform extensions are subgroups of the general linear group. The determination of such groups always reduces to the solution of linear algebraic equations in order to calculate the coefficients in the generators of these groups.

The calculation of linear groups in the cases where they are admitted by the system under study can always be reduced to constructive procedures, as well as the further determination of canonical coordinates for reducing the order of the system.

There are no general methods for finding nonlinear groups of symmetries. Furthermore, even if the generator of a nonlinear group is guessed, there still remains the difficult obstacle that it is generally impossible to find canonical coordinates of a nonlinear group in a closed form.

We consider below three mechanical problems that illustrate the procedure of calculation of linear groups of symmetries.

3.5.1. Falling of a Heavy Homogeneous Thread

A heavy homogeneous thread dangles from the edge of a table and slides down without friction under gravity (see Fig. 3.26). Find the law of motion of the thread.

The forces exerted on the dangling portion of the thread by the portion lying on the table are assumed to be balanced in the horizontal plane by the reaction of the table.

As far as the forces acting vertically are concerned, apart from the force of gravity, a force arises at the upper edge of the table due to continuous involving of new and new elements of the thread in accelerated falling. A force of this kind is referred to as a reactive force. Meshcherskii's equation with a reactive force for a system of varying mass has the form

$$m\frac{d^2x}{dt^2} = F + \frac{dm}{dt}(v - \dot{x}),$$

where v is the velocity at which new elements of mass enter the system. Thus, $v - \dot{x}$ is the relative velocity of the new mass elements and the system.

In our case, the system consists of the dangling portion of the thread. Denote the length of this portion by x. If ρ is the mass density of the thread per unit length, then the mass of the system is $m = x\rho$. Taking into account the fact that $F = \rho g x$ and the vertical velocity of the entering elements of mass is zero ($v = 0$), we obtain the desired equation of motion of the thread in the form

$$x\ddot{x} = gx - \dot{x}^2,$$

where ρ has been canceled.

As usual, rewrite this equation as a system in the Cauchy normal form:

$$\dot{x} = y, \quad \dot{y} = g - \frac{y^2}{x}.$$

The phase flow of this system determines a group of transformations of the phase space into itself with generator (see Section 1.4)

$$A = y\frac{\partial}{\partial x} + \left(g - \frac{y^2}{x}\right)\frac{\partial}{\partial y}.$$

In accordance with Section 1.8, a symmetry group of the system must satisfy the condition

$$[A, U] = \lambda(x, y)\, A,$$

where $\lambda(x, y)$ is an unknown scalar function. We will seek the generator U satisfying this equation in the algebra of generators of the general linear group GL(2); thus,

$$U = (ax + by)\frac{\partial}{\partial x} + (cx + dy)\frac{\partial}{\partial y}.$$

Calculating the commutator yields

$$[A, U] = \left[-cx + (a - d)y + b\left(g - \frac{y^2}{x}\right)\right]\frac{\partial}{\partial x} + \left[dg + 3cy + (d - a)\frac{y^2}{x} - b\frac{y^3}{x^2}\right]\frac{\partial}{\partial y}.$$

Thus, we obtain the following equations for determining the components of U:

$$-cx + (a - d)y + b\left(g - \frac{y^2}{x}\right) = \lambda(x, y)\, y,$$

$$dg + 3cy + (d - a)\frac{y^2}{x} - b\frac{y^3}{x^2} = \lambda(x, y)\left(g - \frac{y^2}{x}\right).$$

Taking $\lambda(x, y) \equiv 1$ and matching the coefficients of the independent functions of x and y, we arrive at the following system of equations for a, b, c, and d:

$$c = 0, \quad a - d = 1, \quad b = 0, \quad d = 1, \quad d - a = -1.$$

Whence,

$$a = 2, \quad b = c = 0, \quad d = 1.$$

Hence, the generator of the desired group of symmetries is expressed as

$$U = 2x\frac{\partial}{\partial x} + y\frac{\partial}{\partial y}.$$

To find canonical coordinates of the group U, one should solve the equation (see Section 1.7)

$$UG \equiv 2x\frac{\partial G}{\partial x} + y\frac{\partial G}{\partial y} = 1.$$

The general solution of this partial differential equation is an arbitrary function of first integrals of the following system of ordinary differential equations:

$$\frac{dx}{2x} = \frac{dy}{y} = \frac{dG}{1}.$$

One of the first integrals can be found from the equation

$$\frac{dx}{2x} = \frac{dy}{y}$$

in the form $G_1 = y/\sqrt{x}$. Another integral follows from the equation

$$\frac{dx}{2x} = \frac{dG}{1};$$

we have $G_2 = G - \frac{1}{2}\ln x$. Thus, the general solution of the equation $UG = 1$ can be represented in the implicit form $F(G_1, G_2) \equiv F\left(\frac{y}{\sqrt{x}}, G - \frac{1}{2}\ln x\right) = 0$, where $F(G_1, G_2)$ is an arbitrary function. Solving this relation for G yields

$$G = \frac{1}{2}\ln x + \Phi\left(\frac{y}{\sqrt{x}}\right),$$

where $\Phi(G_1)$ is an arbitrary function of one argument.

In accordance with Section 1.7, the variables

$$\alpha = \frac{y}{\sqrt{x}}, \quad \beta = \frac{1}{2}\ln x$$

can be taken to be canonical coordinates of the group U. In terms of the new variables, the system becomes

$$\dot{\alpha} = e^{-\beta}\left(g - \tfrac{3}{2}\alpha^2\right),$$
$$\dot{\beta} = \tfrac{1}{2}e^{-\beta}\alpha.$$

Whence we obtain the separable equation

$$\frac{d\alpha}{d\beta} = \frac{2g - 3\alpha^2}{\alpha},$$

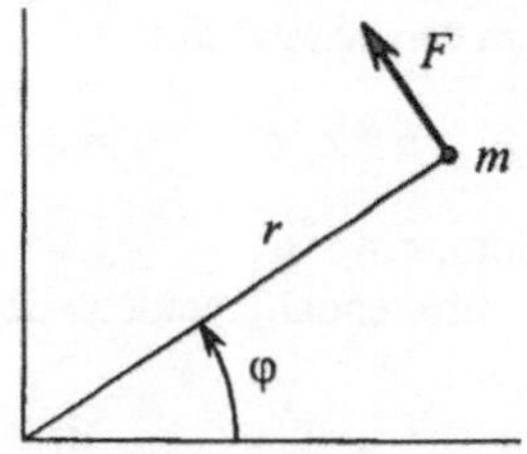

Figure 3.27

whose solution is given by

$$(2g - 3\alpha^2)e^{6\beta} = C,$$

where C is an arbitrary constant.

This solution, being rewritten in terms of the original variables,

$$2gx^3 - 3x^2y^2 = \text{const},$$

is a first integral of the system under study. Solving this integral for y, we obtain

$$y = \frac{\sqrt{2gx^3 - C}}{\sqrt{3}\,x}.$$

This allows us to write out the following first order equation for x:

$$\dot{x} = \frac{\sqrt{2gx^3 - C}}{\sqrt{3}\,x}.$$

Integrating yields

$$t = \sqrt{3}\int_0^x \frac{x\,dx}{\sqrt{2gx^3 - C}} + B,$$

where B and C are arbitrary constants.

For $C = 0$, we obtain a particular solution, which when solved for x becomes

$$x = \frac{g}{6}(t - B)^2.$$

In the general case, the above integral is elliptic and obtaining the explicit dependence $x(t)$ is reduced to the inversion of the integral.

3.5.2. Motion of a Point Particle Under the Action of a Follower Force

Consider the plane motion of point particle of mass m under the action of a force F whose magnitude is constant and direction is perpendicular to the position vector of the particle at any time instant (see Fig. 3.27).

Taking advantage of the arbitrariness in choosing the scale of the variables, we set $m = 1$ and $F = 3\sqrt{2}$. In the polar coordinate system, the motion of the particle obeys the equations

$$\ddot{r} - r\dot{\varphi}^2 = 0, \quad \frac{d}{dt}(r^2\dot{\varphi}) = 3\sqrt{2}\,r.$$

In the Cauchy normal form this system can be rewritten as

$$\dot{r} = v, \quad \dot{v} = \sigma^2 r^{-3}, \quad \dot{\sigma} = 3\sqrt{2}\, r,$$

where σ denotes the angular momentum, $\sigma = r^2\dot{\varphi}$.

The above system of first order differential equations determines a group with infinitesimal generator

$$A = v\frac{\partial}{\partial r} + \frac{\sigma^2}{r^3}\frac{\partial}{\partial v} + 3\sqrt{2}\,\frac{\partial}{\partial \sigma}.$$

Let us proceed to determining a symmetry group of the system. We first search for the infinitesimal generator U of the group, which must satisfy the condition

$$[A, U] = \lambda(r, v, \sigma)\, A,$$

where $\lambda(r, v, \sigma)$ is an arbitrary function of the phase variables. Just as in the previous problem, we seek the generator U in the algebra of generators of the general linear group GL(2):

$$U = (a_{11}r + a_{12}v + a_{13}\sigma)\frac{\partial}{\partial r} + (a_{21}r + a_{22}v + a_{23}\sigma)\frac{\partial}{\partial v} + (a_{31}r + a_{32}v + a_{33}\sigma)\frac{\partial}{\partial \sigma}.$$

The problem thus reduces to determining the constant matrix $\mathcal{A} = \{a_{ij}\}$.

Compute the commutator $[A, U]$:

$$\begin{aligned}
[A, U] = &\left[(3\sqrt{2}\,a_{13} - a_{21})r + (a_{11} - a_{22})v + a_{12}\frac{\sigma^2}{r^3} - a_{23}\sigma\right]\frac{\partial}{\partial r} \\
&+ \left[3\sqrt{2}\,a_{23}r + a_{21}v + (a_{22} - 2a_{33} + 3a_{11})\frac{\sigma^2}{r^3} - 2\frac{\sigma}{r^3}(a_{31}r + a_{32}v) + 3\frac{\sigma^2}{r^4}(a_{12}v + a_{13}\sigma)\right]\frac{\partial}{\partial v} \\
&+ \left[3\sqrt{2}\,(a_{33} - a_{11})r + (a_{31} - 3\sqrt{2}\,a_{12})v + a_{32}\frac{\sigma^2}{r^3} - 3\sqrt{2}\,a_{13}\sigma\right]\frac{\partial}{\partial \sigma}.
\end{aligned}$$

Matching the components of this differential operator with those of λA yields

$$\begin{aligned}
&(3\sqrt{2}\,a_{13} - a_{21})r + (a_{11} - a_{22})v + a_{12}\frac{\sigma^2}{r^3} - a_{23}\sigma = \lambda v, \\
&3\sqrt{2}\,a_{23}r + a_{21}v + (a_{22} - 2a_{33} + 3a_{11})\frac{\sigma^2}{r^3} - 2\frac{\sigma}{r^3}(a_{31}r + a_{32}v) + 3\frac{\sigma^2}{r^4}(a_{12}v + a_{13}\sigma) = \frac{\sigma^2}{r^3}\lambda, \\
&3\sqrt{2}\,(a_{33} - a_{11})r + (a_{31} - 3\sqrt{2}\,a_{12})v + a_{32}\frac{\sigma^2}{r^3} - 3\sqrt{2}\,a_{13}\sigma = 3\sqrt{2}\,r\lambda.
\end{aligned}$$

For these relations to become identities it is necessary and sufficient that the coefficients of all the monomials $r^l v^m \sigma^n$ vanish. This will lead to a system of linear algebraic equations for the unknown coefficients a_{ij}. In this case, the undetermined coefficient $\lambda(r, v, \sigma)$ must generally be sought as a series expansion in such monomials. Since all monomials on the right-hand sides satisfy the condition $m \ge 0$, $n \ge 0$, this condition must obviously be imposed on the expansion of λ. In the general case, a procedure of this sort is quite cumbersome. For this reason, one should begin with simpler assumptions, e.g., the assumption that the function λ is constant or does not contain negative powers.

Consider the first assumption: $\lambda = \text{const}$. Without loss of generality, we can set $\lambda = 1$. This leads to the system

$$\begin{aligned}
&3\sqrt{2}\,a_{13} - a_{21} = 0, \quad a_{11} - a_{22} = 1, \quad a_{12} = 0, \quad a_{23} = 0, \\
&a_{23} = 0, \quad a_{21} = 0, \quad a_{22} - 2a_{33} + 3a_{11} = 1, \quad a_{31} = a_{32} = a_{12} = a_{13} = 0, \\
&a_{33} - a_{11} = 1, \quad a_{31} - 3\sqrt{2}\,a_{12} = 0, \quad a_{32} = a_{13} = 0.
\end{aligned}$$

It turns out that only diagonal entries are nonzero; we have

$$\begin{aligned} a_{11} - a_{22} &= 1, \\ 3a_{11} + a_{22} - 2a_{33} &= 1, \\ a_{11} - a_{33} &= -1. \end{aligned}$$

The solution of this system is $a_{11} = 2$, $a_{22} = 1$, $a_{33} = 3$.

Thus, we obtain a symmetry group with generator

$$U = 2r\frac{\partial}{\partial r} + v\frac{\partial}{\partial v} + 3\sigma\frac{\partial}{\partial \sigma}.$$

To reduce the order of the system of differential equations under study with the generator U just obtained, one should perform a change of variables $(r, v, \sigma) \to (x, y, z)$, where x, y, and z are canonical coordinates of the group, i.e., functions of r, v, and σ which satisfy the conditions $Ux = 1$, $Uy = 0$, and $Uz = 0$. Thus, the functions $y(r, v, \sigma)$ and $z(r, v, \sigma)$ are invariants of the symmetry group defined by the generator U, and $x(r, v, \sigma)$ determines an invariant family of the group. To find these functions, it suffices to solve the partial differential equation

$$2r\frac{\partial G}{\partial r} + v\frac{\partial G}{\partial v} + 3\sigma\frac{\partial G}{\partial \sigma} = 1$$

or, equivalently, the system of ordinary differential equations

$$\frac{dr}{2r} = \frac{dv}{v} = \frac{d\sigma}{3\sigma} = \frac{dG}{1}.$$

This system has the following three first integrals:

$$G_1 = \frac{v}{2\sqrt{r}}, \quad G_2 = \frac{\sqrt{2}\,\sigma}{2\sqrt{r^3}}, \quad G_3 = G - \frac{1}{2}\ln r.$$

Thus, the variables

$$x = \ln\sqrt{r}, \quad y = \frac{v}{2\sqrt{r}}, \quad z = \frac{\sqrt{2}\,\sigma}{2\sqrt{r^3}}$$

can be taken to be canonical coordinates.

Since any function of first integrals is a first integral, the factors $1/2$ and $\sqrt{2}/2$ in y and z can be omitted, but we do retain them for the sake of convenience in what follows.

The inversion of the change of variables is given by

$$r = e^{2x}, \quad v = 2ye^{x}, \quad \sigma = \sqrt{2}\,ze^{3x}.$$

In terms of the new variables, the original equations $\dot r = v$, $\dot v = \sigma^2/r^3$, and $\dot\sigma = 3\sqrt{2}\,r$ become

$$\dot x = ye^{-x}, \quad \dot y = (z^2 - y^2)e^{-x}, \quad \dot z = 3(1 - yz)e^{-x}.$$

On dividing the third equation by the second, we obtain

$$\frac{dz}{dy} = \frac{3(1 - yz)}{z^2 - y^2}.$$

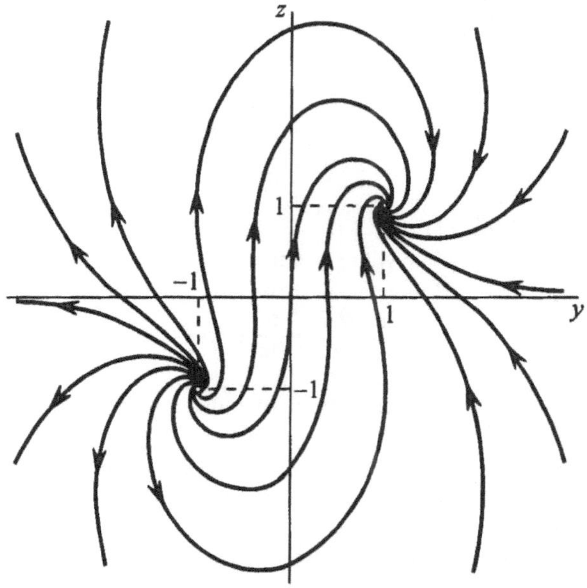

Figure 3.28

Integral curves of this equation are illustrated in Fig 3.28. The equation has two fixed points: $y = z = 1$ and $y = z = -1$. The first point is a stable focus and the second, an unstable focus.

To the stationary solutions $y = z = \pm 1$ there correspond exact particular solutions in terms of the original variables. To find these solutions, one should first integrate the equation $\dot{x} = ye^{-x}$. For $y = \pm 1$ we have $e^x = e^{x_0} \pm t$. Substituting this solution into the expressions of the inverse change of variables, we obtain

$$r = \left(\sqrt{r_0} \pm t\right)^2, \quad v = \pm 2\left(\sqrt{r_0} \pm t\right), \quad \sigma = \pm\sqrt{2}\left(\sqrt{r_0} \pm t\right)^3.$$

The plus sign corresponds to the asymptotically stable particular solution and the minus sign, to the asymptotically unstable solution.

It is of interest to find the equation describing the trajectory of motion of the point particle in the polar coordinates which corresponds the exact particular solutions just obtained.

We proceed from the relation $\dot{\varphi} = \sigma/r^2$. With the above particular solutions, it becomes

$$\dot{\varphi} = \pm\frac{\sqrt{2}}{\sqrt{r_0} \pm t}.$$

Integrating yields

$$\varphi = \sqrt{2}\ln\left(1 \pm \frac{t}{\sqrt{r_0}}\right).$$

On solving this relation for t, we express r, v, and σ in terms of the polar angle:

$$r = r_0 e^{\sqrt{2}\varphi}, \quad v = \pm 2\sqrt{r_0}\, e^{\frac{\sqrt{2}}{2}\varphi}, \quad \sigma = \pm\sqrt{2r_0^3}\, e^{\frac{3\sqrt{2}}{2}\varphi}.$$

The stable trajectory (plus sign) is a logarithmic spiral that goes to infinity. Both the velocity v and the angular momentum σ increase without bound.

The unstable trajectory (minus sign) is also a logarithmic spiral which, however, tends to the origin ($r = 0$) and reaches it in a finite time, while making infinitely many turns about the origin. The velocity v is directed toward the origin and decreases, just as the angular momentum, to zero.

Almost all integral curves in the (y, z) plane wind themselves round the unstable focus as $t \to -\infty$ and round the stable focus as $t \to +\infty$. However, one cannot state that the point $(1, 1)$ is a globally stable singular point of the equation $dz/dy = 3(1 - yz)/(z^2 - y^2)$, since the trajectories do not all tend to the stable focus. There is a single trajectory that winds itself round the unstable focus as $t \to -\infty$ and approaches the y-axis as an asymptote as $t \to +\infty$.

This trajectory can be found by seeking a solution $z(y)$ in the form

$$z = \frac{a}{y} + \frac{b}{y^2} + \cdots.$$

Substituting this expansion into the equation under consideration, we obtain

$$z = \frac{3}{4y} + O\left(\frac{1}{y^5}\right).$$

This solution behaves asymptotically as a hyperbola, and a curve approaching the y-axis as $t \to +\infty$ corresponds to the branch of the hyperbola which is in the third quadrant. The trajectory asymptotic to the hyperbola in the first quadrant approaches it as $t \to -\infty$ and winds itself round the stable focus as $t \to +\infty$.

It is of interest to find out what trajectory in the plane of motion of the point particle there corresponds the trajectory close to the mentioned hyperbola in the (y, z) plane.

Represent the equations of the integral curves in the (y, z) curve in the form

$$\frac{dy}{dx} = \frac{z^2 - y^2}{y}, \quad \frac{dz}{dx} = 3\frac{1 - yz}{y}.$$

Then, on the manifold $z = \frac{3}{4y} + O(y^{-5})$, we have

$$\frac{dy}{dx} = -y + O\left(\frac{1}{y^3}\right).$$

Whence in follows that $y(x)$ behaves asymptotically as

$$y = y_0 e^{-x}.$$

(Recall that we consider the branch for which $y \to -\infty$, in particular, $y_0 < 0$.)

Further the equation $\dot{x} = ye^{-x}$ yields $\dot{x} = y_0 e^{-2x}$ and, hence,

$$e^{2x} \equiv r = 2(C + y_0 t), \quad y = \frac{y_0}{\sqrt{2(C + y_0 t)}}.$$

Whence we obtain the asymptotic behavior of the velocity v and the angular momentum σ,

$$v = 2ye^x = 2y_0, \quad \sigma = \sqrt{2}\, ze^{3x} = \frac{3\sqrt{2}}{y_0}(C + y_0 t)^2,$$

and the derivative of the polar angle φ,

$$\dot{\varphi} = \frac{\sigma}{r^2} = \frac{3\sqrt{2}}{4y_0}.$$

Thus, the singular trajectory in question is asymptotically a spiral of Archimedes which reaches the origin ($r = 0$) in a finite time, $t = -C/y_0$.

To summarize, the overall picture of the plane motion of a point particle acted upon by a follower load is as follows. There are two trajectories that bring the particle to the origin in a finite time; one is a logarithmic spiral with infinitely many turns around the origin and the other is a trajectory close to a spiral of Archimedes with finite number of turns. All the other trajectories go to infinity asymptotically approaching a logarithmic spiral.

3.5.3. The Problem of an Optimal Shape of a Body in an Air Flow

Below we consider a problem on searching for an optimal shape of the forepart of an axisymmetric body in an air flow.[7] The optimality criterion is the minimum air drag in high-speed motion.

Here we use the simplest assumptions about the pressure force and viscous drag in a laminar flow. Specifically, we adopt the following relations for the pressure and the shear stress at the body surface:

$$p - p_\infty = \rho V^2 \delta^2 \left(\frac{dF}{dx}\right)^2 \qquad \text{(pressure)},$$

$$\tau = \frac{k}{\sqrt{x}} \frac{p_\infty V^2}{2} \qquad \text{(shear stress)}.$$

Here, ρ is the air density, v the velocity of the flow at infinity, $F(x)$ a function determining the dependence of the cross-sectional radius on the axial distance of the cross-section from the body vertex, δ a small dimensionless parameter that emphasizes the prolateness of the body, and k is a scaling factor. The function $r = F(x)$ is unknown; it must satisfy the boundary conditions

$$F(0) = 0, \quad F(1) = 1.$$

Under the assumption that $\delta \ll 1$, the drag force can be approximated by

$$Q = 2\pi \rho V^2 \delta^2 \int_0^1 \left[\left(\frac{dF}{dx}\right)^3 + \frac{K^3}{\sqrt{x}}\right] F(x)\, dx,$$

where K is a scaling factor, $K = \left(\frac{1}{2}k/\delta^2\right)^{1/3}$.

Thus, we have obtained a typical problem of variational calculus: find a function $F(x)$ (called an extremal) that minimizes the above integral. For the function $F(x)$ to be an extremal it is necessary that it satisfy the Euler–Lagrange equation

$$\frac{d}{dx} \frac{\partial L}{\partial (dF/dx)} - \frac{\partial L}{\partial F} = 0,$$

where L is the integrand,

$$L = \left[\left(\frac{dF}{dx}\right)^3 + \frac{K^3}{\sqrt{x}}\right] F(x).$$

As a result, we arrive at the following differential equation for $F(x)$:

$$3F \frac{dF}{dx} \frac{d^2F}{dx^2} + \left(\frac{dF}{dx}\right)^3 - \frac{1}{2} \frac{K^3}{\sqrt{x}} = 0; \qquad F(0) = 0, \quad F(1) = 1.$$

Let us reduce this equation to an autonomous system in the Cauchy normal form by introducing the notation $x \equiv \tau$, $F \equiv y$, and $dF/dx \equiv z$. In doing so, we obtain

$$\frac{dx}{d\tau} = 1, \quad \frac{dy}{d\tau} = z, \quad \frac{dz}{d\tau} = -\frac{z^2}{3y} + \frac{K^3}{6\sqrt{x}\, yz}.$$

We consider this system as an example that illustrates how a symmetry group can be sought if one has an a priori suspicion, just as in the two previous examples, that it may be found as a group of extensions (scaling of the variables).

---

[7] A. Miele (ed.), *Theory of Optimum Aerodynamic Slopes. Chapter 15*, Academic Press, 1965.

Indeed, introduce the scaling

$$x = \alpha x', \quad y = \beta y', \quad z = \gamma z',$$

where the coefficients α, β, and γ are to be determined. The system in question becomes

$$\frac{dx'}{d\tau} = \frac{1}{\alpha}, \quad \frac{dy'}{d\tau} = \frac{\gamma}{\beta}z', \quad \frac{dz'}{d\tau} = -\frac{\gamma}{\beta}\frac{z'^2}{3y'} + \frac{1}{\sqrt{\alpha}\,\beta\gamma^2}\frac{K^3}{6\sqrt{x'}\,y'z'}.$$

It is obvious that the system does not change under this transformation, apart from a common scalar factor, if the scaling coefficients satisfy the condition

$$\frac{1}{\alpha} = \frac{\gamma}{\beta} = \frac{1}{\sqrt{\alpha}\,\beta\gamma^2}.$$

Whence it follows that $\alpha = \gamma^{-6}$ and $\beta = \gamma^{-5}$. Thus, the desired symmetry group has the form

$$x' = \gamma^6 x, \quad y' = \gamma^5 y, \quad z' = \gamma^{-1} z.$$

Introduce a new group parameter θ by the formula $\gamma = 1 + \theta$ and retain only the linear terms in θ to obtain

$$x' = (1 + 6\theta)x, \quad y' = (1 + 5\theta)y, \quad z' = (1 - \theta)z.$$

Thus, the generator of the symmetry group is expressed as

$$U = 6x\frac{\partial}{\partial x} + 5y\frac{\partial}{\partial y} - z\frac{\partial}{\partial z}.$$

As usual, the search for canonical coordinates of the group requires the equation $UG = 1$ to be solved. We have

$$UG \equiv 6x\frac{\partial G}{\partial x} + 5y\frac{\partial G}{\partial y} - z\frac{\partial G}{\partial z} = 1.$$

The corresponding characteristic system has the form

$$\frac{dx}{6x} = \frac{dy}{5y} = \frac{dz}{-z} = \frac{dG}{1}.$$

This system has the following three first integrals:

$$G_1 = 6G - \ln x, \quad G_2 = yx^{-5/6}, \quad G_3 = zx^{1/6}.$$

The equations acquire a more convenient form if the canonical coordinates are taken to be

$$u = \ln x, \quad v = \frac{1}{K}yx^{-5/6}, \quad w = \frac{1}{K}zx^{1/6}.$$

In accordance with Section 1.7, the function u is the logarithm of an eigenfunction of the generator U and the functions v and w are invariants of the group, i.e.,

$$Uu = 6, \quad Uv = 0, \quad Uw = 0.$$

To transform the system to the canonical coordinates, we need also the inversion of the change of variables:

$$x = e^u, \quad y = Kve^{\frac{5}{6}u}, \quad y = Kve^{-\frac{1}{6}u}.$$

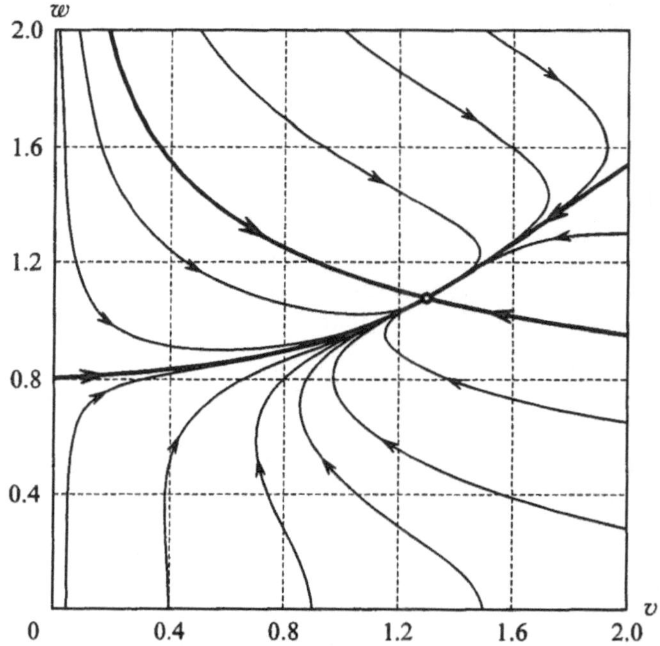

Figure 3.29

In terms of the new variables, the system becomes

$$\dot{u} = e^{-u}, \quad \dot{v} = \frac{1}{6}e^{-u}(6w - 5v), \quad \dot{w} = \frac{e^{-u}}{6vw}(1 - 2w^3 + vw^2).$$

Just as in the previous example, the reduction in the order of the system is accomplished by dividing the third equation by the second. We have

$$\frac{dw}{dv} = \frac{1 - 2w^3 + vw^2}{vw(6w - 5v)}. \qquad (*)$$

The singular points of this differential equation are determined from the algebraic system

$$1 - 2w^3 + vw^2 = 0, \quad vw(6w - 5v) = 0,$$

which splits into two subsystems

$$(1) \quad 1 - 2w^3 + vw^2 = 0, \quad 6w - 5v = 0,$$
$$(2) \quad 1 - 2w^3 = 0, \quad v = 0.$$

Whence we find the singular points:

$$(1) \quad (v, w) = \left(\sqrt[3]{\tfrac{54}{25}}, \sqrt[3]{\tfrac{5}{4}}\right) \approx (1.29;\ 1.08),$$
$$(2) \quad (v, w) = \left(0, \tfrac{1}{\sqrt[3]{2}}\right) \approx (0, 0.79\ldots).$$

The phase portrait of the systems in the variables (v, w) is show in Fig. 3.29.

The first singular point is simple; the corresponding solution in terms of the original variables has the form

$$y = Kve^{5u/6} = K\sqrt[3]{\tfrac{54}{25}}\,x^{5/6},$$
$$z = Kwe^{-u/6} = K\sqrt[3]{\tfrac{5}{4}}\,x^{-1/6}.$$

This solution satisfies the stated boundary conditions, $y(1) = 1$, if the coefficient K, which depends on the viscosity of the incident flow, is taken to be $K = \sqrt[3]{25/54}$. This case and this coefficient are said to be critical. The optimal shape that the forepart of the moving body should have in the critical case is given by $y = x^{5/6}$.

The second singular point is complex. The vector field, which can be represented by the equations

$$\frac{dv}{du} = \frac{6w - 5v}{6}, \qquad \frac{dw}{du} = \frac{1 - 2w^3 + vw^2}{6vw},$$

becomes infinite at this point. For this reason, the second singular point cannot be used for constructing another particular solution of the original system.

A further analysis of the problem for noncritical cases may be performed with the application of approximate methods to equation (*).

Chapter 4

Finite-Dimensional Hamiltonian Systems

4.1. Legendre Transformation

In this chapter, we will consider mechanical systems whose differential equations of motion can be represented in the form of the Lagrange equations

$$\frac{d}{dt}\frac{\partial L}{\partial \dot{q}_i} - \frac{\partial L}{\partial q_i} = 0, \qquad i = 1, \dots, n,$$

where $L(t, q, \dot{q})$ is a scalar descriptive function called the Lagrangian function (or Lagrangian). This function is the difference between the kinetic and potential energies of the system,

$$L(t, q, \dot{q}) = T(t, q, \dot{q}) - \Pi(t, q).$$

It is convenient to reduce the Lagrange equations to the Hamilton equations, a system of $2n$ equations of the first order in a special symmetric form. In the reduction procedure, Legendre transformations are used. These transformations play an important role in mathematics and mechanics by themselves. For this reason, we introduce these transformations in the general form.

Let us pass from some variables $x_1, \dots, x_n$ to other variables $y_1, \dots, y_n$ in accordance with the formulas

$$y_1 = f_1(x_1, \dots, x_n),$$
$$\dots\dots\dots\dots\dots\dots$$
$$y_n = f_n(x_1, \dots, x_n).$$

This change of variables is called a Legendre transformation if the right-hand sides have a potential. This means that there exists a scalar function $V(x_1, \dots, x_n)$ such that

$$f_k = \frac{\partial V}{\partial x_k}, \qquad k = 1, \dots, n.$$

The potential V is called nondegenerate if the condition

$$\det\left(\frac{\partial^2 V}{\partial x_k \partial x_l}\right) \neq 0$$

is satisfied. This potential is said to be strongly nondegenerate if the equations $y_k = f_k(x_1, \dots, x_n)$ can be smoothly and uniquely solved for x_k,

$$x_k = \varphi_k(y_1, \dots, y_n), \qquad k = 1, \dots, n.$$

THEOREM. *If a transformation $y_k = f_k(x_1, \dots, x_n)$, $k = 1, \dots, n$, is potential and the potential $V(x_1, \dots, x_n)$ of this transformation is strongly nondegenerate, then the inverse transformation*

$x_k = \varphi_k(y_1, \dots, y_n)$ is also potential and the corresponding potential $W(y_1, \dots, y_n)$ is also strongly nondegenerate. This potential is related to V by

$$W(y_1, \dots, y_n) = \left[\sum_{i=1}^{n} x_i y_i - V(x_1, \dots, x_n) \right]_{x_k = \varphi_k(y_1, \dots, y_n)}.$$

Proof. We differentiate this relation with respect to y_k to obtain

$$\frac{\partial W}{\partial y_k} = \left[x_k + \sum_{i=1}^{n} \frac{\partial x_i}{\partial y_k} y_i - \sum_{i=1}^{n} \frac{\partial V}{\partial x_i} \frac{\partial x_i}{\partial y_k} \right]_{x_k = \varphi_k(y_1, \dots, y_n)};$$

whence, taking into account the fact that $\partial V / \partial x_i = \varphi_k(y_1, \dots, y_n)$, we find that

$$\frac{\partial W}{\partial y_k} = \varphi_k(y_1, \dots, y_n).$$

This means that the transformation $x_k = \varphi_k(y_1, \dots, y_n)$ is potential. The potentials V and W are said to be conjugate.

4.2. Hamiltonian Systems. Poisson Bracket

We use a Legendre transformation to pass from the phase variables $t, q, \dot{q}$ of a Lagrangian system $L(t, q, \dot{q})$ to new variables t, q, p. We transform only the velocities $\dot{q}$ into the variables p, which are called the generalized momenta. These variables are related to $\dot{q}$ by the Legendre transformation

$$p_k = \frac{\partial L}{\partial \dot{q}_k}, \qquad k = 1, \dots, n,$$

in which the potential is the Lagrangian function. In this case, the variables t and q are treated as parameters.

The conjugate potential is referred to as the Hamiltonian function (or the Hamiltonian). In accordance with the above theorem, this function is determined by the relation

$$H(t, q, p) = \left[\sum_{i=1}^{n} p_i \dot{q}_i - L(t, q, \dot{q}) \right]_{\dot{q} = f(t, q, p)},$$

where the function $\dot{q} = f(t, q, p)$ is the result of the inversion of the equations $p_i = p_i(t, q, \dot{q}) \equiv \partial L / \partial \dot{q}_i$ with respect to $\dot{q}$.

The equations in terms of the new phase variables are referred to as the Hamilton equations. These equations can be obtained as follows. Since $f(t, q, p)$ determines a Legendre transformation with potential $H(t, q, p)$, we have

$$\dot{q} = \frac{\partial H}{\partial p}.$$

One can readily see that $\frac{\partial H}{\partial q} = -\frac{\partial L}{\partial q}$. Then, the Lagrange equations $\frac{d}{dt} \frac{\partial L}{\partial \dot{q}} = \frac{\partial L}{\partial q}$ imply

$$\dot{p} = -\frac{\partial H}{\partial q}.$$

Thus, the Hamilton equations form a system of $2n$ ordinary differential equations of the first order in the Cauchy normal form

$$\dot{q}_k = \frac{\partial H}{\partial p_k}, \quad \dot{p}_k = -\frac{\partial H}{\partial q_k}, \qquad k = 1, \dots, n.$$

If the Hamiltonian of this system is independent of time, then the Hamilton equations are autonomous. In this case, these equations define a one-parameter group of automorphisms of the phase space with infinitesimal generator

$$U = \frac{\partial H}{\partial p}\frac{\partial}{\partial q} - \frac{\partial H}{\partial q}\frac{\partial}{\partial p}.$$

Such a group is called a *Hamiltonian group.* The action of the generator of a Hamiltonian group on a smooth function $G(q,p)$ is called the *Poisson bracket* of the functions $H(q,p)$ and $G(q,p)$:

$$UG \equiv \frac{\partial H}{\partial p}\frac{\partial G}{\partial q} - \frac{\partial H}{\partial q}\frac{\partial G}{\partial p} \equiv \{H,G\}.$$

Properties of Poisson brackets.

PROPERTY 1. A Poisson bracket is linear in each argument,

$$\{\lambda_1 H_1 + \lambda_2 H_2,\, G\} = \lambda_1\{H_1,G\} + \lambda_2\{H_2,G\}.$$
$$\{H,\, \lambda_1 G_1 + \lambda_2 G_2\} = \lambda_1\{H,G_1\} + \lambda_2\{H,G_2\}.$$

where λ_1 and λ_2 are real numbers.

PROPERTY 2. A Poisson bracket possesses skew-symmetry,

$$\{H,G\} = -\{G,H\}.$$

PROPERTY 3. Leibniz's rule holds in the differentiation with respect to any variable, for example,

$$\frac{\partial}{\partial q_k}\{H,G\} = \left\{\frac{\partial H}{\partial q_k}, G\right\} + \left\{H, \frac{\partial G}{\partial q_k}\right\}.$$

The first three properties are obvious.

PROPERTY 4. Any functions $H(q,p)$, $G(q,p)$, and $F(q,p)$ satisfy the Poisson identity

$$\big\{\{H,G\},F\big\} + \big\{\{F,H\},G\big\} + \big\{\{G,F\},H\big\} = 0.$$

Proof. Consider the first term $\big\{\{H,G\},F\big\}$ of this relation. Since a Poisson bracket is linear in its arguments and Property 3 holds, we can put the outer bracket inside to obtain

$$\big\{\{H,G\},F\big\} = \big\{\{H,F\},G\big\} + \big\{H,\{G,F\}\big\}.$$

Rearranging the right-hand side of this relation to the left-hand side and taking into account Property 2 prove the property in question.

PROPERTY 5. The Poisson bracket of two functions H and G is the Hamiltonian of the differential system defined by the commutator of the operators of the systems with Hamiltonians H and G. In other words,

$$[U_H, U_G] = \frac{\partial\{H,G\}}{\partial p}\frac{\partial}{\partial q} - \frac{\partial\{H,G\}}{\partial q}\frac{\partial}{\partial p},$$

where U_H and U_G are the generators of the Hamiltonian groups with Hamiltonians H and G, respectively, i.e.,

$$U_H = \frac{\partial H}{\partial p}\frac{\partial}{\partial q} - \frac{\partial H}{\partial q}\frac{\partial}{\partial p}, \quad U_G = \frac{\partial G}{\partial p}\frac{\partial}{\partial q} - \frac{\partial G}{\partial q}\frac{\partial}{\partial p}.$$

Proof. In this case, the rule of commutation presented in Section 1.3 becomes

$$[U_H,U_G] = \left(U_H\frac{\partial G}{\partial p} - U_G\frac{\partial H}{\partial p}\right)\frac{\partial}{\partial q} - \left(U_H\frac{\partial G}{\partial q} - U_G\frac{\partial H}{\partial q}\right)\frac{\partial}{\partial p}$$
$$= \left(\left\{H,\frac{\partial G}{\partial p}\right\} - \left\{G,\frac{\partial H}{\partial p}\right\}\right)\frac{\partial}{\partial q} - \left(\left\{H,\frac{\partial G}{\partial q}\right\} - \left\{G,\frac{\partial H}{\partial q}\right\}\right)\frac{\partial}{\partial p}.$$

With reference to Properties 2 and 3, the property is proved.

COROLLARY. *The Poisson identity can be represented in the operator form*

$$[U_H, U_G]\,F + [U_F, U_H]\,G + [U_G, U_F]\,H = 0.$$

Consider the functions $H(q,p)$ such that q belongs to the configuration manifold of a mechanical system and p to the tangent bundle of this manifold. The linear space of such functions with the introduced operation of product of two functions, which is specified by the Poisson bracket, is called the Lie algebra of Hamiltonians.

4.3. Nonautonomous Hamiltonian Systems

Consider now the case where the Lagrangian function and, hence, the Hamiltonian function depend on time explicitly, $H = H(t,q,p)$. In this case, the Hamilton equations $\dot{q} = \frac{\partial H}{\partial p}$, $\dot{p} = -\frac{\partial H}{\partial q}$ must be treated in the extended direct product of the configuration manifold and its tangent bundle.

We perform this extension of the manifold as follows. First, we consider t as a new generalized coordinate. It should be found out what quantity plays a role of the conjugate momentum for this coordinate and how the Hamiltonian function must be transformed for this function to depend on the new set of variables.

Let $H(t,p,q)$ be the original Hamiltonian function. We denote the desired transformed Hamiltonian function corresponding to $H(t,q,p)$ by $H^*(t,T,q,p)$, where T is the conjugate momentum for the variable t. The function H^* must define the Hamilton equations for the entire set of variables. We write out these equations separately for the variables (t,T) and (q,p):

$$\dot{t} = \frac{\partial H^*}{\partial T}, \quad \dot{T} = -\frac{\partial H^*}{\partial t},$$
$$\dot{q} = \frac{\partial H^*}{\partial p}, \quad \dot{p} = -\frac{\partial H^*}{\partial q}.$$

For this system to describe the original nonautonomous system with Hamiltonian $H(t,q,p)$ it is sufficient to set

$$H^*(t,T,q,p) \equiv T + H(t,q,p).$$

In this case, the equations for (p,q) coincide with the original equations. The equation $\dot{t} = \partial H^*/\partial T$ permits us to reveal the original meaning of the new generalized coordinate up to an arbitrary additive constant. From the equation $\dot{T} = -\partial H^*/\partial t$, we can find out the physical meaning of the momentum T corresponding to time t.

For this purpose, we differentiate the Hamiltonian function $H(t,q,p)$ with respect to time along the actual trajectories:

$$\frac{dH}{dt} = \frac{\partial H}{\partial t} + \sum_{i=1}^{n}\left(\frac{\partial H}{\partial q_i}\dot{q}_i + \frac{\partial H}{\partial p_i}\dot{p}_i\right) = \frac{\partial H}{\partial t};$$

whence, taking into account the relation $\partial H^*/\partial t \equiv \partial H/\partial t$, we obtain

$$\dot{T} = -\frac{dH}{dt} \quad \Longrightarrow \quad T = -H(t,q,p) + \text{const}.$$

Thus, along the trajectories of the system, the generalized momentum which is the conjugate of time differs from the negative of the Hamiltonian function by a constant.

In the sequel, with reference to the extension presented, we do not distinguish between the autonomous and nonautonomous systems. A nonautonomous system will be treated as a special case of an autonomous system which can be immediately integrated for two variables.

4.4. Integrals of Hamiltonian Groups. Noether's Theorem

The general definition of differential and integral invariants is presented in Section 1.10. In the current section, we discuss the specific forms acquired by these invariants in the Hamiltonian case.

Consider an arbitrary Hamiltonian group with generator

$$U = \frac{\partial H}{\partial p}\frac{\partial}{\partial q} - \frac{\partial H}{\partial q}\frac{\partial}{\partial p}; \qquad H = H(p,q), \quad q = \{q_i\}, \quad p = \{p_i\}, \quad i = 1,\dots,n.$$

A differential invariant of the zeroth order, or a finite invariant, is a function $G(q,p)$ which is a root of the generator of this group, i.e.,

$$UG = 0.$$

If the function $G(q,p)$ is defined in the same domain as the Hamiltonian $H(q,p)$, then this invariant is said to be global. If this function is defined in a subdomain of the domain of definition of H, then this invariant is called local.

In accordance with the definition of a Poisson bracket, an invariant can be equivalently defined by the relation $\{H,G\} = 0$.

The functions H and G occur symmetrically in the Poisson bracket. Thus, if G is an invariant of the Hamiltonian group with Hamiltonian H, then H is an invariant of the group with Hamiltonian G.

Two functions $H(q,p)$ and $G(q,p)$ are said to be *in involution* if the Poisson bracket of these function is zero.

The invariants of the Hamiltonian groups are first integrals of the corresponding Hamiltonian mechanical systems of differential equations, since we have (see Section 1.5)

$$\frac{dG}{dt} = UG.$$

Let us determine the algebraic structure of the family of functionally independent finite invariants (first integrals) of a Hamiltonian group with Hamiltonian $H(q,p)$.

THEOREM. *If $G(q,p)$ and $F(q,p)$ are invariants of a Hamiltonian group, then their Poisson bracket is also an invariant of this group, provided that this bracket is not identically zero.*

Proof. What is to be proved is that the function $\{G,F\}$ is in involution with the function H, i.e., $\{H,\{G,F\}\} = 0$. Using the Poisson identity, we have

$$\{H,\{G,F\}\} = -\{G,\{F,H\}\} - \{F,\{H,G\}\}.$$

Two last brackets are zero, since $\{H,F\} = \{H,G\} = 0$ in accordance with the hypotheses of the theorem. This proves the theorem.

The theorem implies that, since the Poisson bracket of any two invariants also belongs to the family of invariants, this family is a Lie algebra.

Since a Poisson bracket is antisymmetric and, hence, $\{H,H\} = 0$, the Hamiltonian function is also an invariant.

Let us clarify the mechanical meaning of the first integral corresponding to this invariant.

In the general case, the Lagrangian function of a natural mechanical system is the sum of a quadratic function of the generalized velocities, a linear function of these velocities, and a function which is independent of the velocities,

$$L(t,q,\dot{q}) = \frac{1}{2}\sum_{i,j=1}^{n} a_{ij}\dot{q}_i\dot{q}_j + \sum_{i=1}^{n} b_i\dot{q}_i + T_0 - \Pi(t,q).$$

The coefficients a_{ij} and b_i depend on time and the generalized coordinates. The function $T_0(t, q)$ is the component of the kinetic energy which does not depend on the generalized velocities. The function $\Pi(t, q)$ is called the potential energy of the mechanical system.

By using the Legendre transformation, we find the generalized momenta:

$$p_i = \frac{\partial L}{\partial q_i} = \sum_{j=1}^{n} a_{ij}\dot{q}_j + b_i.$$

Since the Lagrange parameters q_i are locally independent, i.e., $\det\{a_{ij}\} \neq 0$, we can solve these equations for $\dot{q}_i$ to obtain

$$\dot{q}_i = \sum_{j=1}^{n} h_{ij}(p_j - b_j),$$

where the h_{ij} are the entries of the inverse of the matrix $\{a_{ij}\}$.

We substitute this expression of $\dot{q}_i$ into the Hamiltonian function

$$H = \sum_{i=1}^{n} p_i\dot{q}_i - L.$$

As a result, we have

$$\begin{aligned} H &= \sum_{i,j=1}^{n} h_{ij}(p_j - b_j)p_i - \frac{1}{2}\sum_{i,j,k,l=1}^{n} a_{ij}h_{ik}h_{jl}(p_k - b_k)(p_l - b_l) - \sum_{i,j=1}^{n} b_i h_{ij}(p_j - b_j) - T_0 + \Pi \\ &= \frac{1}{2}\sum_{i,j=1}^{n} h_{ij}p_ip_j + \frac{1}{2}\sum_{i,j=1}^{n} h_{ij}b_ib_j - \sum_{i,j=1}^{n} h_{ij}b_ip_j - T_0 + \Pi. \end{aligned}$$

If the mechanical system is *conservative*, then its kinetic energy is a quadratic form of the velocities which does not depend on time, i.e., $b_i \equiv 0$ and $T_0 \equiv 0$. In this case, the potential energy is also independent of time. Then the Hamiltonian has the form

$$H = \frac{1}{2}\sum_{i,j=1}^{n} h_{ij}p_ip_j + \Pi,$$

which expresses the *total energy* of the mechanical system.

The knowledge of first integrals of a mechanical system helps us to construct solutions of the differential equations of this system. The first integrals can be also of interest by themselves. The functions which do not change during the motion permit one to reveal important hidden laws of the motion.

In systems of the general form, the existence of a symmetry permits one to decrease the order of the system (see Section 1.8). Similarly, the existence of first integrals of the Hamiltonian systems can be connected with symmetries of special form. One of the results obtained in this direction is Noether's theorem.

THEOREM (EMMY NOETHER). *Let a Hamiltonian system with Hamiltonian $H(t, q, p)$ be given and let a Lie group of this system acting in the extended configuration space $(t, q) \to (t', q')$ in accordance with the relations*

$$t' = t + \xi(t, q)\tau + \cdots,$$

$$q' = q + \eta(t, q)\tau + \cdots$$

be a symmetry group of the Hamiltonian action

$$\int_{t_1}^{t_2} L(t, q, \dot{q})\, dt,$$

where $L(t, q, \dot{q})$ is the Lagrangian function which is the conjugate of the Hamiltonian. Then the system has the first integral

$$\sum_{i=1}^{n} \eta_i(t, q)\, p_i - \xi(t, q)\, H(t, q, p) = \text{const}.$$

Proof. In this case, the invariance condition of the functional J_1 obtained in Section 1.10 has the form

$$\overset{(1)}{U} L + \dot{\xi} L = 0,$$

where $\overset{(1)}{U}$ is the once prolonged generator of the symmetry group in question. In accordance with the prolongation theory (see Section 1.10), this generator has the form

$$\overset{(1)}{U} = \xi \frac{\partial}{\partial t} + \eta \frac{\partial}{\partial q} + (\dot{\eta} - \dot{\xi}\dot{q}) \frac{\partial}{\partial \dot{q}}.$$

Thus, the invariance condition becomes

$$\xi \frac{\partial L}{\partial t} + \sum_{i=1}^{n} \eta_i \frac{\partial L}{\partial q_i} + \sum_{i=1}^{n} (\dot{\eta}_i - \dot{\xi}\dot{q}_i) \frac{\partial L}{\partial \dot{q}_i} + \dot{\xi} L = 0.$$

Using the Lagrange equations

$$\frac{\partial L}{\partial q_i} = \frac{d}{dt} \frac{\partial L}{\partial \dot{q}_i},$$

we represent this condition in the form

$$\xi \frac{\partial L}{\partial t} + \sum_{i=1}^{n} \eta_i \frac{d}{dt} \frac{\partial L}{\partial \dot{q}_i} + \sum_{i=1}^{n} \dot{\eta}_i \frac{\partial L}{\partial \dot{q}_i} - \sum_{i=1}^{n} \dot{\xi} \dot{q}_i \frac{\partial L}{\partial \dot{q}_i} + \frac{d\xi}{dt} L = 0.$$

Since

$$\frac{\partial L}{\partial t} = -\frac{\partial H}{\partial t} = -\frac{dH}{dt}$$

and

$$\frac{\partial L}{\partial \dot{q}_i} = p_i,$$

we have

$$-\xi \frac{dH}{dt} + \frac{d}{dt} \left(\sum_{i=1}^{n} \eta_i \frac{\partial L}{\partial \dot{q}_i} \right) - \frac{d\xi}{dt} \left(\sum_{i=1}^{n} \dot{q}_i p_i \right) + \frac{d\xi}{dt} L = 0;$$

whence it follows that

$$\frac{d}{dt} \left(-\xi H + \sum_{i=1}^{n} \eta_i \frac{\partial L}{\partial \dot{q}_i} \right) = 0.$$

This proves the theorem.

Remark. If the symmetry group does not transform time, i.e., $\xi(t,q) = 0$, then not only the Hamiltonian action but also the Lagrangian function is a differential invariant, i.e.,

$$\overset{(1)}{U} L = 0.$$

The corresponding first integral has the form $\sum_{i=1}^{n} \eta_i p_i = \text{const}$.

The above theorem can be readily generalized to the nonholonomic systems. Let a system with Lagrangian $L(t, q, \dot{q})$ have the nonholonomic constraints

$$a_{k1}\dot{q}_1 + \cdots + a_{kn}\dot{q}_n + a_k = 0 \qquad (k = 1, \ldots, l < n),$$

where the coefficients $a_{ki}(t,q)$ and $a_k(t,q)$ depend on time and the generalized coordinates. In this case, Noether's theorem can be generalized to the nonholonomic systems as follows.

THEOREM. *If the Lagrangian function $L(t, q, \dot{q})$ of a mechanical system with the nonholonomic constraints $a_{k1}\dot{q}_1 + \cdots + a_{kn}\dot{q}_n + a_k = 0$ $(k = 1, \ldots, l < n)$ is an invariant of the Lie group with generator $U = \eta\frac{\partial}{\partial q}$ the coefficients of which coincide with the virtual displacements of the system, i.e., $a_{k1}\eta_1 + \cdots + a_{kn}\eta_n = 0$, then this system has the first integral*

$$\sum_{i=1}^{n} \eta_i \frac{\partial L}{\partial \dot{q}_i} = \text{const}.$$

Proof. We represent the equations of motion of the nonholonomic system in the form of the Lagrange equations with undetermined coefficients,.

$$\frac{\partial L}{\partial q_i} = \frac{d}{dt}\frac{\partial L}{\partial \dot{q}_i} + \sum_{k=1}^{l} a_{ki}\mu_k \qquad (i = 1, \ldots, n),$$

where the μ_k are additional unknowns. We can determine these unknowns by closing the equations presented by the equations of the nonholonomic constrains. We substitute $\partial L/\partial q_i$ into the invariance condition of the Lagrangian to obtain

$$\sum_{i=1}^{n} \eta_i \frac{\partial L}{\partial q_i} + \sum_{i=1}^{n} \dot{\eta}_i \frac{\partial L}{\partial \dot{q}_i} = \sum_{i=1}^{n} \eta_i \frac{d}{dt}\frac{\partial L}{\partial \dot{q}_i} + \sum_{k=1}^{l}\sum_{i=1}^{n} a_{ki}\eta_i\mu_k + \sum_{i=1}^{n} \dot{\eta}_i \frac{\partial L}{\partial \dot{q}_i} = 0.$$

Taking into account the relation $\sum_{i=1}^{n} a_{ki}\eta_i = 0$, we find that

$$\sum_{i=1}^{n} \eta_i \frac{\partial L}{\partial q_i} = \text{const},$$

which proves the theorem.

4.5. Conservation Laws and Symmetries

The symmetries of the differential equations of mechanics are usually associated with certain laws of conservation. Let us consider some examples.

The group of translations in time $t' = t + \tau$, $q' = q$. If the Hamiltonian action is invariant under this group, then, in accordance with Noether's theorem and relations $\eta \equiv 0$ and $\xi \equiv 1$, we have

$$H(q, p) = \text{const}.$$

This means that the law of conservation of energy is a corollary of the invariance of the action under translations in time.

The group of translations in space $t' = t$, $q'_k = q_k + \tau$, $q'_i = q_i$ $(i = k)$. In this case, we have $\xi = 0$ and $\eta_k = 1$. Then, Noether's theorem implies $p_k = \text{const}$ (law of conservation of momentum).

Example. Consider a mechanical system consisting of n point particles,

$$L = \frac{1}{2}\sum_{i=1}^{n} m_i(\dot{x}_i^2 + \dot{y}_i^2 + \dot{z}_i^2) + V(x, y, z).$$

If this system admits translation in a spatial coordinate, $x'_1 = x_1 + \tau, \ldots, x'_n = x_n + \tau$, then we have $\eta_1 = \cdots = \eta_n = 1$ and $\xi = 0$. Then, from Noether's theorem it follows that

$$\sum_{i=1}^{n} m_i \dot{x}_i = \text{const}$$

(law of conservation of momentum).

If this system admits a group of rotations about a spatial axis (e.g., the x-axis),

$$\begin{aligned} x'_k &= x_k, \\ y'_k &= y_k \cos\tau + z_k \sin\tau, \\ z'_k &= -y_k \sin\tau + z_k \cos\tau, \end{aligned}$$

with generator

$$U = \sum_{k=1}^{n}\left(z_k \frac{\partial}{\partial y_k} - y_k \frac{\partial}{\partial z_k}\right),$$

then we have $\xi = 0$, $\eta_k = 0$, $\eta_{k+n} = z_k$, and $\eta_{k+2n} = -y_k$ $(k = 1, \ldots, n)$. In this case, Noether's theorem implies

$$\sum_{i=1}^{n} m_i(\dot{y}_i z_i - y_i \dot{z}_i) = \text{const}$$

(law of conservation of angular momentum).

4.6. Integral Invariants

In Section 1.10, we considered the following three types of integral invariants of arbitrary Lie groups:

$$J_1 = \int_{t_1}^{t_2} \Phi(t, q, \dot{q}, \ldots)\, dt, \quad J_2 = \oint_{\gamma} \Phi_k(q)\, dq_k, \quad J_3 = \int \cdots \int_{v} \Phi(q)\, dq_1 \ldots dq_n.$$

In the special case of Hamiltonian groups, the invariants J_2 and J_3 are of particular interest. In the theory of Hamiltonian systems, these invariants are of fundamental importance. In what follows, we will discuss this question in detail. Unlike the invariants of the J_2 and J_3 types, the invariants of the J_1 type have neither fundamental importance nor any particular interest for the Hamiltonian systems. Nevertheless, invariants of this type occur in some specific problems.

For example, the Lorentz group with generator $U = p\frac{\partial}{\partial q} + q\frac{\partial}{\partial p}$ is a Hamiltonian group with Hamiltonian function $H = \frac{1}{2}(q^2 - p^2)$. For this group, the invariant of the J_1 type is the intrinsic time

$$\int_0^q \left[1 - \left(\frac{dp}{dq}\right)^2\right]^{1/2} dq.$$

Here, we use the conjugate variables q and p which are traditional for the Hamiltonian systems rather than the variables t and x occurring in the Lorentz group, $q \equiv t$ and $p \equiv x$.

The rotation group with generator $U = p\frac{\partial}{\partial q} - q\frac{\partial}{\partial p}$ is also Hamiltonian with $H = \frac{1}{2}(q^2 + p^2)$. For this group, the invariant of the J_1 type is the arc length of the curve

$$\int_{q_1}^{q_2} \left[1 + \left(\frac{dp}{dq}\right)^2\right]^{1/2} dq.$$

4.6.1. Poincaré–Cartan Invariants

Consider integrals of the J_2 type for Hamiltonian groups. It turns out that for these integrals to be invariant their integrands must belong to a quite narrow class. The existence of such invariants is a criterion for the group to be Hamiltonian.

THEOREM (POINCARÉ INTEGRAL INVARIANT). *The contour integral*

$$J_2 = \oint_\Gamma Q(q,p)\,dq + P(q,p)\,dp$$

is an invariant of an arbitrary Hamiltonian group if and only if the relations

$$Q(q,p) \equiv cp, \quad P(q,p) \equiv 0$$

hold, where c is a scalar constant.

Proof. To simplify the calculations, we consider the case where p and q are scalar.

1. *Necessity.* If the group $q' = q + \eta_1(q,p)t + \cdots$, $p' = p + \eta_2(q,p)t + \cdots$ is Hamiltonian, then $Q \equiv cp$ and $P \equiv 0$.

An invariance condition for such an integral was obtained in Section 1.10 in the general case. This condition implies the derivative of this integral along the trajectories of the group,

$$\frac{dJ_2}{dt} = \frac{d}{dt}\oint_{\Gamma'} Q(q,p)\,q + P(q,p)\,dp \qquad (q,p \to q',p'),$$

to be zero. This derivative was calculated in the form (in the notation of Section 1.10)

$$\frac{dJ_2}{dt} = \oint_{\Gamma'} \left(UQ + Q\frac{\partial\eta_1}{\partial q} + P\frac{\partial\eta_2}{\partial q}\right) dq + \left(UP + Q\frac{\partial\eta_1}{\partial p} + P\frac{\partial\eta_2}{\partial p}\right) dp.$$

The derivative is zero if and only if the relation

$$\left(UQ + Q\frac{\partial\eta_1}{\partial q} + P\frac{\partial\eta_2}{\partial q}\right)_p = \left(UP + Q\frac{\partial\eta_1}{\partial p} + P\frac{\partial\eta_2}{\partial p}\right)_q$$

holds. Since the group in question is Hamiltonian, we have

$$UQ = \{H,Q\}, \quad UP = \{H,P\}, \quad \eta_1 = \frac{\partial H}{\partial p}, \quad \eta_2 = -\frac{\partial H}{\partial q}$$

and the invariance condition becomes

$$\{H,Q\}_p + Q_p \frac{\partial^2 H}{\partial q \partial p} - P_p \frac{\partial^2 H}{\partial q^2} = \{H,P\}_q + Q_q \frac{\partial^2 H}{\partial p^2} - P_q \frac{\partial^2 H}{\partial q \partial p}.$$

By using the relations $\{H,Q\}_p = \{H_p,Q\}+\{H,Q_p\}$ and $\{H,P\}_q = \{H_q,P\}+\{H,P_q\}$, we cancel out the Poisson brackets $\{H_q,P\}$ and $\{H_p,Q\}$ and the remaining terms of the last equation to obtain

$$\{H,Q_p\} = \{H,P_q\},$$

or

$$\{H,\ Q_p - P_q\} = 0;$$

whence, taking into account the arbitrariness of the function $H(q,p)$, we obtain

$$Q_p - P_q = c.$$

A particular solution of this equation is given by

$$Q = cp, \quad P = 0.$$

The general solution of this equation is the sum of a particular solution (for example, the solution just presented) and the general solution of the homogeneous equation

$$Q = cp + \frac{\partial M}{\partial q}, \quad P = \frac{\partial M}{\partial p},$$

where $M(q,p)$ is an arbitrary differentiable function. This proves the necessity.

2. *Sufficiency*. If the integral

$$J_2 = \oint_\Gamma p\, dq$$

is an integral invariant of the group

$$q' = q + \eta_1(q,p)t + \cdots, \quad p' = p + \eta_2(q,p)t + \cdots,$$

then this group is Hamiltonian.

In this case, the invariance condition obtained, i.e., the condition for the relation $dJ_2/dt = 0$ to hold, becomes

$$\left(Up + p\frac{\partial \eta_1}{\partial q}\right)_p = \left(p\frac{\partial \eta_1}{\partial p}\right)_q.$$

Since $Up = \eta_2$, this condition implies that

$$\frac{\partial \eta_1}{\partial q} + \frac{\partial \eta_2}{\partial p} = 0;$$

whence it follows that $\eta_1 = \partial H/\partial p$ and $\eta_2 = -\partial H/\partial q$, where $H(q,p)$ is an arbitrary differentiable function. This completely proves the theorem.

As shown in Section 4.3, a Hamiltonian depending explicitly on time can be reduced to an autonomous Hamiltonian by introducing the momentum which is the conjugate of time. The autonomous system thus obtained has a simpler structure than the system of general form. For this reason, we can add some details to the theorem just proved for the nonautonomous case, in which the integral has the form

$$\oint_\Gamma T\,dt + p\,dq.$$

In this case, the contour Γ in the space of the variables (t, T, q, p) can be chosen from a narrower class of closed contours. Instead of considering this contour in the general parametric form

$$t = t(\alpha), \quad T = T(\alpha), \quad q = q(\alpha), \quad p = p(\alpha)$$

with arbitrary functions of the parameter α, we make the class of the curves Γ narrower by using the relation $T = -H(t, q, p)$, which holds along the trajectories of the system. Then we parametrize the contour with respect to the variable T, depending on the parametrization of this contour with respect to the remaining variables

$$T(\alpha) = -H\left(t(\alpha), q(\alpha), p(\alpha)\right).$$

Then the integral Poincaré invariant can be considered in a space of fewer variables, t, q, and p, and has the form

$$\oint_\Gamma \left[-H(t, q, p)\,dt + p\,dq\right].$$

In this special case, the integral Poincaré invariant is called the Poincaré–Cartan invariant.

Remark. The theorem proved above implies that the integral $\int[-H\,dt + p\,dq]$ is an integral invariant of the group defined by the equations

$$\frac{dt}{d\tau} = 1, \quad \frac{dq}{d\tau} = \frac{\partial H}{\partial p}, \quad \frac{dp}{d\tau} = -\frac{\partial H}{\partial q}.$$

However, this integral is also invariant under the wider group

$$\frac{dt}{d\tau} = \pi(t, q, p), \quad \frac{dq}{d\tau} = \pi(t, q, p)\frac{\partial H}{\partial p}, \quad \frac{dp}{d\tau} = -\pi(t, q, p)\frac{\partial H}{\partial q},$$

where $\pi(t, q, p)$ is an arbitrary scalar function.

This group has the same scalar trajectories. But the motion along each trajectory changes. In other words, the Poincaré–Cartan invariant has the same value not only along the curves, into which the initial contour is displaced by the phase flow of the system with Hamiltonian $H(t, q, p)$, but also along any curve Γ enclosing the same tube of trajectories (see Fig. 4.1). This fact follows from the theorem below.

THEOREM. *For the integral $\int[-H\,dt + p\,dq]$ to be an integral invariant of the group $\frac{dp}{d\tau} = \pi$, $\frac{dq}{d\tau} = \pi M$, $\frac{dp}{d\tau} = \pi N$ with an arbitrary function $\pi(t, q, p)$, it is necessary and sufficient that*

$$M = \frac{\partial H}{\partial p}, \quad N = -\frac{\partial H}{\partial q}.$$

Proof. Let us write out the germ of the group,

$$\begin{aligned} t' &= t + \pi\tau + \cdots, \\ q' &= q + \pi M\tau + \cdots, \\ p' &= p + \pi N\tau + \cdots \end{aligned}$$

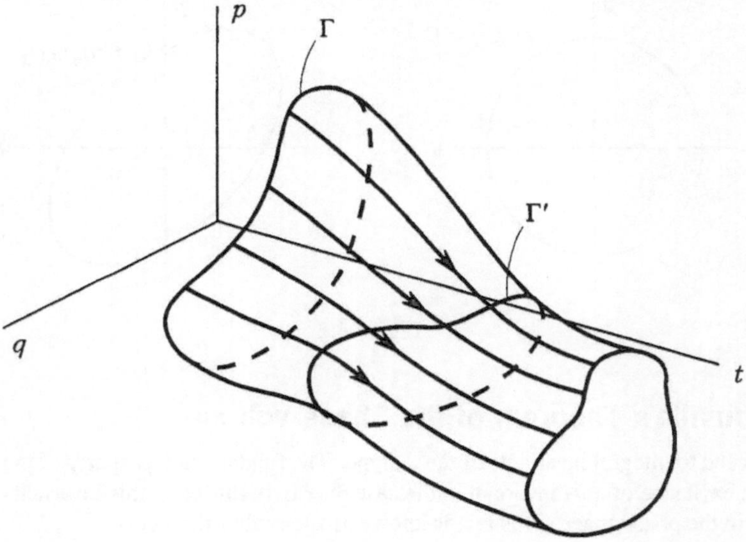

Figure 4.1

and the group generator,

$$U = \pi\left(\frac{\partial}{\partial t} + M\frac{\partial}{\partial q} + N\frac{\partial}{\partial p}\right).$$

Under this group, the integral is transformed as

$$\oint_{\Gamma'} [-H(t', q', p')\,dt' + p'\,dq'] = \oint_{\Gamma} [-H(t, q, p)\,dt + p\,dq] \\ + \tau \oint_{\Gamma} [-UH\,dt + Up\,dq + (-H\,d\pi + p\,d(\pi M))] + \cdots.$$

The necessary and sufficient condition of invariance has the form

$$\oint_{\Gamma} [-UH\,dt + Up\,dq - H\,d\pi + p\,d(\pi M)] = 0.$$

Integrating by parts,

$$\oint H\,d\pi = -\oint \pi\,dH, \quad \oint p\,d(\pi M) = -\oint \pi M\,dp,$$

yields

$$\oint_{\Gamma}\left[\left(-M\frac{\partial H}{\partial q} - N\frac{\partial H}{\partial p}\right)dt + (N + H_q)\,dq + (-M + H_p)\,dp\right]\pi = 0.$$

Since the function π is arbitrary, we have

$$M\frac{\partial H}{\partial q} + N\frac{\partial H}{\partial p} = 0, \quad N + H_q = 0, \quad -M + H_p = 0.$$

This overdetermined system has the unique solution

$$N = -H_q, \quad M = H_p,$$

which proves the theorem.

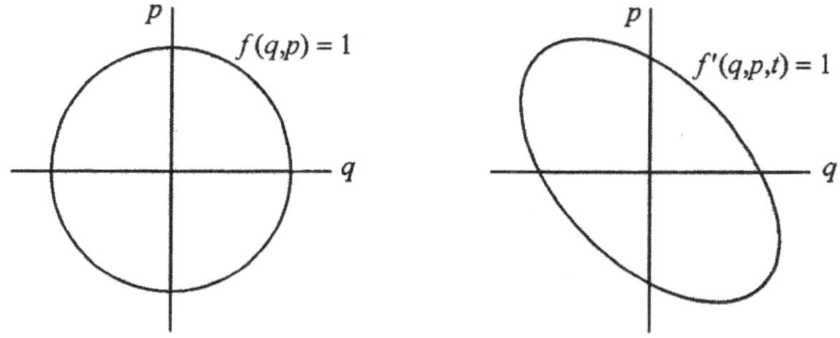

Figure 4.2

4.6.2. Liouville's Theorem of the Phase Volume

We now proceed to integral invariants of the J_3 type. The fundamental property of the Hamiltonian groups is the existence of a J_3 invariant in the case $\Phi \equiv 1$. In this case, this invariant is the volume of a domain in the phase space. This fact is known as Liouville's theorem.

THEOREM (CONSERVATION OF THE PHASE VOLUME). *The integral*

$$J_3 = \int \cdots \int_V dq_1 \ldots dq_n \, dp_1 \ldots dp_n$$

is an invariant of any Hamiltonian group. In other words, if the map $(q, p) \to (q', p')$ changes the domain of integration, $V \to V'$, then the integral over this domain remains unchanged:

$$\int \cdots \int_{V'} dq_1 \ldots dq_n \, dp_1 \ldots dp_n = \int \cdots \int_V dq_1 \ldots dq_n \, dp_1 \ldots dp_n.$$

Proof. Let us use the invariance criterion for such integrals which was established in Section 1.10; this criterion has the form

$$U\Phi + \Phi \operatorname{div} \eta = 0.$$

In our case, we have $\Phi \equiv 1$ and

$$\operatorname{div} \eta = \sum_{i=1}^{n} \left(\frac{\partial^2 H}{\partial q_i \partial p_i} - \frac{\partial^2 H}{\partial p_i \partial q_i} \right) = 0,$$

which proves the theorem.

Example. The Hamiltonian system $\dot{q}' = p'$, $\dot{p}' = q'$ with $H = \frac{1}{2}(p'^2 - q'^2)$ generates a group of transformations of the phase space into itself,

$$q' = q \cosh t + p \sinh t, \quad p' = q \sinh t + p \cosh t \qquad (t \text{ is the group parameter}),$$

where q and p are the initial conditions and q', p' is the solution of the system.

Consider the integral

$$\iint_{q^2+p^2 \leq 1} dq \, dp.$$

The map $(q, p) \to (q', p')$ transforms the boundary of the circle $f = q^2 + p^2 = 1$ into the boundary

$$f = q'^2 + p'^2 = (q \cosh t + p \sinh t)^2 + (q \sinh t + p \cosh t)^2 = f'(t, q, p) = 1.$$

In accordance with Liouville's theorem, we have (see Fig. 4.2)

$$\iint_{f=q^2+p^2 \leq 1} dq \, dp \equiv \iint_{f'(t,q,p) \leq 1} dq \, dp.$$

4.7. Canonical Transformations

We proceed by considering an example.

Example. The Hamiltonian of a one-dimensional linear oscillator has the form $H = \frac{1}{2}(p^2 + q^2)$. The corresponding equations of motion are given by $\dot{q} = p$ and $\dot{p} = -q$.

In these equations of motion, we pass to the polar coordinates $(q, p) \to (\varphi, t) : q = r\cos\varphi$, $p = r\sin\varphi$ to obtain the equation of motion of the oscillator, $\dot{\varphi} = -1$ and $\dot{r} = 0$, in terms of the new variables.

In this case, the Hamiltonian function is transformed as

$$\widetilde{H}(\varphi, r) = H(r\cos\varphi, r\sin\varphi) = \tfrac{1}{2}r^2.$$

If this function were the Hamiltonian of the transformed system, then the relations

$$\dot{\varphi} = \frac{\partial \widetilde{H}}{\partial r} = r, \quad \dot{r} = -\frac{\partial \widetilde{H}}{\partial \varphi} = 0$$

would hold. These relations do not coincide with the performed direct transformation of the equations. Thus, in terms of the new variables, the transformed Hamiltonian function is not related to the transformed equations.

The Lagrange equations are covariant with respect to any admissible (smooth) changes of the generalized coordinates. The Hamilton equations are not covariant with respect to any admissible changes of the phase variables.

The class of changes of the phase variables for which the Hamilton equations are covariant is the class of canonical transformations.

DEFINITION. *A transformation $(q, p) \to (\widetilde{q}, \widetilde{p})$ is called be canonical if it preserves the relation of the transformed equations to the transformed Hamiltonian, whatever this Hamiltonian may be. In other words, the transformation $(q, p) \to (\widetilde{q}, \widetilde{p})$ is canonical if the diagram*

$$\begin{array}{ccc} \boxed{H(q,p)} & \longrightarrow & \boxed{\dot{q} = H_p, \ \dot{p} = -H_q} \\ \downarrow & & \downarrow \\ \boxed{\widetilde{H}(\widetilde{q},\widetilde{p}) \equiv H\big(q(\widetilde{q},\widetilde{p}), p(\widetilde{q},\widetilde{p})\big)} & \longrightarrow & \boxed{\dot{\widetilde{q}} = \widetilde{H}_{\widetilde{p}}, \ \dot{\widetilde{p}} = -\widetilde{H}_{\widetilde{q}}} \end{array}$$

is commutative for any Hamiltonian.

Remark. Sometimes the definition of *valent* canonical transformations is encountered in the literature. These transformations take any Hamiltonian system to a Hamiltonian system, the Hamiltonian of the transformed system being equal to the Hamiltonian of the original system multiplied by a constant, i.e., $\widetilde{H}(\widetilde{q}, \widetilde{p}) = cH\big(q(\widetilde{q}, \widetilde{p}), p(\widetilde{q}, \widetilde{p})\big)$. This constant c is called the valence of the transformation. Such an extension of the class of canonical transformations makes little sense. One can readily show that any valent canonical transformation is the composition of a canonical transformation and a stretching transformation, $q = k_1\widetilde{q}$, $p = k_2\widetilde{p}$.

Canonicity conditions for changes of variables. Let $x = (q, p)$ denote the phase $2n$-vector whose first n coordinates coincide with the generalized coordinates q_i and the other n coordinates coincide with the generalized momenta p_i. With this notation, the Hamilton system of equations can be represented as

$$\dot{x}_i = \sum_{j=1}^{2n} J_{ij} \frac{\partial H}{\partial x_j} \qquad (i = 1, \ldots, 2n)$$

or, in vector form,

$$\dot{x} = J\frac{dH}{dx}$$

where J is the $2n \times 2n$ symplectic matrix,

$$J = \begin{pmatrix} 0 & E \\ -E & 0 \end{pmatrix}$$

and E is the $n \times n$ identity matrix.

The matrix J has the property $J^2 = -E_{2n\times 2n}$ (analogue of the imaginary unit).

Consider a change of variables $x \to \widetilde{x}$,

$$\widetilde{x} = \widetilde{x}(x).$$

We will find the conditions the function $\widetilde{x}(x)$ must satisfy for this change of variables to be canonical. It is obvious that

$$\dot{\widetilde{x}}_i = \sum_{j=1}^{2n} \frac{\partial \widetilde{x}_i}{\partial x_j}\dot{x}_j = \sum_{j,k=1}^{2n} \frac{\partial \widetilde{x}_i}{\partial x_j} J_{jk} \frac{\partial \widetilde{H}}{\partial x_k}.$$

On the other hand,

$$\frac{\partial \widetilde{H}}{\partial x_k} = \sum_{l=1}^{2n} \frac{\partial \widetilde{H}}{\partial \widetilde{x}_l}\frac{\partial \widetilde{x}_l}{\partial x_k}.$$

Finally,

$$\dot{\widetilde{x}}_i = \sum_{j,k,l=1}^{2n} \frac{\partial \widetilde{x}_i}{\partial x_j} J_{jk} \frac{\partial \widetilde{x}_l}{\partial x_k}\frac{\partial \widetilde{H}}{\partial \widetilde{x}_l}$$

or, in vector form,

$$\dot{\widetilde{x}} = \widetilde{J}\frac{d\widetilde{H}}{d\widetilde{x}}, \qquad \widetilde{J} = \frac{d\widetilde{x}}{dx} J \left(\frac{d\widetilde{x}}{dx}\right)^{\mathrm{T}}.$$

Thus, for the change of variables $\widetilde{x}(x)$ to be canonical, it is necessary and sufficient that

$$\frac{d\widetilde{x}}{dx} J \left(\frac{d\widetilde{x}}{dx}\right)^{\mathrm{T}} = J.$$

Since the obtained canonicity condition contains constraints on the derivatives of the equations expressing the change of variables, this condition is called the *local criterion of canonicity*. From among all differentiable transformations of variables, the local criterion of canonicity selects the second-order manifold of canonical transformations.

Example. Separate the manifold of canonical changes of variables from the family of linear transformations

$$\begin{pmatrix} \widetilde{q} \\ \widetilde{p} \end{pmatrix} = \begin{pmatrix} u & v \\ -v & u \end{pmatrix}\begin{pmatrix} q \\ p \end{pmatrix},$$

where u and v are real numbers.

Solution. We take advantage of the criterion of canonicity. In this example, we have

$$\frac{d\widetilde{x}}{dx} = \begin{pmatrix} u & v \\ -v & u \end{pmatrix}.$$

The application of the criterion yields

$$\begin{pmatrix} u & v \\ -v & u \end{pmatrix} \begin{pmatrix} 0 & 1 \\ -1 & 0 \end{pmatrix} \begin{pmatrix} u & -v \\ v & u \end{pmatrix} = \begin{pmatrix} 0 & 1 \\ -1 & 0 \end{pmatrix}.$$

We multiply the matrices on the right-hand side of this equation to obtain

$$\begin{pmatrix} 0 & u^2+v^2 \\ -u^2-v^2 & 0 \end{pmatrix} = \begin{pmatrix} 0 & 1 \\ -1 & 0 \end{pmatrix}.$$

Whence,

$$u^2 + v^2 = 1.$$

The manifold of canonical transformations is a unit circle in the plane (u, v) of all transformations under consideration.

Note that the matrix of the linear transformation in question can be represented in the form

$$\begin{pmatrix} u & v \\ -v & u \end{pmatrix} = uE + vJ, \qquad E = \begin{pmatrix} 1 & 0 \\ 0 & 1 \end{pmatrix}, \quad J = \begin{pmatrix} 0 & 1 \\ -1 & 0 \end{pmatrix}.$$

It is apparent that the family of such transformations is isomorphic to the algebra of complex numbers. Hence, the manifold of the canonical transformations in this family is isomorphic to the rotation group SO(2).

THEOREM. *The map realized by the phase flow of a Hamiltonian system is canonical.*

Proof. Suppose a family of changes of variables $x \to \widetilde{x} : \widetilde{x} = \widetilde{x}(x)$ is generated by the Hamiltonian system

$$\frac{d\widetilde{x}}{d\tau} = J\frac{dH}{d\widetilde{x}}, \qquad \widetilde{x}(0) = x.$$

Introduce the notation $A = d\widetilde{x}/dx$ and $\widetilde{J} = AJA^{\mathrm{T}}$. Each entry of the latter matrix is a function of x and τ, $\widetilde{J}_{ij}(x,\tau)$. Any such function can be represented in the form of the Lie series (Section 1.5)

$$\widetilde{J}_{ij}(x,\tau) = \widetilde{J}_{ij}(x,0) + \tau\big(U\widetilde{J}_{ij}\big)_{\tau=0} + \cdots.$$

If $U\widetilde{J}\big|_{\tau=0} = 0$, then all $U^k\widetilde{J}\big|_{\tau=0} = 0$.

Let us calculate $U\widetilde{J}\big|_{\tau=0}$. We have

$$\widetilde{x} = x + J\frac{dH}{dx}\tau + \cdots, \qquad \frac{d\widetilde{x}}{dx} = E + J\frac{d^2H}{dx^2}\tau + \cdots, \qquad \left(\frac{d\widetilde{x}}{dx}\right)^{\mathrm{T}} = E - \frac{d^2H}{dx}J\tau + \cdots,$$

$$\frac{d\widetilde{x}}{dx}J\left(\frac{d\widetilde{x}}{dx}\right)^{\mathrm{T}} = \left(E + J\frac{d^2H}{dx^2}\tau + \cdots\right)J\left(E - \frac{d^2H}{dx^2}J\tau + \cdots\right) = J + \left(J\frac{d^2H}{dx^2} - J\frac{d^2H}{dx^2}\right)\tau + \cdots.$$

The coefficient of τ is zero. Thus, we obtain $\widetilde{J} = J$, i.e., the phase flow of the Hamiltonian system satisfies the condition of canonicity.

Generating functions. These functions allow one to construct canonical transformations by applying certain simple operations to these functions.

To introduce generating functions, we use the one-to-one correspondence between the Hamilton equations and Poincaré–Cartan integral invariant,

$$\begin{cases} \dfrac{dt}{d\tau} = 1, \\[2mm] \dfrac{dq}{d\tau} = \dfrac{\partial H}{\partial p}, \\[2mm] \dfrac{dp}{d\tau} = -\dfrac{\partial H}{\partial q} \end{cases} \sim \quad \oint_{\Gamma} [-H\,dt + p\,dq].$$

This correspondence was established previously. Consider the transformation $(q, p) \to (\widetilde{q}, \widetilde{p})$. In terms of the new variables, the equations have the form

$$\frac{dt}{d\tau} = 1, \quad \frac{d\widetilde{q}}{d\tau} = M(t, \widetilde{q}, \widetilde{p}), \quad \frac{d\widetilde{p}}{d\tau} = N(t, \widetilde{q}, \widetilde{p}).$$

If the integral invariant has the same form in terms of the new variables,

$$\oint_{\widetilde{\Gamma}} [-\widetilde{H}\, dt + \widetilde{p}\, d\widetilde{q}],$$

then we have $M = \partial \widetilde{H}/\partial \widetilde{p}$ and $N = -\partial \widetilde{H}/\partial \widetilde{q}$ by virtue of that has been proved. Hence, the transformation $(q, p) \to (\widetilde{q}, \widetilde{p})$ is canonical.

For the Poincaré–Cartan invariant to preserve its form after the passage to new variables, it is necessary and sufficient that all terms in the integrand apart from $-\widetilde{H}\, dt + \widetilde{p}\, d\widetilde{q}$ form a total differential,

$$\oint_{\Gamma} [-H\, dt + p\, dq] = \oint_{\widetilde{\Gamma}} [-\widetilde{H}\, dt + \widetilde{p}\, d\widetilde{q} + dF(t, \widetilde{q}, \widetilde{p})].$$

It is convenient for further calculations to represent the arbitrary function $F(t, \widetilde{q}, \widetilde{p})$ in the form

$$F(t, \widetilde{q}, \widetilde{p}) \equiv S\big(t, q(t, \widetilde{q}, \widetilde{p}), \widetilde{p}\big)$$

by using the transformation $(q, p) \to (\widetilde{q}, \widetilde{p})$. We substitute this representation into the total differential to obtain

$$\oint_{\Gamma} (-H\, dt + p\, dq) = \oint_{\widetilde{\Gamma}} \left(-\widetilde{H}\, dt + \widetilde{p}\, d\widetilde{q} + \frac{\partial S}{\partial t}\, dt + \frac{\partial S}{\partial q}\, dq + \frac{\partial S}{\partial \widetilde{p}}\, d\widetilde{p}\right).$$

Whence,

$$\widetilde{H} = \frac{\partial S}{\partial t} + H\big(t, q(t, \widetilde{q}, \widetilde{p}), p(t, \widetilde{q}, \widetilde{p})\big), \quad \widetilde{q} = \frac{\partial S}{\partial \widetilde{p}}, \quad p = \frac{\partial S}{\partial q}.$$

Thus, we have the following algorithm for the construction of a canonical transformation: (i) choose an arbitrary differentiable function of mixed variables $S(t, q, \widetilde{p})$, (ii) construct the equations $\widetilde{q} = \partial S/\partial \widetilde{p}$ and $p = \partial S/\partial q$, and (iii) solve these equations for either new or old variables. The solution obtained is a canonical transformation.

Example. Consider the function $S = q\widetilde{p}^2$. The required partial derivatives have the form

$$\frac{\partial S}{\partial q} = \widetilde{p}^2, \quad \frac{\partial S}{\partial \widetilde{p}} = 2q\widetilde{p}.$$

Hence, the canonical change of variables induced by this generating function has the form

$$p = \widetilde{p}^2, \quad \widetilde{q} = 2q\widetilde{p}.$$

By solving these equations for the new variables, we obtain the change of variables $(q, p) \to (\widetilde{q}, \widetilde{p})$ which is also canonical,

$$\widetilde{q} = 2q\sqrt{p}, \quad \widetilde{p} = \sqrt{p}.$$

This is confirmed by the local criterion of canonicity,

$$\frac{d\widetilde{x}}{dx} = \begin{pmatrix} 2\sqrt{p} & q/\sqrt{p} \\ 0 & 1/(2\sqrt{p}) \end{pmatrix}, \quad \frac{d\widetilde{x}}{dx} J = \begin{pmatrix} -q/\sqrt{p} & 2\sqrt{p} \\ -1/(2\sqrt{p}) & 0 \end{pmatrix},$$

$$\frac{d\widetilde{x}}{dx} J \left(\frac{d\widetilde{x}}{dx}\right)^{\mathrm{T}} = \begin{pmatrix} -q/\sqrt{p} & 2\sqrt{p} \\ -1/(2\sqrt{p}) & 0 \end{pmatrix} \begin{pmatrix} 2\sqrt{p} & 0 \\ q/\sqrt{p} & 1/(2\sqrt{p}) \end{pmatrix} = \begin{pmatrix} 0 & 1 \\ -1 & 0 \end{pmatrix}.$$

The choice of the mixed variables in the generating function is not unique. The following variants are possible:

$$S_1(t,q,\widetilde{p}),\quad S_2(t,\widetilde{q},p),\quad S_3(t,q,\widetilde{q}),\quad S_4(t,p,\widetilde{p}).$$

After similar manipulations, we obtain the relations

$$\widetilde{H}=\frac{\partial S_2}{\partial t}+H,\quad \widetilde{p}=-\frac{\partial S_2}{\partial \widetilde{q}},\quad q=-\frac{\partial S_2}{\partial p},$$
$$\widetilde{H}=\frac{\partial S_3}{\partial t}+H,\quad p=\frac{\partial S_3}{\partial q},\quad \widetilde{p}=-\frac{\partial S_3}{\partial \widetilde{q}},$$
$$\widetilde{H}=\frac{\partial S_4}{\partial t}+H,\quad q=-\frac{\partial S_4}{\partial p},\quad \widetilde{q}=\frac{\partial S_4}{\partial \widetilde{p}}$$

for the three new functions S_2, S_3, and S_4, respectively.

Example. Construct the generating functions for the family of canonical transformations

$$\begin{pmatrix}\widetilde{q}\\ \widetilde{p}\end{pmatrix}=\begin{pmatrix}u & v\\ -v & u\end{pmatrix}\begin{pmatrix}q\\ p\end{pmatrix}\qquad (u^2+v^2=1).$$

Solution. To find the generating function S_1, we solve this system for p and $\widetilde{q}$. As a result, we have

$$p=\frac{\widetilde{p}+vq}{u},\quad \widetilde{q}=\frac{q+v\widetilde{p}}{u}\qquad (u\neq 0).$$

Thus, we have to solve the equations

$$\frac{\partial S_1}{\partial q}=\frac{\widetilde{p}+vq}{u},\quad \frac{\partial S_1}{\partial \widetilde{p}}=\frac{q+v\widetilde{p}}{u}.$$

Whence,

$$S_1=\int_0^1\left[\left(\frac{\widetilde{p}+vq}{u}\right)q+\left(\frac{q+v\widetilde{p}}{u}\right)\widetilde{p}\right]\tau\,d\tau=\frac{1}{2u}\left[2q\widetilde{p}+v(q^2+\widetilde{p}^2)\right].$$

To find the generating function $S_2(\widetilde{q},p)$, we solve the system for $\widetilde{p}$ and q to obtain

$$q=\frac{\widetilde{q}-vp}{u},\quad \widetilde{p}=\frac{p-v\widetilde{q}}{u}\qquad (u\neq 0).$$

Solving the corresponding equations yields

$$S_2=\frac{1}{2u}\left[-2p\widetilde{q}+v(p^2+\widetilde{q}^2)\right].$$

To find the generating function $S_3(q,\widetilde{p})$, we solve the system for p and $\widetilde{p}$ to obtain

$$p=\frac{\widetilde{q}-uq}{v},\quad \widetilde{p}=\frac{u\widetilde{q}-q}{v}\qquad (v\neq 0);$$

whence it follows that

$$S_3=\frac{1}{2u}\left[2q\widetilde{q}-u(q^2+\widetilde{q}^2)\right].$$

Finally, for the function S_4, we have

$$q=\frac{up-\widetilde{p}}{v},\quad \widetilde{q}=\frac{p-u\widetilde{p}}{v}\qquad (v\neq 0)$$

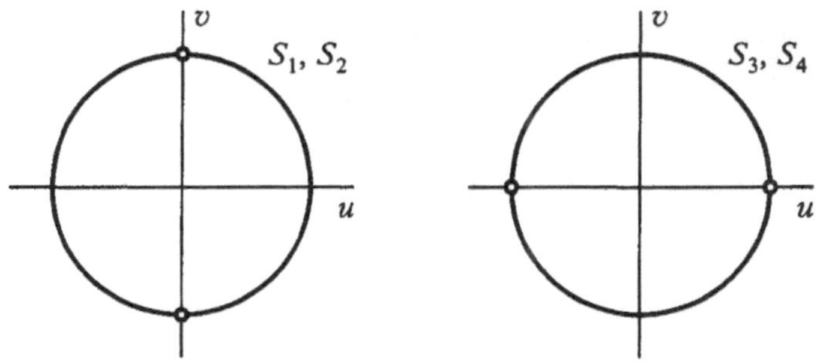

Figure 4.3

and, hence,

$$S_4 = \frac{1}{2u}\left[2p\widetilde{p} - u(p^2 + \widetilde{p}^2)\right].$$

In this example, the generating functions S_1 and S_2 do not exist for the symplectic transformation

$$\begin{pmatrix}\widetilde{q}\\ \widetilde{p}\end{pmatrix} = J\begin{pmatrix}q\\ p\end{pmatrix}$$

and the functions S_3 and S_4 do not exist for the identity transformation (see Fig. 4.3)

$$\begin{pmatrix}\widetilde{q}\\ \widetilde{p}\end{pmatrix} = E\begin{pmatrix}q\\ p\end{pmatrix}.$$

Let us summarize the result of this section. Two methods are presented to test a given transformation of variables $(q, p) \to (\widetilde{q}, \widetilde{p})$ (or $x \to \widetilde{x}$) for canonicity.

First method (local criterion). Construct the matrix of partial derivatives $A = d\widetilde{x}/dx$. After that, check the condition

$$AJA^{\mathrm{T}} = J.$$

Second method. Substitute the change of variables $\widetilde{q} = \widetilde{q}(q, p)$ and $\widetilde{p} = \widetilde{p}(q, p)$ into the differential form $\widetilde{p}\, d\widetilde{q} - p\, dq$. After that, check if the resulting differential form $\widetilde{p}\, d\widetilde{q} - p\, dq = Q(q, p)\, dq + P(q, p)\, dp$ is a total differential, i.e., $Q_p = P_q$.

Two sources of canonical transformations are indicated.

First source. Choose an arbitrary function $S(\widetilde{q}, \widetilde{p})$ as the generating Hamiltonian. Then the solution of the initial-value (Cauchy) problem

$$\frac{d\widetilde{q}}{d\tau} = \frac{\partial S}{\partial \widetilde{p}}, \quad \frac{d\widetilde{p}}{d\tau} = -\frac{\partial S}{\partial \widetilde{q}}, \qquad \widetilde{q}(0) = q, \quad \widetilde{p}(0) = p$$

yields the following one-parameter family of canonical transformations:

$$\widetilde{q} = \widetilde{q}(\tau, q, p), \quad \widetilde{p} = \widetilde{p}(\tau, q, p).$$

Second source. Choose one of the generating functions of mixed variables, for example, $S(q, \widetilde{p})$. Solve the equations of the transformation given in implicit form, $p = S_q$ and $\widetilde{q} = S_{\widetilde{p}}$.

The drawback of the first source is the necessity to solve differential equations. The drawback of the second source is the necessity to invert functions.

In any case, the results can be extended to nonautonomous systems by using the considerations of Section 4.3.

4.8. Hamilton–Jacobi Equation

Consider a Hamiltonian system

$$\dot{q} = H_p, \quad \dot{p} = -H_q, \qquad H = H(t, q, p).$$

We will find a canonical change of variables $(q, p) \to (\widetilde{q}, \widetilde{p})$ reducing the equations of this system to the simplest form. To this end, we use the generating function of the first type,

$$S = S(t, q, \widetilde{p}), \quad p = S_q, \quad \widetilde{q} = S_{\widetilde{p}}.$$

This function is chosen for definiteness. One can also use other functions.

Since the change of variables is canonical, the relation $\widetilde{H} = S_t + H$ between the old and new Hamiltonians, which was established previously, holds.

The equations of motion have the simplest form if $\widetilde{H} = 0$. This condition leads to the partial differential equation

$$\frac{\partial S}{\partial t} + H\left(t, q, \frac{\partial S}{\partial q}\right) = 0$$

for the generating function S. This equation is referred to as the Hamilton–Jacobi equation.

A function $S(t, q_1, \dots, q_n, \alpha_1, \dots, \alpha_n)$ satisfying the Hamilton–Jacobi equation and depending on n arbitrary variables $\alpha_1, \dots, \alpha_n$ is called a *complete integral* of this equation, provided that there are no additive constants in this function and the condition

$$\det\left(\frac{\partial^2 S}{\partial q_i \partial \alpha_j}\right) \neq 0$$

is satisfied.

Assume the arbitrary constants α_i to be the new momenta $\widetilde{p}_i$ in the complete integral obtained. Then, by using this complete integral as the generating function of a canonical transformation, we reduce the original differential Hamilton equations to the form

$$\dot{\widetilde{q}} = 0, \quad \dot{\widetilde{p}} = 0,$$

with the apparent solution $\widetilde{q} = \text{const}$, $\widetilde{p} = \text{const}$.

We solve the equations $\widetilde{q} = S_{\widetilde{p}}$ and $p = S_q$ to obtain the solution of the original Hamilton equations:

$$q = q(t, \widetilde{q}, \widetilde{p}), \quad p = p(t, \widetilde{q}, \widetilde{p}).$$

The established connection between the Hamilton and Hamilton–Jacobi equations can be used to solve the inverse problem: find the complete integral of a partial differential equation of the first order on the basis of the solutions of the corresponding ordinary differential equations.

We proceed as follows. Let the general solution of the Hamiltonian system $\dot{q} = H_p$, $\dot{p} = -H_q$ with $H = H(t, q, p)$ be known,

$$q = q(t, \alpha, \beta), \quad p = p(t, \alpha, \beta),$$

where α and β are arbitrary constants.

We solve these equations for p and α to obtain

$$\alpha = F(t, q, \beta), \quad p = G(t, q, \beta).$$

Then we can find a complete integral of the partial differential equation

$$\frac{\partial S}{\partial t} + H\left(t, q, \frac{\partial S}{\partial q}\right) = 0$$

4.10. The Angle–Action Variables

The canonical angle–action variables are introduced in the case where the manifold of the level of the first integrals H_k is compact. In what follows, these variables will be used to form the procedures of the approximate integration of a system close to Liouville integrable systems.

The angle–action variables are constructed as follows. In the previous section, we found the function $S(q,\alpha) = \int_0^1 \sum_{k=1}^n q_k f_k(\theta q, \alpha)\, d\theta$ and used it as a generating function of the canonical change of variables $(q,p) \to (\widetilde{q}, \widetilde{p})$ after setting $\alpha \equiv \widetilde{p}$.

To obtain the angle–action variables, we do not set $\alpha \equiv \widetilde{p}$ but relate the constants of integration to the new momenta differently. We set

$$\widetilde{p}_n(\alpha) = \frac{1}{2\pi} \int_0^{2\pi} p(\alpha,\theta) \frac{dq(\alpha,\theta)}{d\theta_k}\, d\theta_k.$$

In other words, the kth component of the new momentum is the average of $p(\alpha,\theta)\, dq/d\theta_k$ along a closed trajectory on the torus. This trajectory is obtained by the variation of the parameter θ_k, the remaining parameters being fixed. One should invert the function $\widetilde{p} = \widetilde{p}(\alpha)$ thus obtained and then substitute $\alpha = \alpha(\widetilde{p})$ into $S(q,\alpha)$. After that, the new coordinates can be constructed in the usual way:

$$\widetilde{q} = \frac{\partial}{\partial \widetilde{p}} S\big(q, \alpha(\widetilde{p})\big).$$

The variables $\widetilde{p}$ and $\widetilde{q}$ are referred to as the *action* and *angle*, respectively. In the literature, the special notation is adopted for these variables—$\widetilde{q} \equiv \varphi$ for the angle and $\widetilde{p} \equiv I$ for the action. Thus, we have

$$I_k(\alpha) = \frac{1}{2\pi} \int_0^{2\pi} \sum_{i=1}^n p_i(\alpha, \theta_1, \dots, \theta_n) \frac{d}{d\theta_k} q_i(\alpha, \theta_1, \dots, \theta_n)\, d\theta_k,$$
$$\varphi_k = \frac{\partial}{\partial I_k} S\big(q, \alpha(I)\big), \qquad (k = 1, \dots, n).$$

We assume for definiteness that the Hamiltonian of the system coincides with the first of the integrals which are in involution, i.e., $H(q,p) \equiv H_1(q,p)$. Then, after the passage to the angle–action variables, we obtain a new Hamiltonian in the form

$$\widetilde{H} = H\left(q, \frac{\partial S}{\partial q}\right) \equiv H\Big(q, f\big(q, \alpha(I)\big)\Big) = \alpha_1(I).$$

In terms of these variables, the equations of the original system become

$$\dot{\varphi} = \frac{d\alpha_1(I)}{dI}, \qquad \dot{I} = 0.$$

By using the relations of the transformations, we represent the solution of these equations $\varphi = \omega(I)t + \varphi_0$, $I = \text{const}$, where $\omega(I) = d\alpha_1/dI$, in terms the original variables:

$$q = q\big(\omega(T)t + \varphi_0,\ I\big), \quad p = p\big(\omega(T)t + \varphi_0,\ I\big).$$

The basic property of the angle–action variables is that the canonical transformation $q(\varphi, I)$, $p(\varphi, I)$ is 2π-periodic in all φ_k's. Thus, the solution obtained represents a conditionally periodic oscillatory process.

Example. Integrate the system specified by the Hamiltonian $H = \frac{1}{16}[(p_1^2 - p_2^2 - q_1^2 + q_2^2)^2 + 4(q_1p_1 - q_2p_2)^2]$. This Hamiltonian is an approximation for the problem of oscillations of a compound pendulum.[8]

One can check that the function $G = q_1q_2 + p_1p_2$ is a first integral of the system with the Hamiltonian H, since the Poisson bracket $\{H, G\}$ is zero. The function $E = \frac{1}{2}(q_1^2 + q_2^2 + p_1^2 + p_2^2)$ is also a first integral.

Since the Hamiltonian H is autonomous, it is also a first integral. However, these three integrals are functionally dependent, $4H + G^2 = E^2$.

The intersection of the surfaces defined by the integrals G and E is compact, not simply connected though. In accordance with Liouville's theorem, each connected component is an invariant torus.

By Liouville's theorem, a Hamiltonian system of the forth order having two first integrals in involution can be reduced to quadratures. To this end, it is sufficient to solve H and G for p_1 and p_2 and equate the resulting expression to the derivatives of the desired generating function with respect to q_1 and q_2. From the equations thus obtained, one can find the generating functions.

To simplify the calculations, we slightly modify this procedure. The first integral G can be considered as the Hamiltonian function of a system the phase flow of which generates a symmetry group for the system with Hamiltonian H. By passing to the canonical coordinates of the symmetry group, we reduce the order of the system by two. To calculate the canonical coordinates, we can use Liouville's theorem not for H and G but for G and E, which is much simpler.

From $G = q_1q_2 + p_1p_2$ and $2E = q_1^2 + q_2^2 + p_1^2 + p_2^2$ we find that

$$2p_1p_2 = 2(G - q_1q_2), \quad p_1^2 + p_2^2 = 2E - q_1^2 - q_2^2;$$

whence it follows that

$$p_1 + p_2 = \sqrt{2(E+G) - (q_1+q_2)^2},$$
$$p_1 - p_2 = \sqrt{2(E-G) - (q_1-q_2)^2}$$

and the following relations hold:

$$p_1 = \tfrac{1}{2}\left[\sqrt{2(E+G) - (q_1+q_2)^2} + \sqrt{2(E-G) - (q_1-q_2)^2}\right],$$
$$p_2 = \tfrac{1}{2}\left[\sqrt{2(E+G) - (q_1+q_2)^2} - \sqrt{2(E-G) - (q_1-q_2)^2}\right].$$

We will seek the generating function of the desired canonical transformation in the form $S(q_1, q_2, y_1, y_2)$ taking into account the condition

$$p_1 = 2\frac{\partial S}{\partial q_1}, \quad p_2 = 2\frac{\partial S}{\partial q_2}.$$

We use the so-called valent canonical transformation with valence 2. For this transformation, the transformed and original Hamiltonians are related by the condition $\widetilde{H}(\widetilde{q}, \widetilde{p}) = 2H\big(q(\widetilde{q}, \widetilde{p}), p(\widetilde{q}, \widetilde{p})\big)$. This is done to avoid fractional coefficients in what follows.

Since $\partial p_1/\partial q_2 = \partial p_2/\partial q_1$ (the fundamental point in Liouville's theorem), we have

$$S = \frac{1}{4}(q_1+q_2)\int_0^1 \sqrt{2(E+G) - (q_1+q_2)^2\tau^2}\,d\tau + \frac{1}{4}(q_1+q_2)\int_0^1 \sqrt{2(E-G) - (q_1-q_2)^2\tau^2}\,d\tau.$$

We choose the new momenta y_1 and y_2 so as to satisfy the relations $2y_1 = E$ and $2y_2 = G$. Then the new generalized coordinates can be found from the condition $x_1 = \partial S/\partial y_1$, $x_2 = \partial S/\partial y_2$.

---

[8] V. Ph. Zhuravlev, *Investigation of nonlinear oscillations of a compound pendulum*, Mechanics of Solids, Vol. 31, No. 3, pp. 137–142, 1996.

Differentiating the function S obtained above yields

$$x_1 = \frac{1}{2}(q_1+q_2)\int_0^1 \frac{d\tau}{\sqrt{4(y_1+y_2)-(q_1+q_2)^2\tau^2}} + \frac{1}{2}(q_1-q_2)\int_0^1 \frac{d\tau}{\sqrt{4(y_1-y_2)-(q_1-q_2)^2\tau^2}},$$
$$x_2 = \frac{1}{2}(q_1+q_2)\int_0^1 \frac{d\tau}{\sqrt{4(y_1+y_2)-(q_1+q_2)^2\tau^2}} - \frac{1}{2}(q_1-q_2)\int_0^1 \frac{d\tau}{\sqrt{4(y_1-y_2)-(q_1-q_2)^2\tau^2}}.$$

Computing the integrals, we obtain

$$x_1 = \frac{1}{2}\left[\arcsin\frac{q_1+q_2}{\sqrt{4(y_1+y_2)}} + \arcsin\frac{q_1-q_2}{\sqrt{4(y_1-y_2)}}\right],$$
$$x_2 = \frac{1}{2}\left[\arcsin\frac{q_1+q_2}{\sqrt{4(y_1+y_2)}} - \arcsin\frac{q_1-q_2}{\sqrt{4(y_1-y_2)}}\right];$$

whence we obtain the old coordinates expressed in terms of the new coordinates,

$$q_1 = \sqrt{y_1+y_2}\sin(x_1+x_2) + \sqrt{y_1-y_2}\sin(x_1-x_2),$$
$$q_2 = \sqrt{y_1+y_2}\sin(x_1+x_2) - \sqrt{y_1-y_2}\sin(x_1-x_2).$$

After that, we find the old momenta

$$p_1 = \sqrt{y_1+y_2}\cos(x_1+x_2) + \sqrt{y_1-y_2}\cos(x_1-x_2),$$
$$p_2 = \sqrt{y_1+y_2}\cos(x_1+x_2) - \sqrt{y_1-y_2}\cos(x_1-x_2).$$

The relations obtained define a canonical change of variables $(q_1, q_2, p_1, p_2) \to (x_1, x_2, y_1, y_2)$ of valence $1/2$. This transformation is the passage to canonical coordinates of the group generated by the Hamiltonian $G = q_1q_2 + p_1p_2$.

In accordance with Lie's theorem, the system with the Hamiltonian H represented in terms of these coordinates has a lower order. We use this change of variables to obtain $\widetilde{H} = \frac{1}{2}(y_1^2 - y_2^2)$. In terms of the new variables, the equations of the system have the form

$$\dot{x}_1 = y_1, \quad \dot{x}_2 = -y_2, \quad \dot{y}_1 = 0, \quad \dot{y}_2 = 0.$$

This system can be integrated in closed form. Together with the change of variables, this system defines the solution of the Hamiltonian system with Hamiltonian H in terms of the original variables.

Let us collect the final results together. The system with Hamiltonian H has the form

$$\dot{q}_1 = \tfrac{1}{4}\left[(p_1^2 - p_2^2 - q_1^2 + q_2^2)p_1 + 2(q_1p_1 - q_2p_2)q_1\right],$$
$$\dot{q}_2 = -\tfrac{1}{4}\left[(p_1^2 - p_2^2 - q_1^2 + q_2^2)p_2 + 2(q_1p_1 - q_2p_2)q_2\right],$$
$$\dot{p}_1 = \tfrac{1}{4}\left[(p_1^2 - p_2^2 - q_1^2 + q_2^2)q_1 - 2(q_1p_1 - q_2p_2)p_1\right],$$
$$\dot{p}_2 = -\tfrac{1}{4}\left[(p_1^2 - p_2^2 - q_1^2 + q_2^2)q_2 - 2(q_1p_1 - q_2p_2)p_2\right].$$

The general solution of this system is

$$q_1 = C_1\sin(C_2^2 t + C_3) + C_2\sin(C_1^2 t + C_4),$$
$$q_2 = C_1\sin(C_2^2 t + C_3) - C_2\sin(C_1^2 t + C_4),$$
$$p_1 = C_1\cos(C_2^2 t + C_3) + C_2\cos(C_1^2 t + C_4),$$
$$p_2 = C_1\cos(C_2^2 t + C_3) - C_2\cos(C_1^2 t + C_4),$$

where C_1, C_2, C_3, and C_4 are arbitrary constants of integration.

Pay attention to the typical nonlinear effect—the frequency of oscillations depends on the amplitude.

Chapter 5

Asymptotic Methods of Applied Mathematics

5.1. Introduction

Recall that, in developing their methods, the founders of the theory of continuous groups aimed at constructing solutions to differential equations. The previous chapters give us an idea of how far the capabilities of the group-theoretic analysis go beyond the original aim. The Procrustean bed of the desire to necessarily integrate equations in quadrature turned out to be too short for the potentials of the group analysis.

The object under study is itself the reason for this state of things. The differential equations integrable in quadrature are only a small fraction of the other equations, and the attempt to find quadratures is most frequently the attempt to see the nature of things more simply than it is in reality.

Long before group theory appeared and quite independently of it, a number of approximate methods for the integration of differential equations had been suggested and developed.

The most significant achievement here is the creation of asymptotic methods; the main contribution to this was made by H. Poincaré. The concept of the normal form of nonlinear differential equations which underlies asymptotic methods is as fundamental as, for example, the concept of the Jordan form of a linear operator.

The fundamental asymptotic methods such as the Krylov–Bogolyubov method and the multiscale method are modifications of the Poincaré normal form method. They differ only in the means for attaining the common aim—the construction of the normal form.

For a long time these methods had been developing without any connection with the ideas of group theory. The attempts that appeared subsequently to construct symmetry groups of differential equations approximately turned out to be in the course of the local analysis, where the system under study is treated in a neighborhood of some exactly integrable system, thus adding little to the essence of the basic idea.

It was not until the main drawback of all asymptotic methods—a rapid increase in the amount of elementary operations with the number of the approximation—had become a serious hindrance to the wide application of these methods in practice that a real progress in this issue was made.

For example, the first approximation of the Krylov–Bogolyubov method for Duffing's equation can be constructed quite easily. But the second approximation requires already hundreds of elementary operations, and the third approximation, dozens of thousands.

Of course, the construction of higher approximations for complicated problems is unthinkable without computer algebra. At the moment, software packages that implement basic asymptotic algorithms are available. However, the geometric increase in the amount of computations with the number of the approximation leads to a problem of qualitative nature. It is common to believe that one of the advantages of averaged equations as compared with the original equations is the fact that, after the rapidly oscillating components have been removed, it is much easier to integrate the equations numerically with computers.

Meanwhile, the total amount of computations including the construction of averaged equations and their numerical solution may be larger than that required in direct numerical integration, even with a finer computational mesh. Thus, it may happen that it is senseless to implement a cumbersome asymptotic procedure.

The important problem of optimization of asymptotic procedures with respect to the number of elementary operations arises.

Of course, one cannot avoid a geometric progression in the increase of this number with the number of the approximation, but the geometric ratio can be minimized, thus allowing one to construct higher approximations sensibly.

For solving this problem, it is quite natural to invoke algebra. To notice an algebraic structure in the chain of the operations being performed or to form these operations deliberately, thus determining such a structure in advance, is to straighten things out where things were out of order before. As was established subsequently, the normal form is based on the most important group-theoretic fact that the operator of the linear part of a system in the normal form commutes with the operator of its nonlinear part. This justifies the application of algebraic methods to the construction of asymptotic expansions.

In the current chapter, such an attempt is made. The application of new procedures based on group-theoretic ideas allowed us to obtain one-dimensional recurrence relations for constructing an arbitrary approximation of the normal form. Relations of this sort were absent from the algorithms that existed previously. And this expresses the significant simplifications that we managed to achieve.

The next section of this chapter is devoted to auxiliary issues of the reduction of conservative mechanical coordinates to normal coordinates. In practice, the asymptotic expansions are most frequently constructed in a neighborhood of linear systems, where the normal coordinates are the most convenient.

Further, asymptotic procedure are presented for constructing the Poincaré normal form and for successive averaging on the basis of Hausdorff's formula in the spirit of Krylov and Bogolyubov. Since Hausdorff's formula contains the result of the transformation of an operator under the group generated by another operator, this makes it possible to avoid the calculation of the inverse functions, the most laborious phase in traditional procedures.

Special attention is paid to Hamiltonian systems. Birkhoff's method of normal form is subjected to radical revision; first, a new definition of the normal form is given. The old definition, based on specifying the form of nonlinear terms in terms of special variables, is replaced by a new definition which possesses invariant properties, i.e., does not depend on the form of the variables. In the new definition, there is no difference between the resonant and nonresonant normal forms. The invariant definition of the normal form of the Hamiltonian function is accompanied by an invariant algorithm for reducing the system to the normal form.

For Hamiltonian systems, the asymptotic expansions are constructed for a scalar object—the Hamiltonian function of the system—rather than for a vector. This is more favorable from the computational point of view. Hence, one of the ways to simplify the mathematical manipulations for systems of general from is to artificially hamiltonize these systems. This issue is considered in one of the sections of this chapter.

The method of tangent approximations stands by itself in this chapter. Here, unlike the other methods presented, a solution of the approximate system is constructed rather than the approximate system itself. What is in common with the other methods is the use of groups, their operators, Lie series, and so on.

The chapter ends with the presentation of classical problems of oscillation theory and their treatment with the new methods.

5.2. Normal Coordinates of Conservative Systems

Consider the motion of a mechanical system near an equilibrium. The kinetic energy of the system is defined by a positive definite quadratic form,

$$T = \frac{1}{2}\sum_{j=1}^{n}\sum_{k=1}^{n} A_{jk}\dot{y}_j\dot{y}_k,$$

where y_j, y_k are the generalized coordinates, $\dot{y}_j$, $\dot{y}_k$ are the generalized velocities, and the coefficients A_{jk} are constant, with $A_{jk} = A_{kj}$. The potential energy of the system,

$$\Phi = \frac{1}{2}\sum_{j=1}^{n}\sum_{k=1}^{n} H_{jk}y_j y_k,$$

is also a *positive* definite quadratic form, in which the coefficients H_{jk} are constant and $H_{jk} = H_{kj}$.

Introduce the matrices

$$\underset{(n\times 1)}{y} = \begin{pmatrix} y_1 \\ \vdots \\ y_n \end{pmatrix}, \qquad \underset{(n\times n)}{A} = (A_{jk}), \qquad \underset{(n\times n)}{H} = (H_{jk}).$$

Using the designation $D = \frac{d}{dt}$ for the operator of differentiation with respect to time, we compile the matrix $f(D)$ with entries $f_{jk}(D) = A_{jk}D^2 + H_{jk}$,

$$\underset{(n\times n)}{f(D)} = \big(f_{jk}(D)\big).$$

Lagrange's equations of the second kind for the system in question have the form

$$F(D)y = 0,$$

or

$$(AD^2 + H)y = 0. \tag{2.1}$$

Denoting $D^2 = -\lambda$, we rewrite the matrix relations in the form

$$f(D) = h(\lambda), \quad h(\lambda) = -A\lambda + H, \quad (-A\lambda + H)y = 0. \tag{2.2}$$

The determinant of the matrix $h(\lambda)$ has n roots, which are called characteristic numbers.

To the positive definite kinetic energy T there corresponds the positive definite quadratic form

$$A(y, y) = y^{\mathrm{T}}Ay, \qquad y^{\mathrm{T}} = (y_1, \ldots, y_n).$$

Along with quadratic forms, we consider the bilinear forms

$$A(x, y) = x^{\mathrm{T}}Ay = \sum_{j=1}^{n}\sum_{k=1}^{n} A_{jk}x_j y_k, \qquad x^{\mathrm{T}} = (x_1, \ldots, x_n).$$

A linear transformation $(x, y) \to (\xi, \eta)$,

$$x = \underset{(n\times n)}{v}\,\underset{(n\times 1)}{\xi}, \quad y = \underset{(n\times n)}{w}\,\underset{(n\times 1)}{\eta}, \qquad x^{\mathrm{T}} = \xi^{\mathrm{T}}v^{\mathrm{T}},$$

modifies the bilinear form as follows:

$$x^{\mathrm{T}}Ay = \xi^{\mathrm{T}}v^{\mathrm{T}}Aw\eta = \xi^{\mathrm{T}}a\eta, \qquad a = v^{\mathrm{T}}Aw.$$

Under a linear transformation $x = \underset{(n\times n)}{v}\ \underset{(n\times 1)}{\xi}$ of the quadratic form $A(x, x)$, the matrix $a = v^{\mathrm{T}}Av$ is symmetric. Indeed,

$$a^{\mathrm{T}} = (Av)^{\mathrm{T}}v = v^{\mathrm{T}}Av = a.$$

Return to Eq. (2.1). We will seek its solution in the form

$$y = \underset{(n\times 1)}{v_m}\, e^{\kappa_m t}.$$

Substituting this y into Eq. (2.1) and canceling $e^{\kappa_m t}$, we arrive at the matrix equation

$$(A\kappa_m^2 + H)v_m = 0,$$

which can be rewritten as

$$(-A\lambda_m + H)v_m = 0, \qquad \kappa_m^2 = -\lambda_m. \tag{2.3}$$

The $n \times 1$ column matrix v_m is called a modal column.

In the scalar form, Eq. (2.3) becomes

$$-\sum_{k=1}^{n} A_{jk}v_{km}\lambda_m + \sum_{k=1}^{n} H_{jk}v_{km} = 0 \qquad (j = 1, \dots, n;\ m = 1, \dots, n).$$

Introducing the matrix

$$\Lambda = \begin{pmatrix} \lambda_1 & \cdots & 0 \\ \vdots & \ddots & \vdots \\ 0 & \cdots & \lambda_n \end{pmatrix} \quad \text{with entries} \quad \Lambda_{jk} = \begin{cases} \lambda_k & \text{for } j = k, \\ 0 & \text{for } j \neq k, \end{cases}$$

and using the modal matrix $\underset{(n\times n)}{v} = (v_1, \dots, v_n)$, we rewrite this system of equations in the form

$$-\sum_{k=1}^{n}\sum_{l=1}^{n} A_{jk}v_{kl}\Lambda_{lm} + \sum_{k=1}^{n} H_{jk}v_{km} = 0,$$

or

$$-Av\Lambda + Hv = 0. \tag{2.4}$$

Let us make the change of variables

$$y = v\eta \tag{2.5}$$

in Eq. (2.1). Premultiplying the resulting equation $(AvD^2 + Hv)\eta = 0$ by v^{T}, we obtain

$$(A^*D^2 + H^*)\eta = 0, \qquad A^* = v^{\mathrm{T}}Av, \quad H^* = v^{\mathrm{T}}Hv.$$

It is easy to verify that the matrices A^* and H^* are symmetric.

Premultiplying now Eq. (2.4) by v^{T}, we have

$$-A^*\Lambda + H^* = 0 \tag{2.6}$$

or, in scalar form,

$$-A^*_{ml}\lambda_l + H^*_{ml} = 0 \qquad (m, l = 1, \dots, n), \tag{2.7}$$

where

$$A^*_{lm} = v_l^{\mathrm{T}} A v_m = A(v_l, v_m) = \sum_{j=1}^{n}\sum_{k=1}^{n} A_{jk} v_{jl} v_{km},$$

$$H^*_{lm} = v_l^{\mathrm{T}} H v_m = H(v_l, v_m) = \sum_{j=1}^{n}\sum_{k=1}^{n} H_{jk} v_{jl} v_{km}.$$

Let us demonstrate that all roots λ_l of Eq. (2.7) are real. Suppose a λ_l and a λ_m are complex conjugate roots. The corresponding modal columns are given by

$$v_{jl} = \alpha + i\beta_j, \quad v_{jm} = \alpha - i\beta_j \qquad (j = 1, \dots, n).$$

Compute the entries of A^*; we have

$$A^*_{ml} = \sum_{j=1}^{n}\sum_{k=1}^{n} A_{jk} v_{jm} v_{kl} = \sum_{j=1}^{n}\sum_{k=1}^{n} A_{jk}(\alpha_j - i\beta_j)(\alpha_k + i\beta_k)$$

$$= \sum_{j=1}^{n}\sum_{k=1}^{n} A_{jk}(\alpha_j\alpha_k + \beta_j\beta_k) + i\sum_{j=1}^{n}\sum_{k=1}^{n} A_{jk}(\alpha_j\beta_k - \beta_j\alpha_k).$$

Since the matrix A is symmetric, the last sum is zero. Hence,

$$A^*_{ml} = A(\alpha, \alpha) + A(\beta, \beta).$$

Similarly, we obtain $H^*_{ml} = H(\alpha, \alpha) + H(\beta, \beta)$.

From Eq. (2.7) it follows that

$$\lambda_l = \frac{H(\alpha, \alpha) + H(\beta, \beta)}{A(\alpha, \alpha) + H(\beta, \beta)},$$

which proves that the root λ_l is real. Consequently, $\alpha_j = v_j$ and $\beta_j = 0$ and, hence,

$$\lambda_l = \frac{H(v_l, v_l)}{A(v_l, v_l)}.$$

Rewrite Eq. (2.7) in the form

$$-A^*_{lm}\lambda_m + H^*_{lm} = 0.$$

Since A^* and H^* are both symmetric, this equation can also be represented as

$$-A^*_{lm}\lambda_l + H^*_{ml} = 0.$$

It follows that

$$(\lambda_m - \lambda_l)A^*_{lm} = 0.$$

If $\lambda_m \neq \lambda_l$, then $A^*_{lm} = A(v_l, v_m) = \sum_{j=1}^{n}\sum_{k=1}^{n} A_{jk} v_{jl} v_{km} = 0$, and hence the columns v_l and v_m are orthogonal. Obviously, $H^*_{lm} = H(v_l, v_m) = \sum_{j=1}^{n}\sum_{k=1}^{n} H_{jk} v_{jl} v_{km}$ is also zero.

Compute $A^*_{ll} = A(v_l, v_l) = \sum_{j=1}^{n}\sum_{k=1}^{n} A_{jk} v_{jl} v_{kl}$ and divide each component of the column v_l by $\sqrt{A^*_{ll}}$. For the new modal vector $v'_l = v_l/\sqrt{A^*_{ll}}$, we have $A(v'_l, v'_l) = 1$. Taking this into account and using Eq. (2.7), we obtain

$$H(v'_l, v'_l) = \lambda_l.$$

Thus, under the change of variables $y \to \eta$ of (2.5), the matrices A and H become

$$A^* = \begin{pmatrix} 1 & 0 & \cdots & 0 \\ 0 & 1 & \cdots & 0 \\ \vdots & \vdots & \ddots & \vdots \\ 0 & 0 & \cdots & 1 \end{pmatrix}, \quad H^* = \begin{pmatrix} \lambda_1 & 0 & \cdots & 0 \\ 0 & \lambda_2 & \cdots & 0 \\ \vdots & \vdots & \ddots & \vdots \\ 0 & 0 & \cdots & \lambda_n \end{pmatrix}.$$

Consequently, in terms of η, the kinetic and potential energies of the system acquire the form

$$T^* = \tfrac{1}{2}(\dot\eta_1^2 + \cdots + \dot\eta_n^2), \quad \Pi^* = \tfrac{1}{2}(\lambda_1\eta_1^2 + \cdots + \lambda_n\eta_n^2).$$

For this reason, the coordinates η are referred to as normal coordinates. Lagrange's equations of the second kind become independent and take the form

$$\ddot\eta + \omega_k^2\eta_k = 0, \qquad \omega_k^2 = \lambda_k, \quad k = 1\ldots, n.$$

Recall that the modal columns v_m can be found by the immediate solution of system (2.3) or taken in the form of nonzero columns of the matrix adjoint to $h(\lambda_m)$.

Consider now the case of multiple roots. As was indicated previously, for multiple roots the matrix $h(\lambda)$ has linear elementary divisors. In this case, modal columns may be nonorthogonal. However, a orthogonalization procedure can be applied to these vectors. Such a procedure is presented below.

Suppose we are given a set of modal columns $v_1, v_1, v_3, \ldots$ Any linear combination of modal columns is a modal column. Construct another set of modal columns, $u_1, u_2, u_3, \ldots$, as follows:

$$\begin{aligned} &u_1 = v_1, \quad z_1 = \frac{u_1}{\sqrt{A(u_1, u_1)}}, \\ &u_2 = v_2 - A(v_2, z_1)\, z_1, \quad z_2 = \frac{u_2}{\sqrt{A(u_2, u_2)}}, \\ &u_3 = v_2 - A(v_3, z_1)\, z_1 - A(v_3, z_2)\, z_2, \quad z_3 = \frac{u_3}{\sqrt{A(u_3, u_3)}}, \\ &\cdots\cdots\cdots\cdots\cdots\cdots\cdots\cdots\cdots\cdots \end{aligned}$$

Since

$$\begin{aligned} &A(u_2, z_1) = A(v_2, z_1) - A(v_2, z_1)\, A(z_1, z_1) = 0, \\ &A(u_3, z_2) = A(v_3, z_2) - A(v_3, z_1)\, A(z_1, z_2) - A(v_3, z_2)\, A(z_2, z_2) = 0, \end{aligned}$$

the new modal columns are orthogonal. It is these columns that should be used for constructing the modal matrix in relation (2.5) that defines the transition to normal coordinates.

Example. Consider a system with kinetic energy $A(\dot y, \dot y) = \dot y_1^2 + \dot y_2^2 + \dot y_3^2$ and potential energy $H(y, y) = -y_1^2 - y_2^2 - y_2^2 + 2y_1y_2 + 2y_2y_3 + 2y_3y_1$. Accordingly,

$$A = \begin{pmatrix} 1 & 0 & 0 \\ 0 & 1 & 0 \\ 0 & 0 & 1 \end{pmatrix}, \quad H = \begin{pmatrix} -1 & 1 & 1 \\ 1 & -1 & 1 \\ 1 & 1 & -1 \end{pmatrix};$$

hence,

$$h(\lambda) = -A\lambda + H = \begin{pmatrix} -\lambda - 1 & 1 & 1 \\ 1 & -\lambda - 1 & 1 \\ 1 & 1 & -\lambda - 1 \end{pmatrix}.$$

The characteristic equation

$$\det h(\lambda) = -(\lambda + 1)^3 + 3(\lambda + 1) + 2 = 0$$

has the roots $\lambda_1 = -2$, $\lambda_2 = -2$, and $\lambda_3 = 1$. To construct the adjoint $F(\lambda)$ of the matrix $h(\lambda)$, we replace each entry by its cofactor and transpose the resulting matrix, thus obtaining

$$F(\lambda) = \begin{pmatrix} \lambda(\lambda+2) & \lambda+2 & \lambda+2 \\ \lambda+2 & \lambda(\lambda+2) & \lambda+2 \\ \lambda+2 & \lambda+2 & \lambda(\lambda+2) \end{pmatrix}.$$

The derivative of $F(\lambda)$ is expressed as

$$F'(\lambda) = \begin{pmatrix} 2(\lambda+1) & 1 & 1 \\ 1 & 2(\lambda+1) & 1 \\ 1 & 1 & 2(\lambda+1) \end{pmatrix}.$$

One should take two nonzero columns of the matrix[9]

$$F'(\lambda) = \begin{pmatrix} -2 & 1 & 1 \\ 1 & -2 & 1 \\ 1 & 1 & -2 \end{pmatrix}$$

and one nonzero column of the matrix

$$F'(\lambda) = \begin{pmatrix} 3 & 3 & 3 \\ 3 & 3 & 3 \\ 3 & 3 & 3 \end{pmatrix}.$$

This yields

$$v_1 = \begin{pmatrix} -2 \\ 1 \\ 1 \end{pmatrix}, \quad v_2 = \begin{pmatrix} 1 \\ -2 \\ 1 \end{pmatrix}, \quad v_3 = \begin{pmatrix} 3 \\ 3 \\ 3 \end{pmatrix}.$$

Simplify these columns by compiling linear combinations of v_1 and v_2 and dividing the columns by constant coefficients; we have

$$v_1 = \begin{pmatrix} 1 \\ 0 \\ -1 \end{pmatrix}, \quad v_2 = \begin{pmatrix} 0 \\ 1 \\ -1 \end{pmatrix}, \quad v_3 = \begin{pmatrix} 1 \\ 1 \\ 1 \end{pmatrix}.$$

Calculate the u_i and z_i following the procedure outlined above and normalize the resulting vectors, thus obtaining

$$z_1 = \begin{pmatrix} 1/\sqrt{2} \\ 0 \\ -1/\sqrt{2} \end{pmatrix}, \quad u_2 = \begin{pmatrix} 0 \\ 1 \\ -1 \end{pmatrix} - \frac{1}{\sqrt{2}} \begin{pmatrix} 1/\sqrt{2} \\ 0 \\ -1/\sqrt{2} \end{pmatrix} = \begin{pmatrix} -1/2 \\ 1 \\ -1/2 \end{pmatrix},$$

$$z_2 = \begin{pmatrix} -1/\sqrt{6} \\ \sqrt{2/3} \\ -1/\sqrt{6} \end{pmatrix}, \quad z_3 = \begin{pmatrix} 1/\sqrt{3} \\ 1/\sqrt{3} \\ 1/\sqrt{3} \end{pmatrix}.$$

In accordance with Eq. (2.5), the passage to normal coordinates η_1, η_2, and η_3 is expressed by the relations

$$\begin{aligned} y_1 &= \frac{1}{\sqrt{2}}\eta_1 - \frac{1}{\sqrt{6}}\eta_2 + \frac{1}{\sqrt{3}}\eta_3, \\ y_2 &= \frac{\sqrt{2}}{\sqrt{3}}\eta_2 + \frac{1}{\sqrt{3}}\eta_3, \\ y_3 &= -\frac{1}{\sqrt{2}}\eta_1 - \frac{1}{\sqrt{6}}\eta_2 + \frac{1}{\sqrt{3}}\eta_3. \end{aligned}$$

In terms of the normal coordinates, the kinetic an potential energies acquire, respectively, the form

$$A^*(\dot\eta, \dot\eta) = \dot\eta_1^2 + \dot\eta_2^2 + \dot\eta_3^2, \quad H^*(\eta, \eta) = -2\eta_1^2 - 2\eta_2^2 + \eta_3^2.$$

[9] B. V. Bulgakov, *Oscillations*, GTTI, Moscow, 1954 [in Russian].

5.3. Single-Frequency Method of Averaging Based on Hausdorff's Formula

Consider a system of nonlinear differential equations of the form

$$\frac{dx}{dt} = X(x, y_1, \dots, y_n, \varepsilon),$$
$$\frac{dy_1}{dt} = Y^{(1)}(x, y_1, \dots, y_n, \varepsilon),$$
$$\dots\dots\dots\dots\dots\dots\dots\dots$$
$$\frac{dy_n}{dt} = Y^{(n)}(x, y_1, \dots, y_n, \varepsilon),$$

where $\varepsilon \ll 1$.

Denoting the set of the variables $y_1, \dots, y_n$ and functions $Y^{(1)}, \dots, Y^{(n)}$ by y and Y, respectively, we rewrite the system as

$$\frac{dx}{dt} = X(x, y, \varepsilon), \quad \frac{dy}{dt} = Y(x, y, \varepsilon). \tag{3.1}$$

Suppose the general solution of this system for $\varepsilon = 0$ is known. Assume that the system has a single fast variable, x. Then, by variation of constants, we reduce the system to the form

$$\begin{aligned} \frac{dx}{dt} &= 1 + \varepsilon X_1(x, y) + \varepsilon^2 X_2(x, y) + \varepsilon^3 X_3(x, y) + \cdots, \\ \frac{dy}{dt} &= \varepsilon Y_1(x, y) + \varepsilon^2 Y_2(x, y) + \varepsilon^3 Y_3(x, y) + \cdots, \end{aligned} \tag{3.2}$$

where

$$Y_i(x, y) = \left(Y_i^{(1)}(x, y), \dots, Y_i^{(n)}(x, y)\right)^{\mathrm{T}}, \qquad i = 1, 2, 3, \dots$$

To system (3.2) there corresponds the following standard form of the system operator:

$$\begin{aligned} A = \frac{\partial}{\partial x} &+ \varepsilon \left[X_1(x, y)\frac{\partial}{\partial x} + Y_1(x, y)\frac{\partial}{\partial y}\right] \\ &+ \varepsilon^2 \left[X_2(x, y)\frac{\partial}{\partial x} + Y_2(x, y)\frac{\partial}{\partial y}\right] + \varepsilon^3 \left[X_3(x, y)\frac{\partial}{\partial x} + Y_3(x, y)\frac{\partial}{\partial y}\right] + \cdots, \end{aligned}$$

where

$$Y_i(x, y)\frac{\partial}{\partial y} = Y_i^{(1)}(x, y)\frac{\partial}{\partial y_1} + \cdots + Y_i^{(n)}(x, y)\frac{\partial}{\partial y_n}.$$

We will seek a change of variables simplifying the operator A in the form of a one-parameter Lie group generated by a system of equations

$$\frac{dx}{d\tau} = \xi(x, y), \quad \frac{dy}{d\tau} = \eta(x, y)$$

with operator

$$U = \xi(x, y)\frac{\partial}{\partial x} + \eta(x, y)\frac{\partial}{\partial y} \qquad (n + 1 \text{ terms}).$$

In other words, we seek a change of variables $p = f(x, y, \tau)$, $q = g(x, y, \tau)$ such that $f(x, y, 0) = x$ and $g(x, y, 0) = y$.

Such a change of variables, as well as its inverse, can be expressed by the Lie series

$$\underset{(1\times1)}{p} = x + \tau U x + \frac{1}{2!}\tau^2 U^2 x + \frac{1}{3!}\tau^3 U^3 x + \cdots = e^{\tau U} x,$$
$$\underset{(n\times1)}{q} = y + \tau U y + \frac{1}{2!}\tau^2 U^2 y + \frac{1}{3!}\tau^3 U^3 y + \cdots = e^{\tau U} y,$$
$$\underset{(1\times1)}{x} = p - \tau U p + \frac{1}{2!}\tau^2 U^2 p - \frac{1}{3!}\tau^3 U^3 p + \cdots = e^{-\tau U} p,$$
$$\underset{(n\times1)}{y} = q - \tau U q + \frac{1}{2!}\tau^2 U^2 q - \frac{1}{3!}\tau^3 U^3 q + \cdots = e^{-\tau U} q.$$

Here and in what follows, it is implied that if the generator U acts on the variables x and y, then it is dependent on these variables, i.e., $U = \xi(x,y)\frac{\partial}{\partial x} + \eta(x,y)\frac{\partial}{\partial y}$, and if U acts on p and q, then $U = \xi(p,q)\frac{\partial}{\partial p} + \eta(p,q)\frac{\partial}{\partial q}$.

Thus, by transforming the operator A, we arrive at an operator B. The latter is expressed in terms of p and q by Hausdorff's formula

$$B(p,q) = A(p,q) + \tau[A(p,q), U(p,q)] + \frac{1}{2!}\tau^2\big[[A(p,q), U(p,q)], U(p,q)\big] + \frac{1}{3!}\tau^3\Big[\big[[A(p,q), U(p,q)], U(p,q)\big], U(p,q)\Big] + \cdots . \tag{3.3}$$

Setting $\tau = \varepsilon$, we seek the operators B and U in the form

$$B = B_0 + \varepsilon B_1 + \varepsilon^2 B_2 + \varepsilon^3 B_3 + \cdots ,$$
$$U = U_0 + \varepsilon U_1 + \varepsilon^2 U_2 + \varepsilon^3 U_3 + \cdots .$$

Represent A as a series in powers of ε to obtain

$$A = A_0 + \varepsilon A_1 + \varepsilon^2 A_2 + \varepsilon^3 A_3 + \cdots ,$$

where

$$A_0 = \frac{\partial}{\partial p}, \quad A_1 = X_1(p,q)\frac{\partial}{\partial p} + Y_1(p,q)\frac{\partial}{\partial q},$$
$$A_2 = X_2(p,q)\frac{\partial}{\partial p} + Y_2(p,q)\frac{\partial}{\partial q}, \quad A_3 = X_3(p,q)\frac{\partial}{\partial p} + Y_3(p,q)\frac{\partial}{\partial q}.$$

Matching the coefficients of like powers of ε and taking into account the fact that $[A_0, U_i] = \partial U_i/\partial p$, we find from Eq. (3.3) that

$$B_0 = \frac{\partial}{\partial p},$$
$$B_1 = A_1 + \frac{\partial U_0}{\partial p},$$
$$B_2 = A_2 + [A_1, U_0] + \big[[A_0, U_0], U_0\big] + \frac{\partial U_1}{\partial p},$$
$$B_3 = A_3 + [A_1, U_1] + \frac{1}{2}\big[[A_1, U_0], U_0\big] + \frac{1}{6}\Big[\big[[A_0, U_0], U_0\big], U_0\Big] + \frac{1}{2}\big[[A_0, U_0], U_1\big] + \frac{1}{2}\big[[A_0, U_1], U_0\big] + \frac{\partial U_2}{\partial p}.$$

It is apparent from the relations obtained that B_0 is uniquely defined. To determine B_1 and U_0, we represent the operator A as

$$A_1 = \langle A_1 \rangle + \widetilde{A}, \qquad \langle A_1 \rangle = \lim_{T\to\infty} \frac{1}{T} \int_0^T A_1\, dp.$$

To find U_0, we choose the operator B_1 so that it does not depend on the fast variable p:

$$B_1 = \langle A_1 \rangle.$$

Then we have

$$\widetilde{A}_1 + \frac{\partial U_0}{\partial p} = 0.$$

Hence,

$$U_0 = -\int \widetilde{A}_1\, dp.$$

After U_0 has been found, one can construct the second approximation:

$$A_2 + [A_1, U_0] + \big[[A_0, U_0], U_0\big] = L_2, \qquad L_2 = \langle L_2 \rangle + \widetilde{L}_2,$$
$$B_2 = \langle L_2 \rangle, \quad U_1 = -\int \widetilde{L}_2\, dp.$$

The subsequent approximations can constructed in a similar manner. Thus, we obtain

$$B = \frac{\partial}{\partial p} + \varepsilon\left[P_1(q)\frac{\partial}{\partial p} + Q_1(q)\frac{\partial}{\partial q}\right] + \varepsilon^2\left[P_2(q)\frac{\partial}{\partial p} + Q_2(q)\frac{\partial}{\partial q}\right] + \varepsilon^3\left[P_3(q)\frac{\partial}{\partial p} + Q_3(q)\frac{\partial}{\partial q}\right] + \cdots .$$

The corresponding equations,

$$\begin{aligned} \frac{dp}{dt} &= 1 + \varepsilon P_1(q) + \varepsilon^2 P_2(q) + \varepsilon^3 P_3(q) + \cdots, \\ \frac{dq}{dt} &= \varepsilon Q_1(q) + \varepsilon^2 Q_2(q) + \varepsilon^3 Q_3(q) + \cdots, \end{aligned} \tag{3.4}$$

are simpler than the original equations (3.2), since the former do not depend on p.

After system (3.4) has been solved, we can return to the original variables with the aid of the truncated Lie series

$$x = p - \varepsilon U_0 p + \varepsilon^2 U_1 p, \quad y = q - \varepsilon U_0 q + \varepsilon^2 U_1 q.$$

There is no need to include the term $\varepsilon^3 U_3$ into these formulas, since the integration of system (3.4) leads to an error of order $\varepsilon^4 t$; on a sufficiently long interval, $t \sim 1/\varepsilon$, this error is of the order of ε^3.

5.4. Poincaré Normal Form

Consider a system of differential equations whose right-hand sides are analytic in a neighborhood of the origin,

$$\dot{y}_1 = f_1(y_1, \ldots, y_n),$$
$$\cdots\cdots\cdots\cdots\cdots\cdots$$
$$\dot{y}_n = f_n(y_1, \ldots, y_n),$$

where the y_k are complex variables.

We assume that the right-hand sides vanish at the origin, i.e., $f_k(0,\ldots,0)=0$. This system can be represented as series in the variables,

$$\dot{y}_1 = a_{11}y_1 + \cdots + a_{1n}y_n + \sum_{\sigma\geq 2} f^1_{m_1\ldots m_n} y_1^{m_1}\ldots y_n^{m_n},$$
$$\cdots\cdots\cdots\cdots\cdots\cdots\cdots\cdots\cdots\cdots\cdots\cdots$$
$$\dot{y}_n = a_{n1}y_1 + \cdots + a_{nn}y_n + \sum_{\sigma\geq 2} f^n_{m_1\ldots m_n} y_1^{m_1}\ldots y_n^{m_n}.$$

The summation is assumed over all positive integers m_k satisfying the condition $\sigma = \sum_{k=1}^n m_k \geq 2$. The number σ is referred to as the order of the nonlinear term $y_1^{m_1}\ldots y_n^{m_n}$.

The following problem is set: find an analytic change of variables $(y_1,\ldots,y_n) \to (z_1,\ldots,z_n)$ such that the maximum possible number of the coefficients a_{ij} and $f^i_{m_1\ldots m_n}$, up to any prescribed order σ, become zero.

The resulting form of the system is called a normal form of the system up to order σ inclusive. To give a more specific definition, it is necessary to establish the form of the nonlinear terms which cannot be removed by any analytic changes of variables.

The reduction to the normal form can be carried out successively. One begins with the linear part. After that, one proceeds to simplifying the second order terms. Then the third order terms are simplified an so on, up to the prescribed order.

The problem of simplification of the linear part is well known: by a linear transformation the linear terms can be reduced to the Jordan form. In this case, the matrix of the linear part has only two nonzero diagonals—the leading diagonal, whose components are the eigenvalues of the matrix, and the secondary upper diagonal, whose components are either zeros or units. We assume that the system in question has been already simplified in respect of the linear terms. In addition, we assume tentatively that the Jordan form is purely diagonal. Thus,

$$\dot{y}_k = \lambda_k y_k + \sum_{\sigma\geq 2} f^1_{m_1\ldots m_n} y_1^{m_1}\ldots y_n^{m_n} \qquad (k=1,\ldots,n).$$

The nature of the changes the system undergoes can be revealed, completely and in full detail, in considering a single nonlinear term in one of the n equations of the system (say, the kth equation):

$$\dot{y}_s = \lambda_s y_s, \qquad s \neq k,$$
$$\dot{y}_k = \lambda_k y_k + f^k y_1^{m_1}\ldots y_n^{m_n},$$

where $m_1,\ldots,m_n$ are some fixed integers. This is accounted for by the fact that the transformation modifying the nonlinear term can be chosen so that it changes neither terms of the lower orders nor the other terms of the same order. This transformation, $(y_1,\ldots,y_n)\to(z_1,\ldots,z_n)$, has the form

$$y_s = z_s, \qquad s \neq k,$$
$$y_k = z_k + h^k z_1^{m_1}\ldots z_n^{m_n}.$$

That is, the nonlinear term is taken in the same form as that to be removed in the equations. The inverse of this transformation is given by

$$z_k = y_k - h^k y_1^{m_1}\ldots y_n^{m_n} + \cdots.$$

The trailing dots denote the terms of higher order. If we are interested in only one step, i.e., the removal of the term in question, and the appearance of higher order terms does not interest us, then it suffices to retain only the first two terms in the inverse transformation. If we are going to continue the simplification procedure, then the inversion of the transformation should be carried out with a higher accuracy.

Differentiating the inverse transformation, we have

$$\dot{z}_k = \dot{y}_k - h^k(m_1 y_1^{m_1-1} y_2^{m_2} \dots y_n^{m_n} \dot{y}_1 + \dots + m_n y_1^{m_1} \dots y_{n-1}^{m_{n-1}} y_n^{m_n-1} \dot{y}_n) + \cdots .$$

Substituting $\dot{y}_l$ from the system under transformation into this equation yields

$$\begin{aligned}\dot{z}_k &= \lambda_k y_k + f^k y_1^{m_1} \dots y_n^{m_n} - (m_1\lambda_1 + \dots + m_n\lambda_n) h^k y_1^{m_1} \dots y_n^{m_n} + \cdots \\ &= \lambda_k y_k + [f^k - (y_1\lambda_1 + \dots + m_n\lambda_n) h^k] y_1^{m_1} \dots y_n^{m_n} + \cdots .\end{aligned}$$

Replacing y by its expression in terms of z, we finally obtain

$$\dot{z}_k = \lambda_k z_k + [f^k - (-\lambda_k + m_1\lambda_1 + \dots + m_n\lambda_n) h^k] z_1^{m_1} \dots z_n^{m_n} + \cdots .$$

In order to remove the nonlinear term in question, h^k should be taken in the form

$$h^k = \frac{f^k}{-\lambda_k + m_1\lambda_1 + \dots + m_n\lambda_n}.$$

This is possible if $\lambda_k \neq m_1\lambda_1 + \dots + m_n\lambda_n$.

If the exponents $m_1, \dots, m_n$ in the nonlinear term of the kth equation satisfy the relation

$$\lambda_k = m_1\lambda_1 + \dots + m_n\lambda_n,$$

such a term is said to be resonant. The resonant terms cannot be eliminated by changes of variables; these do not alter under any polynomial changes of variables at all. In other words, the resonant terms are invariants of polynomial transformations of coordinates.

The removal of nonlinear terms having the same order can be generally accomplished for each term independently of the other by a change of variables of the form

$$y_k = z_k + \sum_{\sigma \geq 2} h^k_{m_1 \dots m_n} z_1^{m_1} \dots z_n^{m_n} \qquad (k = 1, \dots, n).$$

The summation is performed over all $m_1, \dots, m_n$ of order $\sigma = m_1 + \dots + m_n \geq 2$. In accordance with the preceding, we have

$$h^k_{m_1 \dots m_n} = \frac{f^k_{m_1 \dots m_n}}{m_1\lambda_1 + \dots + m_n\lambda_n - \lambda_k}.$$

After all nonresonant terms of the order in question have been eliminated, one can proceed to the elimination of the nonresonant terms of the next order.

Thus, the Poincaré normal form of a system is a form where only resonant terms are present in the expansions of the right-hand sides in powers of the variables. If the system contains no nonlinear terms for which $m_1\lambda_1 + \dots + m_n\lambda_n = \lambda_k$ up to some fixed order σ, then such a system can be reduced by an analytic transformation to a linear form, apart from the terms of higher order of smallness.

We have considered the case of diagonal Jordan form for the linear part of the system. If the Jordan form is nondiagonal (the general case), the normal form is the same, but the reduction procedure becomes more complicated.

Consider for simplicity the case where there is only one Jordan cell corresponding to a double root. Without loss of generality we assume $\lambda_1 = \lambda_2$. Thus, the system to be simplified has the form

$$\begin{aligned}\dot{y}_1 &= \lambda_1 y_1 + y_2 + f^1 y_1^{m_1} \dots y_n^{m_n}, \\ \dot{y}_2 &= \lambda_2 y_2, \\ \dot{y}_s &= \lambda_s y_s \qquad (3 \leq s \leq n).\end{aligned}$$

We assume that there is a nonlinear term in the first equation alone. From the discussion that follows it will be clear how to proceed in the general case.

We consider a change of variables $(y_1, \dots, y_n) \to (z_1, \dots, z_n)$ having the same form as before,

$$y_1 = z_1 + h^1 z_1^{m_1} \dots z_n^{m_n}, \qquad z_1 = y_1 - h^1 y_1^{m_1} \dots y_n^{m_n} + \cdots .$$

We proceed just as in the case of simple roots. We have

$$\dot{z}_1 = \lambda_1 z_1 + z_2 + [f^1 - (m_1\lambda_1 + \cdots + m_n\lambda_n - \lambda_1)h^1] z_1^{m_1} \dots z_n^{m_n} - h^1 m_1 z_1^{m_1-1} z_2^{m_2+1} z_3^{m_3} \dots z_n^{m_n} + \cdots .$$

As before, in the nonresonant case,

$$(m, \lambda) \equiv m_1\lambda_1 + \cdots + m_n\lambda_n \neq \lambda_1,$$

the term $z_1^{m_1} \dots z_n^{m_n}$ can be eliminated. But, unlike the case of simple roots, where no other terms of the same order appeared, we now have a new term of the same order,

$$z_1^{m_1-1} z_2^{m_2+1} z_3^{m_3} \dots z_n^{m_n}.$$

If the condition $(m, \lambda) \neq \lambda_1$ is satisfied, then this term is also nonresonant and, hence, can be eliminated by a similar transformation. In this case, however, an new term of the same order appears,

$$z_1^{m_1-2} z_2^{m_2+2} z_3^{m_3} \dots z_n^{m_n}.$$

Repeating this procedure m_1 times, we arrive at a nonresonant term of the form

$$z_2^{m_1+m_2} z_3^{m_3} \dots z_n^{m_n},$$

which can be eliminated with no other terms of the same order appearing.

Now the consideration of an arbitrary collection of Jordan cells is quite easy.

Thus, we have proved the following theorem, with a constructive procedure of reduction to the normal form described.

THEOREM (POINCARÉ–DULAC). *By an analytic change of variables $y \to z$, a system $\dot{y} = Ay + \sum_m f_m y^m$ can be reduced for any finite order to a form where all nonlinear terms are resonant.*

Example. Consider the following second order equation reduced to the normal form with a nonlinearity of the fifth degree:

$$\dot{x}_1 = x_2, \quad \dot{x}_2 = -x_1 - 32x_1^5.$$

We proceed from the normalization of the linear part. Note that the operator of the system has the form

$$A = x_2 \frac{\partial}{\partial x_1} - (x_1 + 32x_1^5) \frac{\partial}{\partial x_2} = A_0 + A_1,$$

where the operator A_0 is linear in x_1 and x_2. The functions $y_1 = x_2 + ix_1$ and $y_2 = x_2 - ix_1$ are eigenfunctions of A_0,

$$A_0 y_1 = iy_1, \quad A_0 y_2 = -iy_2,$$

where $\lambda_1 = i$ and $\lambda_2 = -i$ are the corresponding eigenvalues. Take these functions to be the new variables. Then we have (see Section 1.7)

$$\begin{aligned} \widetilde{A} &= Ay_1 \frac{\partial}{\partial y_1} + Ay_2 \frac{\partial}{\partial y_2} \\ &= (iy_1 - 32x_1^5) \frac{\partial}{\partial y_1} + (-iy_2 - 32x_1^5) \frac{\partial}{\partial y_2} \\ &= \left[iy + i(y - \overline{y})^5\right] \frac{\partial}{\partial y} + \left[-i\overline{y} + i(y - \overline{y})^5\right] \frac{\partial}{\partial \overline{y}}, \end{aligned}$$

where we have used the notation $y_1 \equiv y$ and $y_2 \equiv \overline{y}$. Thus, we obtain the following system whose linear part is reduced to the normal form:

$$\dot{y} = i\left[y + (y - \overline{y})^5\right], \quad \dot{\overline{y}} = -i\left[\overline{y} + (\overline{y} - y)^5\right].$$

The nonlinear terms are expressed as

$$(y - \overline{y})^5 = y^5 - 5y^4\overline{y} + 10y^3\overline{y}^2 - 10y^2\overline{y}^3 + 5y\overline{y}^4 - \overline{y}^5.$$

Determine the resonant terms in the first equation:

$$\begin{array}{lll} y^5: & \sum m_k\lambda_k = 5\lambda_1 = 5i \neq \lambda_1 & \text{nonresonant,} \\ y^4\overline{y}: & \sum m_k\lambda_k = 4\lambda_1 + \lambda_2 = 3i \neq \lambda_1 & \text{nonresonant,} \\ y^3\overline{y}^2: & \sum m_k\lambda_k = 3\lambda_1 + 2\lambda_2 = i = \lambda_1 & \text{resonant,} \\ y^2\overline{y}^3: & \sum m_k\lambda_k = 2\lambda_1 + 3\lambda_2 = -i \neq \lambda_1 & \text{nonresonant,} \\ y\overline{y}^4: & \sum m_k\lambda_k = \lambda_1 + 4\lambda_2 = -3i \neq \lambda_1 & \text{nonresonant,} \\ \overline{y}^5: & \sum m_k\lambda_k = 5\lambda_2 = -5i \neq \lambda_1 & \text{nonresonant.} \end{array}$$

Thus, there is a single resonant term, $y^3\overline{y}^2$; in the second equation, the resonant terms is $y^2\overline{y}^3$. (There is no need to treat the second equation specially, since it is the complex conjugate of the first equation.) Hence, by some change of variables $y \to z$, the first equation can be reduced to the form $\dot{z} = iz + 10i(z\overline{z})^2 z + \cdots$. This is just the normal form of the original system, apart from sixth order terms.

Note that the normal form obtained can easily be integrated. Indeed, it is not difficult to notice that $z\overline{z} = r^2$ (r is the modulus of z) is a first integral of the system in the normal form. Hence, integrating yields

$$z = z_0 e^{i(1+10r^4)t}, \qquad |z_0| = r.$$

It follows that the frequency of this oscillator is related to the modulus r by $\omega = 1 + 10r^4 + \cdots$. If we are interested in the change of variables relating y to z, then, in accordance with procedure presented above, it should be sought in the form

$$y = z + h_{50}z^5 + h_{41}z^4\overline{z} + h_{32}z^3\overline{z}^2 + h_{23}z^2\overline{z}^3 + h_{14}z\overline{z}^4 + h_{05}\overline{z}^5, \quad \overline{y} = \cdots.$$

The f's in the formula $h^k_{m_1 \ldots m_n} = f^k_{m_1 \ldots m_n} / [(m, \lambda) - \lambda_k]$ are given by

$$f_{50} = 1, \quad f_{41} = -5, \quad f_{23} = -10, \quad f_{14} = 5, \quad f_{05} = -1.$$

Hence, the h's are evaluated as

$$h_{50} = \frac{1}{4i}, \quad h_{41} = -\frac{5}{2i}, \quad h_{23} = \frac{5}{i}, \quad h_{13} = -\frac{5}{4i}, \quad h_{05} = \frac{1}{6i}.$$

Construction of the normal form in Hausdorff's asymptotics. Although constructive, the idea of finding changes of variables to reduce a system to the normal form is of secondary importance, helping us to understand the internal structure of the normal form. To carry out practical calculations, especially in the cases where the system is not so simple and several approximations are required, a special algorithm should be developed.

To work out such an algorithm, we proceed from the analysis of quasilinear systems,

$$\dot{y}_k = \lambda_k y_k + \varepsilon \sum_{\sigma \geq 2} f^k_{m_1 \ldots m_n} y_1^{m_1} \ldots y_n^{m_n}, \qquad \sigma = m_1 + \cdots + m_n,$$

where ε is a small parameter.

The general case will be discussed at the end of this topic.

It should be emphasized that restricting the class of systems results in broadening the possibilities of the analysis of these systems. For example, the construction of a normal form with nonresonant terms eliminated requires infinitely many approximations. If there is a small parameter in the system, then the normal form is constructed in all orders at once by one approximation with an accuracy of ε^2, by two approximations with an accuracy of ε^3, etc. In other words, unlike a system without a small parameter, where any finite number of approximations restricts the study of motions to a small neighborhood of the origin, in a system with a small parameter, finite changes of the variables can be investigated.

The system we are currently studying generates a one-parameter Lie group of transformations of the phase space into itself; the generator of the group is given by

$$A = \lambda_1 y_1 \frac{\partial}{\partial y_1} + \cdots + \lambda_n y_n \frac{\partial}{\partial y_n} + \varepsilon \sum_{\sigma \geq 2} y_1^{m_1} \ldots y_n^{m_n} \left(f^1_{m_1 \ldots m_n} \frac{\partial}{\partial y_1} + \cdots + f^n_{m_1 \ldots m_n} \frac{\partial}{\partial y_n} \right).$$

Consider a group of transformations $(y_1, \ldots, y_n) \to (z_1, \ldots, z_n)$ with generator

$$U = \xi_1(z, \varepsilon) \frac{\partial}{\partial z_1} + \cdots + \xi_n(z, \varepsilon) \frac{\partial}{\partial z_n}$$

and a canonical parameter τ.

In accordance with Hausdorff's formula (see Section 1.8), the operator of the transformed system in terms of the new variables acquires the form

$$B = Z_1(z) \frac{\partial}{\partial z_1} + \cdots + Z_n(z) \frac{\partial}{\partial z_n},$$

$$B = A + \tau[A, U] + \frac{\tau^2}{2!}[[A, U], U] + \cdots.$$

In this formulas, the operator A is formally written in terms of the variables z rather than y:

$$A = \lambda_1 z_1 \frac{\partial}{\partial z_1} + \cdots + \lambda_n z_n \frac{\partial}{\partial z_n} + \varepsilon \ldots$$

Let us adopt the notation

$$A_k = A + o(\varepsilon^k), \quad B_k = B + o(\varepsilon^k), \quad U_k = U + o(\varepsilon^k),$$

that is, the operators A_k, B_k, and U_k differ from the respective operators A, B, and U by terms of order higher than ε^k. The operators A_k, B_k, and U_k are referred to as asymptotics of order k of A, B, and U.

Note that the asymptotics of operators in the n-dimensional space form a ring, i.e., the sum of two asymptotics of order k and the product of two asymptotics of order k are also asymptotics of order k. By the product of operators their Poisson bracket is understood. (In general, the asymptotics of order k even form a Lie algebra, but only the properties of the ring will interest us in what follows.)

By setting $\tau = \varepsilon$, we rewrite Hausdorff's formula for the asymptotics as follows:

$$\begin{aligned}
B_0 &= A_0,\\
B_1 &= A_1 + \varepsilon[A_0, U_0],\\
B_2 &= A_2 + \varepsilon[A_1, U_1] + \tfrac{1}{2}\varepsilon^2\big[[A_0, U_0], U_0\big],\\
&\dots\dots\dots\dots\dots\dots\\
B_k &= A_k + \sum_{j=1}^{k} \frac{\varepsilon^j}{j!} \underbrace{\big[\dots[[}_{j} A_{k-j}, U_{k-j}], U_{k-j}], \dots\big],\\
&\dots\dots\dots\dots\dots\dots
\end{aligned}$$

Here the properties of the asymptotics ring have been taken into account.

For the system in question, we have

$$A_0 = \lambda_1 z_1 \frac{\partial}{\partial z_1} + \dots + \lambda_n z_n \frac{\partial}{\partial z_n}, \qquad A_1 = A.$$

In this form, all asymptotics beginning with A_1 coincide. This special case is not necessary at all; more complicated dependences of A on ε are possible.

Rewrite the above formulas for the asymptotics in the form

$$\begin{aligned}
B_0 &= A_0,\\
B_1 &= \varepsilon[A_0, U_0] + L_1,\\
B_2 &= \varepsilon[A_0, U_1] + L_2,\\
&\dots\dots\dots\dots\\
B_k &= \varepsilon[A_0, U_{k-1}] + L_k,\\
&\dots\dots\dots\dots
\end{aligned}$$

which is more convenient in what follows. The operator L_k is defined as

$$L_k = A_k + \varepsilon[A_{k-1} - A_0,\ U_{k-1}] + \sum_{j=2}^{k} \frac{\varepsilon^j}{j!} \underbrace{\big[\dots[[}_{j} A_{k-j}, U_{k-j}], U_{k-j}], \dots\big].$$

An important remark should be made here, which will allow us to convert the above chain of relations for the asymptotics of increasing order into a recursive procedure for the successive calculation of the asymptotics B_k and U_k, provided the asymptotics A_k are known.

The operator L_k contains the term

$$\varepsilon[A_{k-1} - A_0,\ U_{k-1}].$$

Since $A_{k-1} - A_0 \sim \varepsilon$, the asymptotic relation

$$\varepsilon[A_{k-1} - A_0,\ U_{k-1}] \sim \varepsilon[A_{k-1} - A_0,\ U_{k-2}] \qquad (\text{for } k \ge 2)$$

holds. As a result, we have the following asymptotic equivalent for the operator L_k:

$$L_k \sim \varepsilon[A_{k-1} - A_0,\ U_{k-2}] + \sum_{j=2}^{k} \frac{\varepsilon^j}{j!} \underbrace{\big[\dots[[}_{j} A_{k-j}, U_{k-j}], U_{k-j}], \dots\big] \qquad (k \ge 2).$$

Thus, it turns out that the operator L_k depends only on those asymptotics of U which are of a lesser order than that of the asymptotics of U_{k-1}.

This permits us to formulate the following recursive procedure.

Denote by $(L_k)_{\mathrm{R}}$ the collection of resonant terms in the expression of L_k in all orders, i.e., the collection of all terms for which $m_1\lambda_1 + \cdots + m_n\lambda_n = \lambda_j$ $(j = 1, \ldots, n)$. Accordingly, the designation $(L_k)_{\mathrm{N}}$ indicates the nonresonant terms.

Then the recursive algorithm is as follows. The relation $B_0 = A_0$ uniquely defines B_0. In the relation $B_1 = \varepsilon[A_0, U_0] + L_1$ $(L_1 = A_1)$, we choose B_1 so that

$$B_1 = (L_1)_{\mathrm{R}};$$

then the equation

$$\varepsilon[A_0, U_0] + (L_1)_{\mathrm{N}} = 0$$

can serve to determine U_0. For known L_2, on substituting this U_0 into L_2, we find

$$B_2 = (L_2)_{\mathrm{R}}$$

and obtain an equation for U_1,

$$\varepsilon[A_0, U_1] + (L_2)_{\mathrm{N}} = 0,$$

and so on.

An arbitrary kth approximation is given by the relations

$$B_k = (L_k)_{\mathrm{R}}, \quad \varepsilon[A_0, U_{k-1}] + (L_k)_{\mathrm{N}} = 0.$$

Consider the issue of solving the equation

$$\varepsilon[A_0, U_{k-1}] + (L_k)_{\mathrm{N}} = 0$$

for the unknown operator U_{k-1}. This equation is called *homological.*

Let the operator L_k have the form

$$L_k = \lambda_1 z_1 \frac{\partial}{\partial z_1} + \lambda_n z_n \frac{\partial}{\partial z_n} + \varepsilon \sum_{\sigma \geq 2} \left(g^1_{m_1 \ldots m_n} \frac{\partial}{\partial z_1} + \cdots + g^n_{m_1 \ldots m_n} \frac{\partial}{\partial z_n} \right) z_1^{m_1} \ldots z_n^{m_n}.$$

The summation is performed over all positive m_s such that $\sum_s m_s \geq 2$. We seek the solution of the operator equation for U_{k-1} in the form

$$U_{k-1} = \sum_{\sigma \geq 2} \left(h^1_{m_1 \ldots m_n} \frac{\partial}{\partial z_1} + \cdots + h^n_{m_1 \ldots m_n} \frac{\partial}{\partial z_n} \right) z_1^{m_1} \ldots z_n^{m_n}. \qquad (*)$$

This operator contains nonlinear terms in z of the same form as those in the operator $(L_k)_{\mathrm{N}}$.

Calculate the commutator:

$$[A_0, U_{k-1}] = \sum_{\sigma \geq 2} \left[h^1_{m_1 \ldots m_n} (\lambda_1 m_1 + \cdots + \lambda_n m_n - \lambda_1) \frac{\partial}{\partial z_1} + \cdots \right.$$
$$\left. + h^n_{m_1 \ldots m_n} (\lambda_1 m_1 + \cdots + \lambda_n m_n - \lambda_n) \frac{\partial}{\partial z_n} \right] z_1^{m_1} \ldots z_n^{m_n}.$$

Thus, the calculation of the commutator is reduced to the multiplication of each lth component of U_{k-1} by the factor $(\lambda_1 m_1 + \cdots + \lambda_n m_n - \lambda_l)$. This permits us to represent the solution of the equation $\varepsilon[A_0, U_{k-1}] + (L_k)_{\mathrm{N}} = 0$ in the form of Eq. $(*)$ in which

$$h^l_{m_1 \ldots m_n} = \frac{g^l_{m_1 \ldots m_n}}{\lambda_l - m_1 \lambda_1 - \cdots - \lambda_n m_n}.$$

The operator A can be reduced to the form of B_k with the normalizing transformation

$$z_l = y_l + \varepsilon U_{k-1} y_l + \frac{1}{2}\varepsilon^2 U_{k-2}^2 y_l + \cdots + \frac{\varepsilon^k}{k!} U_0^k y_l.$$

Properties of the normal form. Recall that, in constructing the algorithm of reduction of a differential system to the normal form, we established the rule of calculation of the Poisson bracket for the operator of the linear system, A_0, and an arbitrary operator, M. If the operator of the linear system is given by

$$A_0 = \lambda_1 z_1 \frac{\partial}{\partial z_1} + \cdots + \lambda_n m_n \frac{\partial}{\partial m_n}$$

and the arbitrary operator has the form

$$M = \sum_{\sigma \ge 1} \left(q^1_{m_1 \dots m_n} \frac{\partial}{\partial z_1} + \cdots + q^n_{m_1 \dots m_n} \frac{\partial}{\partial z_n} \right) z_1^{m_1} \dots z_n^{m_n}, \qquad \sigma = m_1 + \cdots + m_n,$$

then the relation

$$[U_0, M] = \sum_{\sigma \ge 1} \left\{ q^1_{m_1 \dots m_n} [(\lambda, m) - \lambda_1] \frac{\partial}{\partial z_1} + \cdots + q^n_{m_1 \dots m_n} [(\lambda, m) - \lambda_n] \frac{\partial}{\partial z_n} \right\} z_1^{m_1} \dots z_n^{m_n}$$

holds, where $(\lambda, m) = \lambda_1 m_1 + \cdots + \lambda_n m_n$.

It follows that if the operator M contains only resonant terms with respect to the operator A_0, for which $(\lambda, m) - \lambda_l = 0$, $[U_0, M] = 0$.

Since, by construction, the normal form of order k contains only resonant terms, we have $[U_0, B_k] = 0$. This property can be used as the basic relation for determining the normal form.

DEFINITION. *A differential system $\dot{y} = \Lambda y + F(y)$ with $F(0) = 0$ is in the Poincaré normal form if the operators of the linear and nonlinear parts of the system commute,*

$$[A_0, A] = 0, \qquad A_0 = \Lambda y \frac{\partial}{\partial y}, \quad A = \left(\Lambda y + F(y)\right) \frac{\partial}{\partial y}.$$

Since the commutator $[A_0, A]$ is invariant under any change of the variables in terms of which A_0 and A are represented, the above definition of the normal form is also invariant. This means that the linear part A_0 is not necessarily diagonal.

In Section 1.9, the principle of superposition of solutions to nonlinear systems in the case where their separate parts commute was proved. This principle is also applicable to the normal form: to construct the general solution of a system in the normal form, it suffices to construct the solution of the linear part separately and then substitute the solution of the system with the linear part dropped for the constants of integration. The property that the operator of the normal form commutes with the operator of its linear part can be used to reduce the order of the system in the normal form. To this end, it suffices to pass in the normal form to canonical coordinates of the group generated by the linear part.

Thus, we have proved the following property: the order of the normal form can always be reduced. It is no mere chance that the example discussed above admitted full integration.

Consider a mechanical system whose linear part corresponds to the system studied in Section 5.2. All roots of the former are pure imaginary an complex conjugate pairwise, $\lambda_1 = \bar{\lambda}_{n+1}, \dots, \lambda_n = \bar{\lambda}_{2n}$ (the system has an even order $2n$). The quantity $|\lambda_s| = |\lambda_{s+n}| = \omega_s$ is called a natural frequency $(s = 1, \dots, n)$. Then the system under study in this section becomes

$$\dot{y}_k = i\omega_k y_k + \varepsilon \sum_{\substack{m_1 + \cdots + m_n \ge 1 \\ p_1 + \cdots + p_n \ge 1}} f^k_{m_1 \dots m_n p_1 \dots p_n} y_1^{m_1} \dots y_n^{m_n} \bar{y}_1^{p_1} \dots \bar{y}_n^{p_n} \qquad (k = 1, \dots, n),$$

$$\dot{\bar{y}}_k = -i\omega_k \bar{y}_k + \cdots.$$

The condition $(\lambda, m) = \lambda_k$, which selects the resonant terms, becomes

$$m_1\omega_1 + \cdots + m_n\omega_n - p_1\omega_1 - \cdots - p_n\omega_n = \omega_k,$$

or

$$(m_1-p_1)\omega_1+\cdots+(m_{k-1}-p_{k-1})\omega_{k-1}+(m_k-p_k-1)\omega_k+(m_{k+1}-p_{k+1})\omega_{k+1}+\cdots+(m_n-p_n)\omega_n=0.$$

If the natural frequencies are rationally incommensurable, i.e., if

$$\mu_1\omega_1 + \cdots + \mu_n\omega_n \neq 0$$

for any integers μ, not all zero, then from the condition that selects the resonant terms it follows that

$$m_1 = p_1, \quad \ldots, \quad m_{k-1} = p_{k-1}, \quad m_k = p_k + 1, \quad m_{k+1} = p_{k+1}, \quad \ldots, \quad m_n = p_n.$$

This implies that only the terms

$$(y_1\bar{y}_1)^{m_1} \ldots (y_k\bar{y}_k)^{p_k} \ldots (y_n\bar{y}_n)^{m_1} y_k$$

can be resonant in this case.

Thus, the normal form of a conservative mechanical system in the nonresonant case is given by

$$\dot{z}_k = i\omega_k z_k + \varepsilon\left(\sum_{\sigma\geq 1} q^k_{m_1\ldots m_n}(z_1\bar{z}_1)^{m_1}\ldots(z_k\bar{z}_k)^{m_k-1}\ldots(z_n\bar{z}_n)^{m_n}\right)z_k, \qquad \sigma = m_1 + \cdots + m_n,$$

$$\dot{\bar{z}}_k = -i\omega_k z_k + \varepsilon\left(\sum_{\sigma\geq 1} \bar{q}^k_{m_1\ldots m_n}(z_1\bar{z}_1)^{m_1}\ldots(z_k\bar{z}_k)^{m_k-1}\ldots(z_n\bar{z}_n)^{m_n}\right)\bar{z}_k, \qquad k = 1, \ldots, n.$$

This system admits reduction in order by n. To this end, one should pass to the variables $r_k = z_k\bar{z}_k$ $(k = 1, \ldots, n)$ to obtain

$$\dot{r}_k = \varepsilon \sum_{\sigma\geq 1}(q^k_{m_1\ldots m_n} + \bar{q}^k_{m_1\ldots m_n})r_1^{m_1}\ldots r_n^{m_n}, \qquad k = 1, \ldots, n.$$

The algorithm just presented can easily be extended to the general procedure of the method of normal form, i.e., to the case where no presence of a small parameter in the system is assumed.

Rewrite the system of differential equation under study in the form

$$\dot{y}_k = \lambda_k y_k + \sum_{\sigma=2} f^k_{m_1\ldots m_n} y_1^{m_1}\ldots y_n^{m_n} + \sum_{\sigma=3} f^k_{m_1\ldots m_n} y_1^{m_1}\ldots y_n^{m_n} + \cdots,$$

where, just as previously, $\sigma = m_1 + \cdots + m_n$.

Thus, the first sum collects only second-order terms (homogeneous quadratic form), the second sum contains only third-order terms (trilinear form), and so on. Rewrite this system once again by introducing a parameter ε:

$$\dot{y}_k = \lambda_k y_k + \varepsilon\sum_{\sigma=2} f^k_{m_1\ldots m_n} y_1^{m_1}\ldots y_n^{m_n} + \varepsilon^2\sum_{\sigma=3} f^k_{m_1\ldots m_n} y_1^{m_1}\ldots y_n^{m_n} + \cdots \qquad (\varepsilon = 1).$$

For the system represented in this form, Hausdorff's formula for asymptotics written out previously can be used. The subsequent manipulations remain the same. In the final formulas, one must set $\varepsilon = 1$, which will result in the normal form of the nonlinear system and the normalizing transformation.

The generalization of the above algorithm to the case of nondiagonal Jordan forms is of no interest for mechanics, since the linear conservative systems always have a diagonal Jordan form. Nevertheless, the algorithm is applicable in this case as well. Everything remains the same, except for the formula specifying the form of the operator to be sought in the homological equation. In the case of a nondiagonal Jordan form, the operator must be completed by the term obtained by shifting the exponents to the right, just as was demonstrated in the proof of the Poincaré–Dulac theorem.

5.5. The Averaging Principle

In the practice of solving problems of nonlinear mechanics, the averaging method (also known as the Krylov–Bogolyubov method) has gained the most acceptance. The majority of the other methods known thus far, which are quite numerous, are either another form of presentation of the averaging method (e.g., the multiscale method) or methods particularly adapted to solving narrower problems than those dealt with in the averaging method. The connection between the averaging method and the method of normal form will be discussed after the algorithm of the former has been outlined.

The averaging method is not a method of solving nonlinear systems of differential equations, but a method of reducing them to some more simple form, for which one of the problems that can be stated for the original system can also be stated. The same holds true for the method of normal form.

The approach where exact solutions of an approximate system are analyzed instead of the direct construction of approximate solutions to the exact system is much more flexible, since it contains the latter as a special case. In addition, this approach leaves room for qualitative analysis, usage of computer for numerical solution, and so on.

The averaging method falls into the so-called local methods of analysis of differential equations. This means that the subject of study is a system which is, in a sense, little different from another system for which the general exact solution can be constructed.

If the behavior of the solutions of little different systems were also little different from each other, then such a statement would be low-informative. However, infinitesimal perturbations of the system can lead to qualitatively different solutions. To establish these qualitative changes is just the most interesting aim of the approximate analysis of nonlinear systems.

Below we discuss the basic ideas of the averaging method in the general statement.

Consider a system of differential equations

$$\frac{dz_i}{dt} = F_i(t, z_1, \dots, z_n) + \varepsilon f_i(t, z_1, \dots, z_n) \qquad (i = 1, \dots, n).$$

Introduce the column matrices

$$z = \begin{pmatrix} z_1 \\ \vdots \\ z_n \end{pmatrix}, \quad F = \begin{pmatrix} F_1 \\ \vdots \\ F_n \end{pmatrix}, \quad f = \begin{pmatrix} f_1 \\ \vdots \\ f_n \end{pmatrix}.$$

Then the above system of equations acquires the form

$$\frac{dz}{dt} = F(t, z) + \varepsilon f(t, z).$$

The variable z occurring in F and f indicates that these functions depend on the collection of variables $z_1, \dots, z_n$. The scalar parameter ε is assumed to be small and nonnegative.

Suppose the degenerate system (with $\varepsilon = 0$) has the exact general solution

$$z = G(t, C),$$

where C is the collection of n arbitrary constants of integration. Hence, the function $G(t, C)$ satisfies the identities

$$\frac{d}{dt}G(t, C) \equiv \frac{\partial}{\partial t}G(t, C) \equiv F\big(t, G(t, C)\big).$$

Our task is to construct an approximate system whose solutions would represent, with a guaranteed accuracy, the solutions of the complete, nondegenerate system.

To solve this problem, we reduce the original system to a special form known as the *standard form* of the averaging method. This reduction is accomplished with a change of variables $z \to x$ defined as

$$z = g(t, x),$$

where $g(t, x)$ is some function to be defined below. Differentiating yields

$$\frac{dz}{dt} = \frac{\partial g}{\partial t} + \frac{\partial g}{\partial x}\dot{x},$$

where

$$\frac{\partial g}{\partial x} = \begin{pmatrix} \dfrac{\partial g_1}{\partial x_1} & \cdots & \dfrac{\partial g_1}{\partial x_n} \\ \vdots & \ddots & \vdots \\ \dfrac{\partial g_n}{\partial x_1} & \cdots & \dfrac{\partial g_n}{\partial x_n} \end{pmatrix}.$$

Substituting these relations into the original system, we obtain

$$\frac{dx}{dt} = \left[\frac{\partial g}{\partial x}\right]^{-1} \left[F\big(t, g(t, x)\big) - \frac{\partial g}{\partial t} + \varepsilon f\big(t, g(t, x), \varepsilon\big)\right].$$

It is apparent from this relation that if $g(t, x) \equiv G(t, x)$, then, by virtue of the identity $G_t \equiv F(t, G)$, this equation in terms of x becomes simpler and takes the form

$$\frac{dx}{dt} = \varepsilon \left[\frac{\partial G}{\partial x}\right]^{-1} f\big(t, G(t, x), \varepsilon\big).$$

In addition, it is evident that the new variables are slow functions of time. This equation, which is often represented as

$$\frac{dx}{dt} = \varepsilon X(t, x, \varepsilon),$$

is referred to as the standard form of a single-frequency system. The meaning of the term "single-frequency" will be explained later.

A specific feature of the system obtained is the presence of the small factor ε on the right-hand side. Note that any systems with perturbations can be reduced to this form, provided that the general solution of the unperturbed system is known. The reduction method is known as the Lagrange method of variation of arbitrary constants.

The next stage involves replacing the exact system in the standard form with an approximate system of equations. This replacement must satisfy the following two conditions: (i) the corresponding solutions of the exact and approximate systems must be close to each other and (ii) the approximate system must be easier analyzable mathematically.

The simplification that is performed in the averaging method is based on the idea of separation of motions explained below.

Let the right-hand side of the system $\frac{dx}{dt} = \varepsilon X(t, x, \varepsilon)$ be periodic in t with period T, i.e.,

$$X(t + T, x, \varepsilon) \equiv X(t, x, \varepsilon).$$

If there is no periodicity in t, then we assume that $T = \infty$. Let us make the change of time $t \to \tau$ by the formula $\tau = \varepsilon t$. The system under study becomes

$$\frac{dx}{d\tau} = X\Big(\frac{\tau}{\varepsilon}, x, \varepsilon\Big).$$

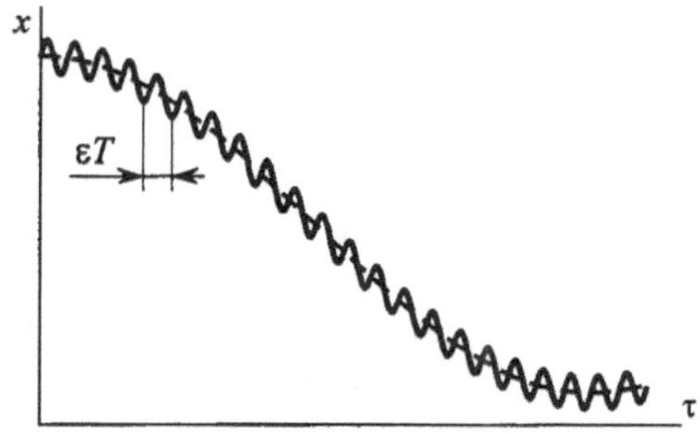

Figure 5.1

The right-hand side of this system is periodic in τ with period εT, i.e., it contains periodic functions of τ varying with a high frequency. A typical form of a particular solution to this system is shown in Fig. 5.1.

This solution can be represented as the sum of a slowly varying component $u(\tau)$, called the *system evolution*, and a rapidly oscillating component $v(\tau)$, called the oscillation, as follows:

$$x = u(\tau) + v(\tau).$$

This vague statement can be made more precise by considering the evolution to mean the result of the action of the following smoothing operator on $x(\tau)$:

$$u(\tau) = \frac{1}{\varepsilon T}\int_0^{\varepsilon T} x(\tau + \xi)\, d\xi.$$

Usually, the system evolution is of more interest. The evolution obeys the equation

$$\frac{du}{d\tau} = \frac{1}{\varepsilon T}\int_0^{\varepsilon T} \frac{dx(\tau+\xi)}{d\tau}\, d\xi = \frac{1}{\varepsilon T}\int_0^{\varepsilon T} X\left(\frac{\tau+\xi}{\varepsilon},\, x(\tau+\xi),\, \varepsilon\right) d\xi.$$

This relation is exact. In integrating with respect to ξ, one should take into account that X depends on ξ not only explicitly but also implicitly, since x also depends on ξ. However, the integration interval is small and x does not change much on this interval. Assuming that $x(\tau + \xi) \approx u(\tau)$, we have

$$\frac{du}{d\tau} \approx \frac{1}{\varepsilon T}\int_0^{\varepsilon T} X\left(\frac{\tau+\xi}{\varepsilon},\, u,\, \varepsilon\right) d\xi = \frac{1}{\varepsilon T}\int_0^{\varepsilon T} X\left(\frac{\xi}{\varepsilon},\, u,\, \varepsilon\right) d\xi.$$

Returning to the original time t, we obtain

$$\frac{du}{dt} \approx \varepsilon X_0(u, \varepsilon),$$

where

$$X_0 = \frac{1}{T}\int_0^{T} X(t, u, \varepsilon)\, dt$$

if X is a periodic function of t and

$$X_0 = \lim_{T\to\infty} \frac{1}{T}\int_0^{T} X(t, u, \varepsilon)\, dt$$

if X is not a periodic function of time.

This is the main technique of the averaging method. The idea of this technique goes back to Gauss. In brief, it involves the replacement of the exact equation

$$\dot{x} = \varepsilon X(t, x, \varepsilon)$$

by an approximate equation

$$\dot{x} = \varepsilon X_0(x, \varepsilon).$$

In the averaging method, the approximate equation is obtained by averaging the exact equation with respect to time occurring explicitly, under the assumption that x is a parameter.

Example. Consider an oscillator with cubic damping,

$$\ddot{z} + \varepsilon \dot{z}^3 + z = 0, \qquad 0 < \varepsilon \ll 1.$$

Rewrite this equation as a first-order perturbed system,

$$\dot{z}_1 = z_2, \quad \dot{z}_2 = -z_1 - \varepsilon z_2.$$

The general solution of the unperturbed system ($\varepsilon = 0$), by

$$z_1 = C_1 \cos t + C_2 \sin t,$$
$$z_2 = -C_1 \sin t + C_2 \cos t,$$

determines the change of variables $(z_1, z_2) \to (x_1, x_2)$:

$$z_1 = x_1 \cos t + x_2 \sin t,$$
$$z_2 = -x_1 \sin t + x_2 \cos t.$$

With this change of variables, the original system becomes

$$\dot{x}_1 \cos t + \dot{x}_2 \sin t = 0,$$
$$-\dot{x}_1 \sin t + \dot{x}_2 \cos t = -\varepsilon(-x_1 \sin t + x_2 \cos t)^3;$$

whence

$$\dot{x}_1 = -\varepsilon(x_1 \sin t - x_2 \cos t)^3 \sin t,$$
$$\dot{x}_2 = \varepsilon(x_1 \sin t - x_2 \cos t)^3 \cos t.$$

These are just the equations of the oscillator under study in the standard form.

The averaging with respect to the explicit time results in the following equations for the evolution:

$$\dot{u}_1 = -\tfrac{3}{8}\varepsilon u_1(u_1^2 + u_2^2), \quad \dot{u}_2 = -\tfrac{3}{8}\varepsilon u_2(u_1^2 + u_2^2).$$

These equations can readily be integrated by passing to the polar coordinates (r, φ), $u_1 = r \cos \varphi$ and $u_2 = r \sin \varphi$. In terms of (r, φ), the above equations become

$$\frac{dr}{dt} = -\frac{3}{8}\varepsilon r^3, \quad \frac{d\varphi}{dt} = 0.$$

Integrating yields

$$r = r_0\left(1 + \tfrac{3}{4}\varepsilon r_0^2 t\right)^{-1/2}, \quad \varphi = \varphi_0.$$

Returning to the original variables, we find an approximate solution of the original problem in the form $z = z_1 = r_0\left(1 + \tfrac{3}{4}\varepsilon r_0^2 t\right)^{-1/2} \cos(t - \varphi_0)$.

The above reasoning, based on the introduction of the notion of the evolution with the aid of a smoothing operator, is euristic and requires a justification. This is dictated not only by a search for a formal mathematical completeness of the solution. If the rigorous conditions of applicability of a method are not satisfied, errors may occur which lead to a crude distortion of our perception of the essence of the phenomenon under investigation. In the example just considered, the approximate solution describes correctly the damping of the oscillations in the system if $\varepsilon > 0$. If there is the minus sign before ε in the equation of the oscillator, then the solution becomes $z = r_0\left(1 - \frac{3}{4}\varepsilon r_0^2 t\right)^{-1/2}\cos(t - \varphi_0)$. It goes to infinity in a finite time for any initial conditions. But actually there are a lot of nonzero initial conditions for which the solution of the exact equation goes to infinity in infinite time. Thus, for $\varepsilon < 0$ ($|\varepsilon| \ll 1$), the approximate solution does not agree with the exact solution even qualitatively.

The theorem below sets the clear bounds of applicability of the averaging method and permits one to avoid disagreements of this sort.

The theorem compares two Cauchy problems:

$$\begin{array}{ll} \text{I} & \text{II} \\ \dot{x} = \varepsilon X(t, x, \varepsilon), & \dot{u} = \varepsilon X_0(u, \varepsilon), \\ x(0) = x_0, & u(0) = x_0. \end{array}$$

THEOREM (BOGOLYUBOV). *Suppose system I satisfies the following conditions:*

(a) $X(t, x, \varepsilon)$ is defined in some domain D of the n-dimensional space of the variables $x = (x_1, \ldots, x_n)$ and on $0 \le t < \infty$ for $0 \le \varepsilon \le \varepsilon_1$ and is measurable in D with respect to t;

(b) independent constants M and λ can be found such that $\|X(t, x, \varepsilon)\| \le M$ and $\|X(t, x, \varepsilon) - X(t, x', \varepsilon)\| \le \lambda \|x - x'\|$ for any values of the variables from the domain of definition; here $\|x\|$ stands for the Euclidean norm of x, $\|x\| = \left(\sum_{i=1}^n x_i^2\right)^{1/2}$;

(c) the limit $\lim_{T\to\infty} \frac{1}{T}\int_0^T X(t, x, \varepsilon)\, dt$ exists uniformly for $u \in D$ and $\varepsilon \in [0, \varepsilon_1]$.

Then if the solution $u(t)$ of the Cauchy problem for system II belongs to D together with its ρ-neighborhood for $0 \le t < \infty$, then for any prescribed number L there exists an ε_0 such that

$$\|x(t) - u(t)\| \to 0$$

for $0 \le t \le L/\varepsilon$ and $0 \le \varepsilon < \varepsilon_0 \le \varepsilon_1$.

A complete proof of this theorem can be found in Besjes (1969).[10] The proof can be substantially simplified by requiring that the function $X(t, x, \varepsilon)$ be continuously differentiable with respect to x.

Proof. Let us make the following change of variables in system I:

$$x \to y: \quad x = y + \varepsilon\varphi(t, x, \varepsilon), \qquad \varphi(t, x, \varepsilon) = \int_0^t [X(t, y, \varepsilon) - X_0(y, \varepsilon)]\, dt.$$

Since X_0 is the time average of X, the function $\varphi(t, x, \varepsilon)$ possesses the property

$$\lim_{T\to\infty} \frac{1}{T}\varphi(T, y, \varepsilon) = 0.$$

For this reason,

$$\|x(t) - y(t)\| \le \varepsilon\|\varphi(t, y, \varepsilon)\| = l\frac{\varepsilon}{l}\left\|\varphi\left(\frac{l}{\varepsilon}, y, \varepsilon\right)\right\| \qquad \left(t = \frac{l}{\varepsilon}\right).$$

[10] I. G. Besjes, *On the asymptotic methods for nonlinear differential equations*, J. Mech., Vol. 8, No. 3, pp. 357–372, 1969.

Since

$$\lim_{\varepsilon\to 0}\frac{\varepsilon}{l}\left\|\varphi\left(\frac{l}{\varepsilon},y,\varepsilon\right)\right\|=0$$

for any fixed l, we have

$$\sup_{l\in[0,L]}\frac{\varepsilon}{l}\left\|\varphi\left(\frac{l}{\varepsilon},y,\varepsilon\right)\right\|\to 0 \quad \text{as} \quad \varepsilon\to 0.$$

With the above change of variables, system I becomes

$$\dot{y}=\varepsilon X_0(y,\varepsilon)+\varepsilon^2 Q(t,y,\varepsilon),$$

where the function $Q(t,y,\varepsilon)$ is bounded in the domain of definition, $\|Q\|\le N$, if the derivative $\frac{\partial X}{\partial x}$ is bounded.

Let us estimate the difference $\|y(t)-u(t)\|$. We have

$$\|y(t)-u(t)\|\le\varepsilon\int_0^t\|X_0(y,\varepsilon)-X_0(u,\varepsilon)\|\,dt+\varepsilon^2\int_0^t\|Q(t,y,\varepsilon)\|\,dt,$$

or

$$\|y(t)-u(t)\|\le\varepsilon\lambda\int_0^t\|y-u\|\,dt+\varepsilon LN.$$

Using the familiar Gronwall inequality,

$$v(t)<C+\int_0^t f(\xi)v(\xi)\,d\xi \quad\Longrightarrow\quad v(t)<C\exp\left[\int_0^t f(\xi)\,d\xi\right] \qquad (v,C,f \text{ are positive}),$$

we obtain

$$\|y(t)-u(t)\|\le\varepsilon LNe^{\lambda L}\to 0 \qquad (\varepsilon\to 0).$$

Since

$$\|x(t)-u(t)\|\le\|x(t)-y(t)\|+\|y(t)-u(t)\|,$$

we have

$$\|x(t)-u(t)\|\to 0 \qquad (\varepsilon\to 0,\ t\in[0,L/\varepsilon]).$$

This proves the theorem.

5.5.1. Averaging of Single-Frequency Systems

The principle of averaging described above may be treated as a first approximation of an asymptotic method for reducing a nonautonomous system in the standard form to an autonomous form. To formulate such an asymptotic method, we proceed, as in the case of the method of normal form, from Hausdorff's formula, which relates three operators—the generator of the transformation group, the operator of the original differential system, and the operator of the differential system averaged in an arbitrary approximation.

Rewrite the standard form introduced above to make it autonomous:

$$\frac{d\varphi}{dt}=1,\qquad \frac{dx}{dt}=\varepsilon X(\varphi,x,\varepsilon).$$

We state the problem: find a change of variables $(\varphi,x)\to(\psi,y):\varphi=\psi,\ y=y(\varphi,x,\varepsilon)$ so as to reduce the above autonomous system with any prescribed accuracy to the form

$$\frac{d\psi}{dt}=1,\qquad \frac{dy}{dt}=\varepsilon Y(y,\varepsilon),$$

where the equations for the slow variables y do not depend on the fast variable ψ.

The original system has the operator

$$A = \frac{\partial}{\partial \varphi} + \varepsilon X(\varphi, x, \varepsilon)\frac{\partial}{\partial x}$$

and the operator of the desired (transformed) system is expressed as

$$B = \frac{\partial}{\partial \psi} + \varepsilon Y(y, \varepsilon)\frac{\partial}{\partial y}.$$

We seek the change of variables as a group

$$\begin{cases} \psi = \varphi, \\ y = y(\varphi, x, \varepsilon, \tau), \end{cases}$$

where τ is the group parameter.

The generator of this group is sought in the form

$$U = \xi(\varphi, x, \varepsilon)\frac{\partial}{\partial x} \equiv \xi_1 \frac{\partial}{\partial x_1} + \cdots + \xi_n \frac{\partial}{\partial x_n}.$$

Just as in Section 1.3, we use Hausdorff's formula to relate the three operators:

$$B = A + \tau[A, U] + \frac{\tau^2}{2!}\big[[A, U], U\big] + \cdots .$$

Introduce into consideration the asymptotics of A, U, and B,

$$A_k = A + o(\varepsilon^k), \quad U_k = U + o(\varepsilon^k), \quad B_k = B + o(\varepsilon^k),$$

and identify τ with ε. Following the same reasoning as that in deriving the asymptotic procedure of constructing the normal form, we arrive at the same sequence of finite series that combine these asymptotics into a one-dimensional recursion scheme:

$$\begin{aligned}
B_0 &= A_0, \\
B_1 &= \varepsilon[A_0, U_0] + A_1, \\
B_2 &= \varepsilon[A_0, U_1] + A_2 + \varepsilon[A_1 - A_0, U_0] + \tfrac{1}{2}\varepsilon^2\big[[A_0, U_0], U_0\big], \\
&\cdots\cdots\cdots\cdots\cdots\cdots \\
B_k &= \varepsilon[A_0, U_{k-1}] + L_k, \\
&\cdots\cdots\cdots\cdots\cdots\cdots
\end{aligned}$$

where

$$L_k = \begin{cases} A_1 & \text{for } k = 1, \\ A_k + \varepsilon[A_{k-1} - A_0, U_{k-2}] + \displaystyle\sum_{j=2}^{k} \frac{\varepsilon^j}{j!} \underbrace{\big[\ldots[[A_{k-j}, U_{k-j}], U_{k-j}], \ldots\big]}_{j} & \text{for } k > 1. \end{cases}$$

As soon as we study a system in the standard form, the operator A_0 is rather simple; specifically,

$$A_0 = \frac{\partial}{\partial \psi},$$

and hence,

$$[A_0, U_k] = \frac{\partial U_k}{\partial \psi}.$$

As far as the aim of the recursive procedure is concerned, all asymptotics B_k of the desired operator B must be independent of φ. It suffices to set

$$B_k = \langle L_k \rangle,$$

where $\langle L \rangle$ stands for the average of L over φ. Then we obtain the following equation for the transformation operator:

$$\varepsilon \frac{\partial U_{k-1}}{\partial \varphi} + \widetilde{L}_k = 0 \quad \Longrightarrow \quad U_{k-1} = -\frac{1}{\varepsilon} \int \widetilde{L}_k \, d\psi,$$

where $\widetilde{L}_k$ is the correction to the average, $L_k = \langle L_k \rangle + \widetilde{L}_k$.

The last two relations completely exhaust the recursive procedure for determining B and U asymptotically.

The first step of the scheme (the first approximation involves the calculation of the average of A_1) is as follows:

$$B_1 = \langle A_1 \rangle = \frac{\partial}{\partial \psi} + \langle \xi(\psi, y, \varepsilon) \rangle \frac{\partial}{\partial y}.$$

After that the transformation operator U_0 is calculated by the formula

$$U_0 = -\frac{1}{\varepsilon} \int \widetilde{A}_1 \, d\psi = -\left(\int \widetilde{\xi} \, d\psi \right) \frac{\partial}{\partial y}.$$

This stage is completely equivalent to the principle of averaging presented above.

The aim of the construction of a B independent of the fast variable could be attained in another way rather than by setting $B_k = \langle L_k \rangle$. However, the average of $\widetilde{\xi}$ would be nonzero in this case and the operator U would be unbounded in the fast variable. To construct transformations bounded in time and reducing the system to an autonomous form, there is no choice other than to set $B_k = \langle L_k \rangle$.

The second step of the scheme is to calculate L_2. We have

$$L_2 = \varepsilon[A_1 - A_0, U_0] + \tfrac{1}{2}\varepsilon^2 [[A_0, U_0], U_0].$$

The second approximation of the averaged system is determined by

$$B_2 = \langle L_2 \rangle.$$

and the second approximation of the transformation operator is evaluated as

$$U_1 = -\frac{1}{\varepsilon} \int \widetilde{L}_2 \, d\psi.$$

And so on.

After the kth transformation operator U_k has been found, the transformation itself can be determined as the Lie series.

Direct transformation ($x \to y$):

$$y = x + \varepsilon U_k x + \frac{\varepsilon^2}{2!} U_k^2 x + \cdots.$$

Inverse transformation ($y \to x$):

$$x = y - \varepsilon U_k y + \frac{\varepsilon^2}{2!} U_k^2 y - \cdots .$$

In the direct transformation, the operator U is represented in terms of the old variables,

$$U_k = \xi(\varphi, x, \varepsilon)\frac{\partial}{\partial x};$$

in the inverse transformation, U is expressed in terms of the new variables:

$$U_k = \xi(\psi, y, \varepsilon)\frac{\partial}{\partial y}.$$

The functions ξ are the same in both cases.

There in no point in taking into account infinitely many terms. Furthermore, the order of the asymptotics of U_k may be reduced by one in each term as compared with the previous term.

In this case, as follows from Bogolyubov's theorem, the difference between the exact and approximate solutions is of the order of ε in the first approximation of the averaging method. For this reason, it suffices to retain only the identity part $y = x$ of the change of variables. This conclusion is suggested by the form of transformations and it makes sense to take this form into account.

The first approximation:

$$B_1 = \langle A_1 \rangle,$$
$$y = x.$$

Since any corrections to the identity transformation make no sense in the this approximation, there is no need to calculate U_0 here.

The second approximation:

$$U_0 = -\frac{1}{\varepsilon}\int \widetilde{A}_1 \, d\psi, \quad B_2 = \left\langle \varepsilon[A_1 - A_0, U_0] + \tfrac{1}{2}\varepsilon^2\big[[A_0, U_0], U_0\big]\right\rangle,$$
$$y = x + \varepsilon U_0 x.$$

The kth approximation:

$$U_{k-2} = -\frac{1}{\varepsilon}\int \widetilde{L}_{k-1} \, d\psi, \quad B_k = \langle L_k \rangle,$$
$$y = x + \varepsilon U_{k-2} x + \frac{\varepsilon^2}{2!} U_{k-3}^2 x + \cdots + \frac{\varepsilon^{k-1}}{(k-1)!} U_0^{k-1} x.$$

Example. Reduce the system

$$\frac{d\varphi}{dt} = 1, \quad \frac{dx}{dt} = -\varepsilon x^3 \cos^2\varphi$$

by a change of variables $(\varphi, x) \to (\psi, y)$ to a form in which the right-hand sides would be independent of ψ.

Write out the operator of the system in terms of the new variables:

$$A = \frac{\partial}{\partial \psi} - \varepsilon y^3 \cos^2\psi \, \frac{\partial}{\partial y}.$$

The first approximation:

$$B_1 = \langle A \rangle = \frac{\partial}{\partial \psi} - \frac{\varepsilon}{2} y^3 \frac{\partial}{\partial y}.$$

In the first approximation, the change of variables is given by

$$\varphi = \psi, \quad x = y.$$

The second approximation:

$$U_0 = -\frac{1}{\varepsilon} \int \widetilde{A}\, d\psi = \left(\frac{y^3}{2} \int \cos 2\psi \, d\psi \right) \frac{\partial}{\partial y} = \frac{1}{4} y^3 \sin 2\psi \, \frac{\partial}{\partial y}.$$

Then we calculate L_2 by the formula

$$L_2 = A + \varepsilon [A - A_0, U_0] + \tfrac{1}{2}\varepsilon^2 [[A_0, U_0], U_0].$$

Compute the commutators:

$$[A - A_0, U_0] = \left[-\varepsilon y^3 \cos^2 \psi \, \frac{\partial}{\partial y}, \, \frac{1}{4} y^3 \sin 2\psi \, \frac{\partial}{\partial y} \right] = 0,$$

$$[[A_0, U_0], U_0] = \left[\frac{\partial U_0}{\partial \psi}, U_0 \right] = \left[\frac{1}{2} y^3 \cos 2\psi \, \frac{\partial}{\partial y}, \, \frac{1}{4} y^3 \sin 2\psi \, \frac{\partial}{\partial y} \right] = 0.$$

Hence,

$$B_2 = \langle L_2 \rangle = \langle A \rangle = B_1 = \frac{\partial}{\partial \psi} - \frac{\varepsilon}{2} y^3 \frac{\partial}{\partial y},$$

$$U_1 = U_0 = \frac{1}{4} y^3 \sin 2\psi \, \frac{\partial}{\partial y}.$$

In the second approximation, the change of variables becomes

$$\varphi = \psi, \quad x = y - \varepsilon U_0 y = y - \varepsilon \left(\frac{1}{4} y^3 \sin 2\psi \, \frac{\partial}{\partial y} \right) y = y - \frac{\varepsilon}{4} y^3 \sin 2\psi.$$

We see that the second approximation for B and U coincides with the first. It is not difficult to see that this will hold true for any approximation in this example. Hence, the operators B and U are found exactly:

$$B = \frac{\partial}{\partial y} - \frac{\varepsilon}{2} y^3 \frac{\partial}{\partial y}, \quad U = \frac{1}{4} y^3 \sin 2\psi \, \frac{\partial}{\partial y}.$$

It follows that the exact transformation $(\psi, y) \to (\varphi, x)$ can also be found. We have

$$\begin{aligned} \varphi &= \psi, \\ x &= y - \varepsilon \left(\frac{1}{4} y^3 \sin 2\psi \, \frac{\partial}{\partial y} \right) y + \frac{1}{2} \varepsilon^2 \left(\frac{1}{4} y^3 \sin 2\psi \, \frac{\partial}{\partial y} \right)^2 y + \cdots \\ &= y + \sum_{n=1}^{\infty} (-1)^n \varepsilon^n \frac{(2n-1)!!}{n!\, 4^n} y^{2n+1} \sin^n 2\psi. \end{aligned}$$

Under this transformation, the original system acquires the form

$$\dot{\psi} = 1, \quad \dot{y} = -\frac{\varepsilon}{2} y^3.$$

Since the algorithm presented contains the averaging procedure, the asymptotic estimates for this algorithm are equivalent to ordinary estimates of the accuracy for the Krylov–Bogolyubov method. However, the actual accuracy provided by this algorithm in specific examples can be higher than that provided by the Krylov–Bogolyubov method. The point is that transformations of coordinates are constructed in the Krylov–Bogolyubov method, whereas in the method under consideration the operators of these transformations are determined. Therefore, if one succeeds in finding an exact expression of the operator in a finite number of steps, as in the example just considered, then the formula $y = e^{\varepsilon U}x$ determines an exact expression of the transformation $(x \to y)$, thus represented as an infinite series. But in the Krylov–Bogolyubov method an infinite series cannot be obtained with a finite number of approximations in principle.

Discuss one more example.

Example. Consider the equation[11]

$$\ddot{q} + q - aq^2 = \mu \cos t.$$

There is a main resonance in this equation. The periodic solution can be found as a series in fractional powers of the parameter μ. This series was calculated in Malkin (1956; see footnote 11), which states that such series cannot be obtained within the framework of the theory of quasilinear systems. We will demonstrate that this is not so. The fact that fractional powers appear in a series is only related to the choice of the scale of variables and is in no way determined by the essence of the asymptotic procedure for constructing solutions.

To apply the quasilinear approach, we introduce into the equation a small scale in accordance with the formula $q = \varepsilon z$ (ε is a small parameter). Then the equation under study can be rewritten as a quasilinear equation,

$$\ddot{z} + z = \varepsilon a z^2 + \frac{\mu}{\varepsilon}\cos t.$$

For the nonlinear and nonhomogeneous terms to have the same order of influence on the oscillator, one should set $\mu/\varepsilon = \varepsilon^2$.

Let us pass to new phase variables, $(t, z, \dot{z}) \to (\varphi, x_1, x_2)$, in accordance in the formulas

$$\begin{aligned} t &= \varphi, \\ z &= x_1 \sin\varphi + x_2 \cos\varphi, \\ \dot{z} &= x_1 \cos\varphi - x_2 \sin\varphi. \end{aligned}$$

In terms of the new variables, the original equation can rewritten as a system in the standard form of the averaging method as follows:

$$\begin{aligned} \dot{\varphi} &= 1, \\ \dot{x}_1 &= \{\varepsilon a(x_1 \sin\varphi + x_2\cos\varphi)^2 + \varepsilon^2\cos\varphi\}\cos\varphi, \\ \dot{x}_2 &= -\{\varepsilon a(x_1 \sin\varphi + x_2\cos\varphi)^2 + \varepsilon^2\cos\varphi\}\sin\varphi. \end{aligned}$$

Applying the above asymptotic method to this system, we seek a change of variables $(\varphi, x_1, x_2) \to (\psi, y_1, y_2)$ reducing the system to a form in which the right-hand sides do not depend on ψ.

In the new variables, the operator of this system is expressed as

$$\begin{aligned} A = A_2 &= \frac{\partial}{\partial\psi} + \{\varepsilon a(y_1\sin\psi + y_2\cos\psi)^2 + \varepsilon^2\cos\psi\}\left(\cos\psi\frac{\partial}{\partial y_1} - \sin\psi\frac{\partial}{\partial y_2}\right), \\ A_1 &= \frac{\partial}{\partial\psi} + \varepsilon a(y_1\sin\psi + y_2\cos\psi)^2\left(\cos\psi\frac{\partial}{\partial y_1} - \sin\psi\frac{\partial}{\partial y_2}\right). \end{aligned}$$

[11] I. G. Malkin, *Some Problems of the Theory of Nonlinear Oscillations*, Gostekhizdat, Moscow, 1956 [in Russian].

The first approximation:

$$B_1 = \langle A_1 \rangle = A_0 = \frac{\partial}{\partial \psi},$$

$$U_0 = -\frac{1}{\varepsilon}\int \widetilde{A}_1\, d\psi = a\{\tfrac{1}{3}(y_2^2 - y_1^2)\sin^3\psi + \tfrac{2}{3}y_1y_2\cos^3\psi - y_2^2\sin\psi\}\frac{\partial}{\partial y_1}$$
$$+ a\{\tfrac{1}{3}(y_2^2 - y_1^2)\cos^3\psi + \tfrac{2}{3}y_1y_2\sin^3\psi - y_1^2\cos\psi\}\frac{\partial}{\partial y_2}.$$

Compute the second approximation. Since $A_1 - A_0 = -\varepsilon[A_0, U]$, we have

$$B_2 = \langle A \rangle + \frac{\varepsilon}{2}\langle [A_1 - A_0,\, U] \rangle = \frac{\partial}{\partial \psi} + \frac{1}{2}\varepsilon^2\left\{1 + \frac{5}{6}a^2 y_2(y_1^2 + y_2^2)\right\}\frac{\partial}{\partial y_1} - \frac{5}{12}\varepsilon^2 a^2 y_1(y_1^2 + y_2^2)\frac{\partial}{\partial y_2}.$$

Since we are not going to construct the third approximation, there is no need to calculate U_1.

In accordance with B_2, the second approximation system acquires the form

$$\frac{d\psi}{dt} = 1,$$
$$\frac{dy_1}{dt} = \frac{\varepsilon^2}{2}\left\{1 + \frac{5}{6}a^2 y_2(y_1^2 + y_2^2)\right\},$$
$$\frac{dy_2}{dt} = -\frac{5}{12}\varepsilon^2 a^2 y_1(y_1^2 + y_2^2).$$

In accordance with U_0, the change of variables is expressed as

$$\varphi = \psi,$$
$$x_1 = y_1 - \varepsilon U_0 y_1 = y_1 - \frac{\varepsilon a}{3}\{(y_2^2 - y_1^2)\sin^3\psi + 2y_1y_2\cos^3\psi - 3y_2^2\sin\psi\},$$
$$x_2 = y_2 - \varepsilon U_0 y_2 = y_2 - \frac{\varepsilon a}{3}\{(y_1^2 - y_2^2)\cos^3\psi + 2y_1y_2\sin^3\psi - 3y_1^2\cos\psi\}.$$

Solving the second approximation system and substituting the solution obtained into this change of variables, we arrive at the solution in terms of φ, x_1, and x_2. In particular, if we are only interested, as in Malkin (1956; see footnote 11), in the periodic solution, then the stationary point should be found from the second approximation system; we have

$$y_1 = 0, \quad y_2 = -\sqrt[3]{\frac{6}{5a^2}}.$$

Thus, for the original variable q we obtain

$$q = -\sqrt[3]{\frac{6\mu}{5a^2}}\cos t + \left(\sqrt[3]{\frac{6\mu}{5a^2}}\right)^2 \frac{a}{2}\left(1 - \frac{1}{3}\cos 2t\right),$$

which exactly coincides with the result obtained in Malkin (1956; see footnote 11) by a different method. Indeed, this solution contains fractional powers of μ, but this have not prevented us from obtaining it from the standpoint of a quasilinear system.

Comparison of the method of normal form with the averaging method. As was established previously in describing the method for calculating the asymptotics of the normal form with Hausdorff's formula, the Poincaré normal form satisfies the condition $[A_0, A] = 0$, where A is the operator of a system in the normal form and A_0 is the operator of the linear part of the system. In the averaging method, the operator of the averaged system does not depend on the fast variable, $\frac{\partial A}{\partial \psi} = 0$. Alternatively, this condition can be represented as $[A_0, A] = 0$, since $A_0 = \frac{\partial}{\partial \psi}$.

Thus, the averaging method is just the method of normal form which is however applied to a system whose operator A_0 was first reduced to the simple form of the translation operator by a passage to canonical coordinates of the group generated by A_0.

In the method of normal form, the passage to canonical coordinates is carried out after constructing the normal form, provided that we wish to reduce the order of the system.

To summarize, in the averaging method, one first passes to canonical coordinates of the linear part, i.e., obtains the standard form, and then seeks normalizing transformations. Conversely, in the method of normal form, one first normalizes the system and then passes to canonical coordinates. In both cases, the final result is the same. Nevertheless, the procedures are different. One method may turn out to be more convenient than the other in specific problems.

Since the procedures are different, this results in the fact that the classes of systems to which these procedures are applicable do not coincide.

The method of normal form is applied to systems with polynomial right-hand sides. This is not necessary for the averaging method; the right-hand sides can be discontinuous or be of a more complicated nature (delay, integro-differential, etc.).

To illustrate this, we will analyze a problem of an oscillator with dry friction.

Example. Consider the system $\ddot{q} + \varepsilon \operatorname{sgn} \dot{q} + q = 0$. In the Cauchy normal form this system is equivalently represented as

$$\dot{q} = p, \quad \dot{p} = -q - \varepsilon \operatorname{sgn} p.$$

The right-hand side of the second equation is discontinuous, and hence the method of normal form is inapplicable.

To apply the averaging method, we reduce the system to the standard form by the change of variables

$$(q, p) \to (\varphi, x): \quad q = x \sin \varphi, \quad p = x \cos \varphi.$$

(The degenerate system determines the group of rotations in the phase plane, and x and φ are canonical coordinates of this group.)

In terms of the canonical coordinates, the system becomes

$$\dot{\varphi} = 1 + \frac{\varepsilon}{x} \sin \varphi \operatorname{sgn} \cos \varphi,$$
$$\dot{x} = -\varepsilon |\cos \varphi|,$$

where the relation $\operatorname{sgn} x \cos \varphi = \operatorname{sgn} \cos \varphi$ $(x \geq 0)$ was taken into account.

The operator of the system is given by

$$A = \frac{\partial}{\partial \varphi} + \varepsilon \left(\frac{1}{x} \sin \varphi \operatorname{sgn} \cos \varphi \frac{\partial}{\partial \varphi} - |\cos \varphi| \frac{\partial}{\partial x} \right).$$

Thus, $A_0 = \frac{\partial}{\partial \varphi}$ and there is nothing to prevent us from calculating the first approximation following the algorithm of the method. We have

$$B_1 = \langle A \rangle = \frac{\partial}{\partial \psi} - \varepsilon \frac{2}{\pi} \frac{\partial}{\partial y}.$$

Hence, the averaged system in the first approximation becomes

$$\dot{\psi} = 1, \quad \dot{y} = -\varepsilon\frac{2}{\pi}.$$

Its general solution is $\psi = \psi_0 + t$, $y = y_0 - \varepsilon\frac{2}{\pi}t$; it exists on the time interval $t \in \left[0, \frac{\pi}{2}y_0/\varepsilon\right]$. The solution in terms of the original variable,

$$q = \left(x_0 - \frac{2\varepsilon}{\pi}t\right)\cos(\varphi_0 + t),$$

shows dying out of oscillations according to a linear law in a finite time.

The construction of the second approximation meets with difficulties owing to the necessity to calculate commutators of nonsmooth functions.

5.5.2. Multifrequency Systems. Resonance

The standard form of a multifrequency system can be obtained by a formal generalization of the above standard form, $\dot{\varphi} = 1$, $\dot{x} = \varepsilon X(\varphi, x)$, where φ is a scalar variable that changes at a rate of 1. If the function $X(\varphi, x)$ is periodic in φ with period $2\pi/\omega$, then, by scaling, the rate of change of φ can be made equal to ω and the function $X(\varphi, x)$ having the period 2π. Then the system can be rewritten in the form $\dot{\varphi} = \omega$, $\dot{x} = \varepsilon X(\varphi, x)$, where ω is said to be the frequency of variation of the right-hand side in time. If now the variable φ is treated as a vector variable, $\varphi = (\varphi_1, \dots, \varphi_m)$, to which there corresponds the vector of frequencies $\omega = (\omega_1, \dots, \omega_m)$, then the resulting system is referred to as a multifrequency system. The dimensions m and n of the vectors φ and x are independent of each other. In addition, a small nonuniformity in the variation of φ is allowed. Thus, we arrive at the following standard form of a multifrequency system:

$$\frac{d\varphi}{dt} = \omega(x) + \varepsilon\Phi(\varphi, x, \varepsilon),$$
$$\frac{dx}{dt} = \varepsilon X(\varphi, x, \varepsilon).$$

The functions $\Phi(\varphi, x, \varepsilon)$ and $X(\varphi, x, \varepsilon)$ are assumed to be periodic in all φ_k's with period 2π. The variables x are said to be slow, and the variables φ fast. The rate of change of the fast variables can depend on the slow variables.

There are a lot of oscillatory systems of various type which can be reduced to this form. For example, consider a quasilinear oscillatory system of the form

$$A\ddot{q} + Bq = \varepsilon f(t, q, \dot{q}),$$

where $q = (q_1, \dots, q_n)$, A and B are $n \times n$ symmetric positive definite matrices, and the vector function f is periodic in t. By a linear change of variables, the matrix A can be reduced to the identity matrix and B to a diagonal form (see reduction to normal coordinates in Section 5.2). Suppose the system has already been reduced to this form, so that $A = E = \operatorname{diag}\{1, \dots, 1\}$ and $B = \operatorname{diag}\{\omega_1^2, \dots, \omega_n^2\}$.

Rewrite the quasilinear system in the form

$$\dot{q}_k = p_k, \qquad k = 1, \dots, n,$$
$$\dot{p}_k = -\omega_k^2 q_k + \varepsilon f_k(t, q_1, \dots, q_n, p_1, \dots, p_n).$$

In each one-dimensional oscillatory subsystem, the passage to canonical coordinates (x_k, φ_k) of the rotation groups in the corresponding phase subspaces can be carried out in accordance with the relations

$$q_k = x_k \sin\varphi_k, \quad p_k = x_k \cos\varphi_k.$$

In terms of the new variables, the system is expressed as

$$\begin{aligned}
\dot\varphi_k &= \omega_k - \frac{\varepsilon}{x_k\omega_k} f_k(t, x_1\sin\varphi_1, \dots, x_n\sin\varphi_n, x_1\cos\varphi_1, \dots, x_n\cos\varphi_n)\sin\varphi_k,\\
\dot x_k &= \frac{\varepsilon}{\omega_k} f_k(t, x_1\sin\varphi_1, \dots, x_n\sin\varphi_n, x_1\cos\varphi_1, \dots, x_n\cos\varphi_n)\cos\varphi_k.
\end{aligned}$$

Denote the period of the right-hand sides in t by $2\pi/\omega_{n+1}$ and introduce the variable $\varphi_{n+1} = \omega_{n+1}t$ instead of t. The functions f_k are 2π-periodic in φ_{n+1}. Also we introduce the notation

$$\begin{aligned}
-\frac{1}{x_k\omega_k} f_k(t, x_1\sin\varphi_1, \dots, x_n\sin\varphi_n, x_1\cos\varphi_1, \dots, x_n\cos\varphi_n)\sin\varphi_k &\equiv \Phi_k(\varphi_1, \dots, \varphi_m, x_1, \dots, x_n),\\
\frac{1}{\omega_k} f_k(t, x_1\sin\varphi_1, \dots, x_n\sin\varphi_n, x_1\cos\varphi_1, \dots, x_n\cos\varphi_n)\cos\varphi_k &\equiv X_k(\varphi_1, \dots, \varphi_m, x_1, \dots, x_n),
\end{aligned}$$

where $m = n + 1$.

In this notation the equations of the quasilinear oscillatory system have the standard form of a multifrequency system. If the function f is conditionally periodic in t (i.e., there are several periods $2\pi/\omega_{n+1}, \dots, 2\pi/\omega_{n+s}$), then additional fast variables $\psi_{n+1} = (2\pi/\omega_{n+1})t, \dots \psi_{n+s} = (2\pi/\omega_{n+s})t$ must be introduced.

Remark 1. For the quasilinear system in question, the frequency vector ω turned out to be constant, and hence, independent of the slow variables. The independence of ω on x is an indication of quasilinearity of the original system. The case where ω depends on x is essentially nonlinear.

Remark 2. By the change of variables

$$(q, p) \to (x, y): \quad q_k = x_k\cos\omega_k t + y_k\sin\omega_k t, \quad p_k = -x_k\sin\omega_k t + y_k\cos\omega_k t$$

the system $\dot q_k = p_k$, $\dot p_k = -\omega_k^2 q_k + \varepsilon f_k$ could be reduced to the standard form of a single-frequency system, since the method of variation of arbitrary constants, used in Subsection 5.5.1, can be applied here as well. This indicates that the terms "single-frequency" and "multifrequency" are, in a sense, a matter of convention. Any system with evident "physical" multifrequenciness can be written in a mathematically single-frequency form. However, the representation of a multifrequency system with explicit separation of the frequency vector—making the main contribution to the rate of change of the vector of fast variables (phase variables)—possesses a number of advantages. This is especially so in studying resonance phenomena, which will be demonstrated in what follows.

The averaging procedure for multifrequency systems involves the passage from the system $\dot\varphi = \omega + \varepsilon\Phi(\varphi, x, \varepsilon)$, $\dot x = \varepsilon X(\varphi, x, \varepsilon)$ to an approximate system $\dot\varphi = \omega + \varepsilon\Phi_0(x, \varepsilon)$, $\dot x = \varepsilon X_0(x, \varepsilon)$, where

$$\begin{aligned}
\Phi_0(x, \varepsilon) &= \frac{1}{(2\pi)^m}\int_0^{2\pi}\cdots\int_0^{2\pi}\Phi(\varphi_1, \dots, \varphi_m, x, \varepsilon)\,d\varphi_1\dots d\varphi_m,\\
X_0(x, \varepsilon) &= \frac{1}{(2\pi)^m}\int_0^{2\pi}\cdots\int_0^{2\pi}X(\varphi_1, \dots, \varphi_m, x, \varepsilon)\,d\varphi_1\dots d\varphi_m.
\end{aligned}$$

The functions Φ_0 and X_0 are the averages of the right-hand sides $\Phi(\varphi, x, \varepsilon)$ and $X(\varphi, x, \varepsilon)$ over all fast variables $\varphi_1, \dots, \varphi_m$ and are referred to as space averages.

The averaged equations are considerably simpler than the original equations, since the equations for x are separated from those for φ. After the equations for x have been integrated, the determination of φ is reduced to quadratures.

The justification of the accuracy of the solutions that follow from the system averaged over the fast variables is given by Bogolyubov's theorem, which was formulated previously for single-frequency systems, provided that there is no resonance in the multifrequency system. If there is a resonance, then one cannot pass on to the averaged equations given by Bogolyubov's theorem without loss of accuracy. The averaging procedure in the resonant case will be considered below.

Before proceeding to the solution of a multifrequency system, we must first define the notion of a resonance and indicate how the presence of a resonance in the standard form of a system can be established. To this end, we define the time averages of the functions $\Phi(\varphi, x, \varepsilon)$ and $X(\varphi, x, \varepsilon)$ as follows:

$$\Phi_0^*(x,\varepsilon,\omega) = \lim_{T\to\infty} \frac{1}{T}\int_0^T \Phi(\omega_1 t+\theta_1, \ldots, \omega_m t+\theta_m, x, \varepsilon)\, dt,$$

$$X_0^*(x,\varepsilon,\omega) = \lim_{T\to\infty} \frac{1}{T}\int_0^T X(\omega_1 t+\theta_1, \ldots, \omega_m t+\theta_m, x, \varepsilon)\, dt.$$

These expressions are the averages of the right-hand sides of the system over time along the trajectory

$$x = \text{const}, \quad \varphi = \omega t + \theta$$

of the degenerate system ($\varepsilon = 0$). The functions $\Phi_0^*(x, \varepsilon, \omega)$ viewed as functions of ω can have discontinuities. The values of the frequencies $\omega = (\omega_1, \ldots, \omega_m)$ at which these functions are discontinuous are just called resonances.[12]

The resonant frequencies satisfy relations of the form

$$\omega_1\lambda_1 + \cdots + \omega_m\lambda_m = 0, \qquad \lambda_1^2 + \cdots + \lambda_m^2 \neq 0,$$

where $\lambda_1, \ldots, \lambda_m$ are integers. Such relations are said to be resonance relations.

It should be emphasized that a resonance relation is not any relation of this form but only a relation that defines a discontinuity of time averages.

Example. Find a resonance in the system defined by Mathieu's equation $\ddot{q}+\nu^2(1-\varepsilon\cos 2t)q=0$. Rewrite this equation in the form of a first order system to obtain

$$\dot{q} = p, \quad \dot{p} = -\nu^2(1-\varepsilon\cos 2t)q.$$

By the change of variables $q = x\sin\varphi_1$, $q_2 = x\nu\cos\varphi_1$, $2t = \varphi_2$, this system can be reduced to the standard form

$$\begin{aligned} \dot{\varphi}_1 &= \nu - \varepsilon\nu\sin^2\varphi_1\cos\varphi_1, \\ \dot{\varphi}_2 &= 2, \\ \dot{x} &= \varepsilon\nu x\sin\varphi_1\cos\varphi_1\cos\varphi_2 \end{aligned}$$

with $\omega_1 = \nu$ and $\omega_2 = 2$. Compute the time average

$$X_0^*(x,\omega_1,\omega_2) = \lim_{T\to\infty}\frac{1}{T}\int_0^T x\omega_1\sin(\omega_1 t+\theta)\cos(\omega_1 t+\theta)\cos\omega_2 t\, dt$$

to obtain

$$X_0^*(x,\omega_1,\omega_2) = \begin{cases} 0 & \text{if } 2\omega_1 \neq \omega_2, \\ \frac{1}{4}x\omega_1\sin 2\theta & \text{if } 2\omega_1 = \omega_2. \end{cases}$$

Thus, in the first approximation, Mathieu's equation has a single resonance determined by the resonance relation $2\omega_1 - \omega_2 = 0$. It is called a main parametric resonance.

[12] Obviously, the definition of a resonance must take into account the order of the approximation. For example, within the framework of the first approximation, discontinuities of order of ε^2 must be disregarded.

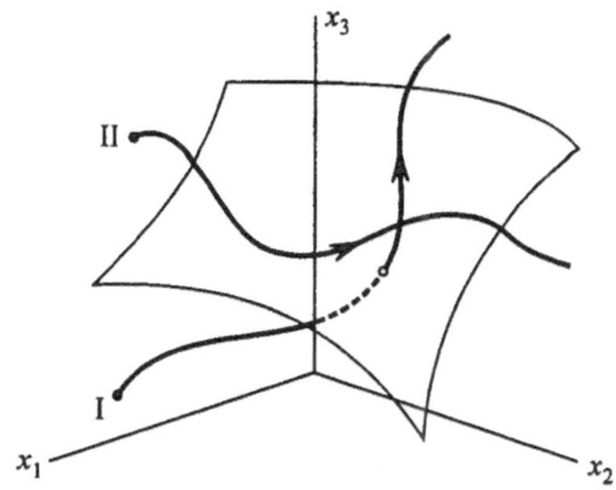

Figure 5.2

The time average may have a discontinuity if the frequency vector ω satisfies several resonance relations with integer coefficients,

$$\omega_1\lambda_{11} + \cdots + \omega_m\lambda_{1m} = 0,$$
$$\cdots\cdots\cdots\cdots\cdots\cdots$$
$$\omega_1\lambda_{s1} + \cdots + \omega_m\lambda_{sm} = 0.$$

The number of such linearly independent relations is referred to as the *multiplicity* of the resonance and the sum of the moduli of the coefficients λ_{ki} in the linearly independent relations is called the *order* of the resonance . The main resonance in Mathieu's equation is a resonance of the third order.

If we deal with a quasilinear system and the frequency vector does not depend on the slow variables, then the fact of the presence or absence of resonances does not depend of the system evolution. If the system is essentially nonlinear, then the resonance relation

$$\lambda_1\omega_1(x) + \cdots + \lambda_m\omega_m(x) = 0$$

determines a surface in the space of slow variables (see Fig. 5.2), which is called a resonance surface. In this case, a resonance occurs when the trajectory of motion of the system intersects the resonance surface and may disappear after that.

The method of asymptotic averaging of quasilinear systems based on Hausdorff's formula. Turn back to a quasilinear multifrequency system in the standard form:

$$\frac{d\varphi}{dt} = \omega + \varepsilon\Phi(\varphi, x, \varepsilon),$$
$$\frac{dx}{dt} = \varepsilon X(\varphi, x, \varepsilon), \qquad \varphi = (\varphi_1, \ldots, \varphi_m), \quad x = (x_1, \ldots, x_n).$$

The algorithm of asymptotic averaging aims, just as in the single-frequency case, at constructing a change of variables $(\varphi, x) \to (\psi, y)$ that would transform the equations of the system to a form which is independent of the fast variables ψ,

$$\frac{d\psi}{dt} = \omega + \varepsilon\Psi(y, \varepsilon),$$
$$\frac{dy}{dt} = \varepsilon Y(y, \varepsilon).$$

The basic recursion scheme that we previously used two times (in the method of normal form and in studying single-frequency systems) remains unchanged:

$$
\begin{aligned}
B_0 &= A_0, \\
B_1 &= \varepsilon[A_0, U_0] + A_1, \\
&\dots\dots\dots\dots\dots\dots \\
B_k &= \varepsilon[A_0, U_{k-1}] + L_k, \\
&\dots\dots\dots\dots\dots\dots
\end{aligned}
$$

where

$$
L_k = \begin{cases} A_1 & \text{for } k = 1, \\ A_k + \varepsilon[A_{k-1} - A_0, U_{k-2}] + \sum_{j=2}^{k} \frac{\varepsilon^j}{j!} \underbrace{[\dots[[}_{j} A_{k-j}, U_{k-j}], U_{k-j}], \dots] & \text{for } k > 1. \end{cases}
$$

Here, A is the operator of the system written in terms of the new variables,

$$
A = (\omega + \varepsilon\Phi)\frac{\partial}{\partial\psi} + \varepsilon X\frac{\partial}{\partial y}, \quad A_0 = \omega\frac{\partial}{\partial\psi}.
$$

The operator of the desired (averaged) system is expressed as

$$
B = (\omega + \varepsilon\Psi)\frac{\partial}{\partial\psi} + \varepsilon Y\frac{\partial}{\partial y}.
$$

The operator of the transformation that brings the given system to the averaged one is sought in the form

$$
U = \xi(\psi, y, \varepsilon)\frac{\partial}{\partial\psi} + \eta(\psi, y, \varepsilon)\frac{\partial}{\partial y}.
$$

The operators A_k, B_k, and U_k are, as before, the kth order asymptotics of the operators A, B, and U, respectively.

The asymptotic solution of this problem is given by the recurrence relations

$$
B_k = \langle L_k \rangle, \quad \varepsilon[A_0, U_{k-1}] + \widetilde{L}_k = 0,
$$

where $\langle L_k \rangle$ stands for the space average of L_k over all phases $\psi_1, \dots, \psi_m$, and $\widetilde{L}_k$ is the complement of the average.

Unlike the single-frequency case, the space average is calculated in this case and the homological equation is modified,

$$
[A_0, U_{k-1}] \equiv \omega_1\frac{\partial U_{k-1}}{\partial\psi_1} + \dots + \omega_m\frac{\partial U_{k-1}}{\partial\psi_m} = -\frac{1}{\varepsilon}\widetilde{L}_k.
$$

The homological equation for the unknown operator U_{k-1} is a system of partial differential equations for each of the components of this operator. For example, if U_{k-1} is given by

$$
U_{k-1} = \xi(\psi, y, \varepsilon)\frac{\partial}{\partial\psi} + \eta(\psi, y, \varepsilon)\frac{\partial}{\partial y}
$$

and the operator $\widetilde{L}_k$ is expressed as

$$
\widetilde{L}_k = Q(\psi, y, \varepsilon)\frac{\partial}{\partial\psi} + P(\psi, y, \varepsilon)\frac{\partial}{\partial y},
$$

then the equation for the operator splits into the following equations for its components:

$$\omega_1 \frac{\partial \xi_i}{\partial \psi_1} + \cdots + \omega_m \frac{\partial \xi_i}{\partial \psi_m} = -\frac{1}{\varepsilon} Q_i \qquad (i = 1, \ldots, m),$$
$$\omega_1 \frac{\partial \eta_j}{\partial \psi_1} + \cdots + \omega_m \frac{\partial \eta_j}{\partial \psi_m} = -\frac{1}{\varepsilon} P_j \qquad (j = 1, \ldots, n).$$

The solution of these equations is given by

$$\xi_i(\psi, y, \varepsilon) = -\frac{1}{\varepsilon} \int Q_i(C_1 + \omega_1 \theta, \ \ldots, \ C_m + \omega_m \theta, \ y, \ \varepsilon)\, d\theta,$$
$$\eta_j(\psi, y, \varepsilon) = -\frac{1}{\varepsilon} \int P_j(C_1 + \omega_1 \theta, \ \ldots, \ C_m + \omega_m \theta, \ y, \ \varepsilon)\, d\theta.$$

After the integrals have been computed, the constants C_i must be substituted by $C_i = \psi_i - \omega_i \theta$.

These quadratures exhaust the procedure of asymptotic averaging for quasilinear systems in the case where no resonances are present in the system.

Resonant case. Consider the case of a resonance of multiplicity s. We have the resonance relations

$$\sum_{i=1}^{m} \lambda_{ki} \omega_i = 0, \qquad k = 1, \ldots, s; \quad s = 1, \ldots, m; \quad s < m.$$

By the neighborhood of a resonance we understand a domain where the resonance relations are zero not exactly but with an accuracy of ε, i.e.,

$$\Lambda \omega = \Delta, \qquad \|\Delta\| \sim \varepsilon, \quad \Lambda = \{\lambda_{ki}\}.$$

The vector of small quantities Δ is called the frequency detuning. In the product $\Lambda\omega$, the vector ω is understood as a column matrix. Without loss of generality, we assume that the rank of the detuning matrix Λ is equal to s. Represent Λ in the block form

$$\Lambda = (\Lambda_0, \Lambda_1),$$

where Λ_1 is the $s \times s$ nondegenerate matrix composed of the last s columns of Λ, and the matrix Λ_0 completes Λ_1 to Λ. Divide the fast variables $\varphi = (\varphi_1, \ldots, \varphi_m)$ into two groups, $\varphi = (\alpha, \beta)$, where $\alpha_i = \varphi_i$ for $i = 1, \ldots, m-s$ and $\beta_j = \varphi_{m-s+j}$ for $j = 1, \ldots, s$. In what follows, we retain the notation $\Phi = (\Phi_1, \ldots, \Phi_{m-s})$ and $\omega = (\omega_1, \ldots, \omega_{m-s})$ for the shortened vectors, while introducing the notation $\Phi' = (\Phi_1, \ldots, \Phi_m)$ and $\omega' = (\omega_1, \ldots, \omega_m)$ for the complete vectors.

Transform the β-part of the fast variables, $(\beta_1, \ldots, \beta_s) \to (\theta_1, \ldots, \theta_s)$, in accordance with the formulas

$$\theta = \Lambda \varphi = (\Lambda_0, \Lambda_1) \begin{pmatrix} \alpha \\ \beta \end{pmatrix} = \Lambda_0 \alpha + \Lambda_1 \beta.$$

Since Λ_1 is nondegenerate, we obtain

$$\beta = \Lambda_1^{-1}(\theta - \Lambda_0 \alpha).$$

In terms of the new variables, the original multifrequency system in the standard form $\dot{\varphi} = \omega + \varepsilon \Phi$, $\dot{x} = \varepsilon X$ becomes

$$\dot{\alpha} = \omega + \varepsilon \Phi\big(\alpha, \ \Lambda_1^{-1}(\theta - \Lambda_0 \alpha), \ x\big),$$
$$\dot{\theta} = \Delta + \varepsilon \Lambda \Phi'\big(\alpha, \ \Lambda_1^{-1}(\theta - \Lambda_0 \alpha), \ x\big),$$
$$\dot{x} = \varepsilon X\big(\alpha, \ \Lambda_1^{-1}(\theta - \Lambda_0 \alpha), \ x\big).$$

The variables θ in this system should be treated as slow variables, since $\dot{\theta} \sim \varepsilon$. Thereby, the number of slow variables increases by the multiplicity of the resonance, s, and the number of fast variables decreases by the same number s. The remaining fast variables, α_i, are not related to each other, and hence the resonance is eliminated. Thus, the procedure of asymptotic averaging described above for the nonresonant case can be applied to this system.

5.6. Asymptotic Integration of Hamiltonian Systems

5.6.1. Birkhoff Normal Form

The reduction of systems of ordinary differential equations to the Poincaré normal form in the neighborhood of a singular point is simplified if the equations have the Hamiltonian form, since the Hamiltonian function can be subjected to necessary transformations rather than the system of differential equations itself. Birkhoff[13] was the first to suggest a definition for a normal form of a Hamiltonian function and an algorithm of reduction to this form. Today the algorithm suggested by Birkhoff is only of methodical significance and used for explaining the essence of the normal form of Hamiltonians. Subsequently, Birkhoff's reduction algorithm were considerably improved by a passage from the generating function of the desired canonical transformation to the generating Hamiltonian with the aid of Lie series.[14] The next step was the formation of closed recursive algorithms that permit one to develop computer programs of symbolic computations for the reduction of Hamiltonians to the normal form.[15]

The algorithm presented in this section is a further considerable simplification, since this algorithm is based on a one-dimensional recursion scheme in contrast to currently available branching multidimensional schemes.

The simplification is attained due to abandoning the search for successive coefficients of Lie series for canonical transformations in favor of using a ring of asymptotics based on these series. This approach has already been applied in the previous sections when constructing the Poincaré normal form; Hausdorff's formula was used as the basis for the ring of asymptotics.

The method of solving the homological equation based on the direct integration of the right-hand sides along the phase flow of the degenerate system leads to substantial simplifications as well.

Consider a Hamiltonian system determined by an autonomous analytic Hamiltonian

$$H(q,p) = H_0(q,p) + H_*(q,p),$$

where $H_0(q,p)$ is the quadratic part of the Hamiltonian, which determines the linear part of the system, and $H_*(q,p)$ is a finite or infinite polynomial with only terms of the third of higher degree. In the theory of quasilinear systems, the Hamiltonian $H_0(q,p)$ is said to be unperturbed (or degenerate), and the function $H_*(q,p)$ is called a perturbation.

In defining the notion of the *Birkhoff normal form* and in reducing the system to it, it is conventional to use the complex combinations $x = p + iq$ and $y = p - iq$ instead of the generalized coordinates and momenta q and p. These relations can be treated as a canonical change of variables with valence $2i$.

In what follows, we retain the old names for the functions depending on the new variables and assume that the valence has been taken into account. Thus, we proceed from the following representation of the Hamiltonian as a function of x and y:

$$H(x,y) = H_0(x,y) + H_*(x,y).$$

By a linear canonical change of variables, the quadratic part of the Hamiltonian of a conservative mechanical system is reducible to the simplest form which can be represented as

$$H_0 = \frac{1}{2}\sum_{k=1}^{n} \lambda_k (q_k^2 + p_k^2) \qquad \text{in terms of } (q,p),$$

$$H_0 = i\sum_{k=1}^{n} \lambda_k x_k y_k \qquad \text{in terms of } (x,y).$$

[13] G. D. Birkhoff, *Dynamical systems*, Amer. Math. Soc. Colloq. Publ., No. 9, p. 296, 1927.

[14] G.-I. Hori, *Theory of general perturbation with unspecified canonical variables*, Publ. Astron. Soc. Jap., Vol. 18, No. 4, pp. 287–296, 1966.

[15] A. P. Markeev and A. G. Sokol'skii, *Some Computational Algorithms for Normalization of Hamiltonian Systems*, Preprint No. 31, Moscow Inst. Appl. Math., Acad. Sci. USSR, 1976.

A similar problem of reducing the perturbed part of the Hamiltonian, $H_*(x,y)$, to the simplest form can be posed. In this case, the transformation of variables which solves this problem must (i) be canonical, (ii) not change the quadratic part of the Hamiltonian, H_0, and (iii) belong to the class of polynomial functions. The last condition implies that the desired change of variables must not have a singularity at the origin.

This form of the Hamiltonian, which cannot be simplified by any polynomial changes of variables, is just called the Birkhoff normal form.

It is not difficult to establish the general structure of the Birkhoff normal form. Suppose the term

$$x_1^{l_1}\dots x_n^{l_n}y_1^{s_1}\dots y_n^{s_n}$$

must be eliminated from $H_*(x,y)$. We will demonstrate that this can be done by means of a canonical transformation $(x,y)\to(u,v)$ with generating function

$$S(x,v)=\sum_{k=1}^{n}x_kv_k+hx_1^{l_1}\dots x_n^{l_n}v_1^{s_1}\dots v_n^{s_n}.$$

The terms $\sum_{k=1}^{n}x_kv_k$ generate an identity transformation.

For this change of variables, we have

$$\begin{aligned}y&=\frac{\partial S}{\partial x}=v+h\frac{\partial}{\partial x}(x^lv^s)=v+h\frac{\partial}{\partial u}(u^lv^s)+\cdots,\\ u&=\frac{\partial S}{\partial v}=x+h\frac{\partial}{\partial v}(x^lv^s)=x+h\frac{\partial}{\partial y}(x^ly^s)+\cdots.\end{aligned}$$

The last relation is obtained by inverting the transformation, and the higher order terms that appear are denoted by dots.

The inverse transformation is given by

$$\begin{aligned}v&=y-h\frac{\partial}{\partial x}(x^ly^s)+\cdots,\\ x&=u-h\frac{\partial}{\partial v}(u^lv^s)+\cdots.\end{aligned}$$

By substituting the old variables expressed in terms of the new ones into the Hamiltonian function, we express the Hamiltonian in terms of the new variables:

$$\widetilde{H}=i\sum_{k=1}^{n}\lambda_ku_kv_k+ih\sum_{k=1}^{n}\lambda_k\left(u_k\frac{\partial}{\partial u_k}-v_k\frac{\partial}{\partial v_k}\right)u^lv^s+H_*.$$

The term proportional to h has the same structure as that to be eliminated. Note that neither terms of lower orders nor other terms of the same order have been changed. This permits one to eliminate nonlinear terms successively, from lower to higher orders.

It is not difficult to see that the resulting expression we plan to use to compensate for the chosen nonlinear term is the following Poisson bracket (see Section 4.2):

$$ih\sum_{k=1}^{n}\lambda_k\left(u_k\frac{\partial}{\partial u_k}-v_k\frac{\partial}{\partial v_k}\right)u^lv^s\equiv h\{H_0,u^lv^s\}.$$

Therefore, if the term u^lv^s is a first integral of the linear part of the system, the Poisson bracket vanishes, and hence this term cannot be compensated for.

Thus, by polynomial changes of variables, all terms but first integrals of the linear part of the system can be eliminated from the Hamiltonian of the nonlinear part, i.e., from H_*.

Just as any system of order $2n$, the system with Hamiltonian H_0,

$$\dot{u}_k = i\lambda_k u_k, \quad \dot{v}_k = -i\lambda_k v_k \qquad (k = 1, \dots, n),$$

always has $2n-1$ independent first integrals, $G_1(u, v), \dots, G_{2n-1}(u, v)$. Obviously, this system has n first integrals of the polynomial form

$$G_k = u_k v_k \qquad (k = 1, \dots, n).$$

The system is so simple that there are no other polynomial first integrals (since it is easy to write out all $2n-1$ independent first integrals), provided that the system is nonresonant. This is the case if

$$k_1\lambda_1 + \dots + k_n\lambda_n \neq 0 \qquad (k_1^2 + \dots + k_n^2 \neq 0)$$

for all integer $k_1, \dots, k_n$.

If there is a resonance, then the number of independent polynomial first integrals exceeds by n the number of resonance relations.

For example, if there is a resonance $\lambda_1 = 3\lambda_2$, then in addition to the n first integrals $G_k = u_k v_k$ $(k = 1, \dots, n)$, the function $G_{n+1} = u_1 v_2^3$ is also a first integral. In this case, the Birkhoff normal form depends on the following arguments: $u_1 v_1, \dots, u_n v_n$, and $u_1 v_2^3$.

DEFINITION. *A Hamiltonian $H = H_0 + H_*$ is said to have the Birkhoff normal form if it depends only on polynomial first integrals (invariants) of the unperturbed part H_0, i.e., if $\{H_0, H_*\} = 0$.*

The above algorithm that made it possible to establish the structure of the normal form is inconvenient for practical calculations.

An invariant algorithm for constructing the Birkhoff normal form. As to Birkhoff himself, he gave a different definition of the normal form. He stated that it must depend on $u_k v_k$ in the (u, v) description and on $q^2 + p^2$ in the (q, p) description.

The above definition is given in an invariant form; it is valid if the Hamiltonian is represented in terms of any variables, not only in terms of (x, y) or (q, p). Furthermore, this definition covers both resonant and nonresonant cases.

The reduction algorithm that follows is also invariant.

We seek canonical transformations without using the generating function $S(x, v)$, which depends on mixed variables. Instead, we seek a generating Hamiltonian, i.e., the Hamiltonian of some auxiliary system whose phase flow determines the group of the transformations used for the normalization of the original Hamiltonian.

Let $Q(x, y)$ be the desired Hamiltonian of this auxiliary system, which means that the equations

$$\frac{dx}{d\tau} = \frac{\partial Q}{\partial y}, \quad \frac{dy}{d\tau} = -\frac{\partial Q}{\partial x}$$

subject to the initial conditions $x(0) = u$ and $y(0) = v$ determine the desired transformation

$$x = x(\tau, u, v), \quad y = y(\tau, u, v),$$

which is the solution of this Cauchy problem.

Such transformations and their inversions can be represented with Lie series as (see Section 1.5)

$$x = u + \tau\{u, Q\} + \frac{\tau^2}{2!}\{\{u, Q\}, Q\} + \cdots,$$
$$y = v + \tau\{v, Q\} + \frac{\tau^2}{2!}\{\{v, Q\}, Q\} + \cdots,$$

$$u = x - \tau\{x, Q\} + \frac{\tau^2}{2!}\{\{x, Q\}, Q\} + \cdots,$$
$$v = y - \tau\{y, Q\} + \frac{\tau^2}{2!}\{\{y, Q\}, Q\} + \cdots.$$

Here, $Q = Q(u, v)$ for the direct transformations and $Q = Q(x, y)$ for the inverse transformations, whereas the function Q itself is the same in both cases.

The transformed Hamiltonian is related to the original one by the Lie series

$$\widetilde{H}(u, v) = H(u, v) + \tau\{H, Q\} + \frac{\tau^2}{2!}\{\{H, Q\}, Q\} + \cdots.$$

This is the basic formula for the reduction algorithm presented below.

Let us formally define the order of nonlinear terms as follows. Introduce a small parameter ε and perform the scaling of variables $x \to \varepsilon x$, $y \to \varepsilon y$. Then, taking into account the valence $1/\varepsilon^2$, we rewrite the Hamiltonian $H(x, y) = H_0(x, y) + H_*(x, y)$ in the form

$$H(x, y, \varepsilon) = H_0(x, y) + \frac{1}{\varepsilon^2}H_*(\varepsilon x, \varepsilon y) \equiv H_0(x, y) + H_*(x, y, \varepsilon).$$

We will refer to any function

$$H_k = H + O(\varepsilon^{k+1})$$

as a kth order asymptotics of the Hamiltonian H.

We seek the Hamiltonian of the auxiliary generating system in the form of an asymptotics

$$H_k = Q + O(\varepsilon^{k+1}).$$

For asymptotics, the Lie series representing the new Hamiltonian can be written in the form

$$\widetilde{H}_0(u, v) = H_0(u, v),$$
$$\widetilde{H}_1(u, v) = H_1(u, v) + \tau\{H_0, Q_0\},$$
$$\widetilde{H}_2(u, v) = H_2(u, v) + \tau\{H_1, Q_1\} + \frac{\tau^2}{2!}\{\{H_0, Q_0\}, Q_0\},$$
$$\cdots\cdots\cdots\cdots\cdots\cdots\cdots\cdots\cdots\cdots\cdots\cdots\cdots\cdots\cdots\cdots\cdots\cdots$$
$$\widetilde{H}_k(u, v) = H_k(u, v) + \tau\{H_{k-1}, Q_{k-1}\} + \cdots + \frac{\tau^k}{k!}\underbrace{\{\ldots\{\{}_{k\ \text{times}} H_0, Q_0\}, Q_0\}, \ldots\},$$
$$\cdots\cdots\cdots\cdots\cdots\cdots\cdots\cdots\cdots\cdots\cdots\cdots\cdots\cdots\cdots\cdots\cdots\cdots$$

where ε has been identified with τ.

Rewrite the Poisson bracket $\{H_{k-1}, Q_{k-1}\}$ in the expression of $\widetilde{H}_k(u, v)$ in the asymptotically equivalent form

$$\{H_{k-1}, Q_{k-1}\} = \{H_0, Q_{k-1}\} + \{H_{k-1} - H_0, Q_{k-1}\} = \{H_0, Q_{k-1}\} + \{H_{k-1} - H_0, Q_{k-2}\}.$$

This equivalence is valid because $H_{k-1} - H_0 \sim O(\tau)$.

With reference to this representation, we rewrite the expressions of the asymptotics of $\widetilde{H}$ as

$\widetilde{H}_0(u, v) = H_0,$

$\widetilde{H}_1(u, v) = H_1 + \tau\{H_0, Q_0\},$

. . . .

$$\widetilde{H}_k(u, v) = H_k + \tau\{H_{k-1}, Q_{k-1}\} + \tau\{H_{k-1} - H_0, Q_{k-2}\} + \sum_{j=2}^{k} \frac{\tau^j}{j!} \underbrace{\{\ldots\{\{H_{k-j}, Q_{k-j}\}, Q_{k-j}\}, \ldots\}}_{j \text{ times}},$$

. . . .

Introduce the notation

$$L_1 = H_1,$$

$$L_k = H_k + \tau\{H_{k-1} - H_0, Q_{k-2}\} + \sum_{j=2}^{k} \frac{\tau^j}{j!} \underbrace{\{\ldots\{\{H_{k-j}, Q_{k-j}\}, Q_{k-j}\}, \ldots\}}_{j \text{ times}} \qquad (k \geq 2).$$

Then the desired kth-order asymptotics of the transformed Hamiltonian can be expressed as

$$\widetilde{H}_k = \tau\{H_0, Q_{k-1}\} + L_k \qquad (k \geq 1).$$

This equation is called *homological*. We will use it to successively increase the order of the asymptotics of the desired Hamiltonian in the normal form, $\widetilde{H}$. The zeroth order asymptotics coincides with the degenerate Hamiltonian. Further on, the function L_k that occurs in the expression of the kth approximation Hamiltonian $\widetilde{H}_k$ turns out to be a known function of u and v, since L_k depends on the asymptotics found at the previous steps. The homological equation serves to determine the unknown functions $\widetilde{H}_k(u, v)$ and $Q_{k-1}(u, v)$.

Prior to solving the homological equation for these functions, we note that the Poisson bracket $\{H_0, Q_{k-1}\}$ is the total derivative of $Q_{k-1}(u, v)$ with respect to t along the trajectories of the degenerate system, i.e., along the family of maps $u \to ue^{i\lambda t}$, $v \to ve^{-i\lambda t}$. Thus,

$$UQ_{k-1} \equiv \{H_0, Q_{k-1}\} = \frac{dQ_{k-1}}{dt}.$$

Furthermore, by definition, $\widetilde{H}_k$ consists only of invariants of the same linear system. Integrating the homological equation along the trajectories of the degenerate system, we obtain

$$\int_0^t \widetilde{H}_k \, dt = \tau \int_0^t \{H_0, Q_{k-1}\} \, dt + \int_0^t L_k \, dt$$

or, taking into account the preceding,

$$\int_0^t L_k\left(ue^{i\lambda t}, ve^{-i\lambda t}\right) dt = t\widetilde{H}_k + \tau\left\{Q_{k-1}(u, v) - Q_{k-1}\left(ue^{i\lambda t}, ve^{-i\lambda t}\right)\right\}.$$

Thus, on integrating the known function $L_k(u, v)$ with respect to t along the trajectories of the degenerate system, we arrive at the desired asymptotics of the normal form of the Hamiltonian $\widetilde{H}_k$ as the coefficient of t, and the desired asymptotics of the generating Hamiltonian Q_{k-1} as the coefficient of τ which does not depend on t.

The above formula completes the algorithm of reducing a Hamiltonian system to the Birkhoff normal form. Thus, the algorithm involves the following steps: (a) compute the function $L_k(u, v)$, (b) take the integral of it with respect to time along the trajectories of the degenerate system, and (c) find the coefficients of t and τ.

The first step of this recursion scheme involves calculating the integral of the first asymptotics of the original Hamiltonian along the trajectories of the degenerate system, specifically,

$$\int_0^t H_1(ue^{i\lambda t}, ve^{-i\lambda t}) \, dt = t\widetilde{H}_1 + \tau Q_0(0) - \tau Q_0(t).$$

Properties of the Birkhoff normal form.

PROPERTY 1. The asymptotics of each particular order of the normal form contains a minimum number of nonlinear terms of this order. This number cannot be reduced by any polynomial transformations. Therefore, any analysis of the system by considering its normal form is the simplest.

PROPERTY 2. Since the perturbation H_* and unperturbed component H_0 of the Hamiltonian H in the normal form commute, the principle of superposition holds for the system (see Section 1.9). By virtue of this principle, to construct the solution of the system with Hamiltonian $H_0 + H_*$, it suffices to know the solution of the systems with Hamiltonians H_0 and H_* separately. The solution of the system with Hamiltonian H_0 is known. It remains to find the solution of the system with Hamiltonian H_*.

PROPERTY 3. The Hamilton equations in the normal form always admit reduction in order, since a solution of the degenerate system is always a symmetry group for the entire system.

Example. Consider Duffing's equation $\ddot{q} + q + q^3 = 0$.

The corresponding Hamiltonian is expressed as $H = \frac{1}{2}\left(p^2+q^2+\frac{1}{2}q^4\right)$. By the change of variables $x = q - ip$, $y = q + ip$, the Hamiltonian is reduced to the form $H = i\left[xy + \frac{\varepsilon}{32}(x+y)^4\right]$.

The first approximation of the normal form. The trajectories of the degenerate system with Hamiltonian $H_0 = ixy$ are given by $x = ue^{it}$, $y = ve^{-it}$. Integrating the Hamiltonian along these trajectories yields

$$\begin{aligned}\int_0^t H_1\,dt = \int_0^t H\,dt &= i\int_0^t \left[uv + \frac{\tau}{32}(ue^{it} + ve^{-it})^4\right] dt\\ &= ituv + \frac{\tau}{32}\left(\frac{1}{4}u^4e^{4it} + 2u^3ve^{2it} + 6iu^2v^2t - 2uv^3e^{-2it} - \frac{1}{4}v^4e^{-4it}\right)\\ &\quad - \frac{\tau}{32}\left(\frac{1}{4}u^4 + 2u^3v - 2uv^3 - \frac{1}{4}v^4\right).\end{aligned}$$

The first order asymptotics of the normal form is the coefficient of t; we have

$$\widetilde{H}_1 = i\left(uv + \frac{3\tau}{16}u^2v^2\right).$$

The first approximation of the generating Hamiltonian is given by the time-invariant coefficient of τ:

$$Q_0 = -\frac{1}{32}\left(\frac{1}{4}u^4 + 2u^3v - 2uv^3 - \frac{1}{4}v^4\right).$$

The second approximation of the normal form. Compute

$$L_2 = \tau\{H_1 - H_0, Q_0\} + \frac{\tau^2}{2}\left\{\{H_0, Q_0\}, Q_0\right\} + H_2.$$

In our case,

$$H_1 - H_0 = \frac{i\tau}{32}(x+y)^4, \quad \{H_0, Q_0\} = \frac{1}{\tau}\left(\widetilde{H}_1 - H_1\right) = -\frac{i}{32}(x^4 + 4x^3y + 4xy^3 + y^4), \quad H_2 = H.$$

On computing the Poisson brackets, substituting $x = ue^{it}$ and $y = ve^{-it}$, and integrating L_2 with respect to t, we find the coefficient of t, thus obtaining

$$\widetilde{H}_2 = i\left(uv + \frac{3\tau}{16}u^2v^2 - \frac{17\tau^2}{256}u^3v^3\right).$$

We confine ourselves to this approximation and, therefore, do not write out the function Q necessary for constructing the third approximation.

The Birkhoff normal form in the nonautonomous case . Consider now the case where the perturbation of a conservative Hamiltonian system depends explicitly on time,

$$H(t,q,p,\varepsilon) = H_0(q,p) + \varepsilon H_*(t,q,p,\varepsilon), \qquad |\varepsilon| \ll 1,$$

where $H_0(q,p)$ is of the same form as in the autonomous case considered above and $H(t,q,p,\varepsilon)$ is an arbitrary smooth function of the variables specified. Unlike the autonomous case, we do not assume here that H_* does not contain terms of order < 3 in q and p.

Any Hamiltonian system with a Lyapunov reducible linear part can be reduced to this form. Special cases of such systems are nonlinear systems in which the small parameter ε does not occur explicitly; then any terms of order ≥ 2 in canonical variables can be treated as perturbations. In this case, the small parameter can be formally introduced by the similarity transformation.

DEFINITION. *In accordance with the above definition of the normal form for an autonomous system, the Hamiltonian $H = H_0 + \varepsilon H_*(t,p,q,\varepsilon)$ is said to have the Birkhoff normal form if the perturbation H_* is a first integral of the linear part of the system, which is determined by the Hamiltonian function $H_0(q,p)$.*

In other words, an nonautonomous Hamiltonian is in the normal form if and only if

$$\frac{\partial H_*}{\partial t} + \{H_*, H_0\} = 0,$$

where, as before, $\{\cdot,\cdot\}$ is a Poisson bracket.

All other known definitions of the Birkhoff normal form are exhausted by this definition. It does not matter whether there are resonances in the linear part of the system. The definition is invariant in the sense that it appeals not to the dependence of H_* on specially selected variables but to the geometric properties of the phase flows of the perturbed and unperturbed systems—these flows must commute.

Actually, such a definition immediately follows from the definition given above for autonomous systems. This can be seen if, following Section 4.3, one treats time t in the nonautonomous Hamiltonian as a generalized coordinate and denote it by φ and then introduces its conjugate, the momentum J. Then the Hamiltonian of this system is expressed as

$$H(\varphi,J,q,p,\varepsilon) = H_0(q,p) + J + \varepsilon H_*(\varphi,q,p,\varepsilon).$$

The system with this Hamiltonian is equivalent to the system with the original Hamiltonian, provided that the initial condition is $\varphi(0) = 0$.

In this case, the condition of the normal form coincides with that introduced above for autonomous systems, $\{H_0 + J, H_*\} = 0$.

Properties of the nonautonomous normal form. The main property of the nonautonomous normal form is that it can be directly reduced to a system with constant coefficients. This property is rather convenient in studying nonlinear oscillations and stability.

THEOREM. *If a system with Hamiltonian $H = H_0 + \varepsilon H_*(t,q,p,\varepsilon)$ satisfies the condition of the normal form, then to construct the general solution of the corresponding Hamilton equations,*

$$\dot{q} = \frac{\partial H}{\partial p}, \quad \dot{p} = -\frac{\partial H}{\partial q},$$

it suffices:

(1) to find the general solution $q = f(t,q_0',p_0')$, $p = g(t,q_0',p_0')$ of the degenerate system

$$\dot{q}_0 = \frac{\partial H_0}{\partial p}, \quad \dot{p}_0 = -\frac{\partial H_0}{\partial q};$$

(2) to find the general solution $q = h(t, q_0'', p_0'')$, $p = l(t, q_0'', p_0'')$ of the system determined by the Hamiltonian H_*,

$$\dot{q}_0 = \varepsilon \frac{\partial H_*}{\partial p}, \quad \dot{p}_0 = -\varepsilon \frac{\partial H_*}{\partial q},$$

provided that time in this equation is treated as zero, i.e., $H_* = H_*(0, q, p, \varepsilon)$.

Then the general solution of the original system is the composition of these solutions taken in any order,

$$q = f\big(t,\ h(t, q_0'', p_0''),\ l(t, q_0'', p_0'')\big) \equiv h\big(t,\ f(t, q_0', p_0'),\ g(t, q_0', p_0')\big),$$
$$p = g\big(t,\ h(t, q_0'', p_0''),\ l(t, q_0'', p_0'')\big) \equiv l\big(t,\ f(t, q_0', p_0'),\ g(t, q_0', p_0')\big).$$

Proof. Take advantage of the representation of the Hamiltonian in the autonomous form, $H(\varphi, J, q, p, \varepsilon) = H_0(q, p) + J + \varepsilon H_*(\varphi, q, p, \varepsilon)$. Then we arrive at the Hamilton equations

$$\dot{\varphi} = 1, \quad \dot{J} = -\varepsilon \frac{\partial H_*}{\partial \varphi}, \quad \dot{q} = \frac{\partial H_0}{\partial p} + \varepsilon \frac{\partial H_*}{\partial p}, \quad \dot{p} = -\frac{\partial H_0}{\partial q} - \varepsilon \frac{\partial H_*}{\partial q}$$

subject to the initial condition $\varphi(0) = 0$.

By virtue of the condition $\{H_0 + J, H_*\} = 0$, the general solution of this system is the composition of the general solutions of the following systems (see Section 1.9):

$$(1) \quad \dot{\varphi} = 1, \quad \dot{J} = 0, \quad \dot{q} = \frac{\partial H_0}{\partial p}, \quad \dot{p} = -\frac{\partial H_0}{\partial q}, \qquad \varphi(0) = 0;$$
$$(2) \quad \dot{\varphi} = 0, \quad \dot{J} = -\varepsilon \frac{\partial H_*}{\partial \varphi}, \quad \dot{q} = \varepsilon \frac{\partial H_*}{\partial p}, \quad \dot{p} = -\varepsilon \frac{\partial H_*}{\partial q}, \qquad \varphi(0) = 0.$$

Integrating the first two equations in both systems, substituting the thus obtained φ and J into the remaining equations, and performing further the necessary composition, we arrive at the property stated in the theorem.

An asymptotic algorithm of reduction to the normal form. The normalization algorithm described above for autonomous systems is applicable in this case, too, if the original Hamiltonian is represented in the extended phase space as follows:

$$H(\varphi, J, q, p, \varepsilon) = H_0(q, p) + J + \varepsilon H_*(\varphi, q, p, \varepsilon).$$

To reduce this Hamiltonian to the normal form, we seek a canonical change of variables $(\varphi, J, q, p) \to (\widetilde{\varphi}, \widetilde{J}, \widetilde{q}, \widetilde{p})$ of the form

$$\varphi = \widetilde{\varphi}, \quad J = J(\tau, \widetilde{\varphi}, \widetilde{J}, \widetilde{q}, \widetilde{p}), \quad q = q(\tau, \widetilde{\varphi}, \widetilde{J}, \widetilde{q}, \widetilde{p}), \quad p = p(\tau, \widetilde{\varphi}, \widetilde{J}, \widetilde{q}, \widetilde{p}).$$

This change of variables is a one-parameter group (with parameter τ) of canonical transformations generated by the phase flow of the Hamiltonian system with Hamiltonian $Q(\varphi, q, p)$, so that the functions that determine the change of variables are solutions of the Cauchy problem

$$\frac{d\varphi}{d\tau} = \frac{\partial Q}{\partial J} = 0, \quad \frac{dJ}{d\tau} = -\frac{\partial Q}{\partial \varphi}, \quad \frac{dq}{d\tau} = \frac{\partial Q}{\partial p}, \quad \frac{dp}{d\tau} = -\frac{\partial Q}{\partial q},$$
$$\varphi(0) = \widetilde{\varphi}, \quad J(0) = \widetilde{J}, \quad q(0) = \widetilde{q}, \quad p(0) = \widetilde{p}.$$

Since the artificially autonomized Hamiltonian has a slightly more specific form than an arbitrary Hamiltonian with the same variables, some simplifications turn out to be possible here as compared with the general case. For example, the generating Hamiltonian can be chosen as a function independent of the momentum J, so that there is no need to do any manipulations with time.

Except for this, the recursive procedure for constructing successive asymptotics $\widetilde{H}_k$ and $\widetilde{Q}_k$ coincides with that described previously.

It is necessary to compute the functions $L_k(\widetilde{\varphi}, \widetilde{J}, \widetilde{q}, \widetilde{p}, \tau)$, $k > 1$, in accordance with the formulas

$$L_1 = H_1,$$

$$L_k = H_k + \tau\{H_{k-1} - H_0, Q_{k-2}\} + \sum_{j=2}^{k} \frac{\tau^j}{j!} \underbrace{\{\ldots\{}_{j \text{ times}} H_{k-j}, Q_{k-j}\}, Q_{k-j}\}, \ldots\} \qquad (k \geq 2, \ \tau = \varepsilon).$$

After that, the equation for the desired normal form $\widetilde{H}_k$ and the generating Hamiltonian Q_{k-1} should be written out:

$$\widetilde{H}_k - \tau\{H_0, Q_{k-1}\} = L_k, \qquad \{H_0, \widetilde{H}_k\} = 0.$$

As a result, we obtain a system of two first-order linear partial differential equations for determining $\widetilde{H}_k$ and Q_{k-1}. These functions can be found, i.e., the system can be integrated, in the following manner.

It is necessary to solve the linear system

$$\dot{\varphi} = 1, \quad \dot{J} = 0, \quad \dot{q} = \frac{\partial H_0}{\partial p}, \quad \dot{p} = -\frac{\partial H_0}{\partial q}.$$

Its solution is given by

$$\varphi = 1 + \widetilde{\varphi}, \quad J = \widetilde{J}, \quad q = q(t, \widetilde{q}, \widetilde{p}), \quad p = p(t, \widetilde{q}, \widetilde{p}).$$

After that, one must solve $\widetilde{\varphi}$, $\widetilde{J}$, $\widetilde{q}$, and $\widetilde{p}$ for φ, J, q, and p and substitute the resulting expressions into the homological system $\widetilde{H}_k - \tau\{H_0, Q_{k-1}\} = L_k$, $\{H_0, \widetilde{H}_k\} = 0$. Then the second equations is satisfied identically and the Poisson bracket in the first equation is found to be equal to the total derivative of Q_{k-1} with respect to time,

$$\{H_0, Q_{k-1}\} = \frac{dQ_{k-1}}{dt}.$$

Therefore, integrating the first equation from 0 to t yields

$$\int_0^t L_k \, dt = \widetilde{H}_k t + \tau Q_{k-1}(0) - \tau Q_{k-1}(t).$$

As before, the desired normal form is the coefficient of t and the generating Hamiltonian is the time-invariant coefficient of τ.

The first step, which corresponds to the first approximation with $k = 1$, involves the integration of the first order asymptotics of the original Hamiltonian along the trajectories of the linear system,

$$\int_0^t H_1 \, dt = \widetilde{H}_1 t + \tau Q_0(0) - \tau Q_0(t).$$

Example. Consider Mathieu's equation $\ddot{x} + [1 + \varepsilon(a + \cos 2t)]x = 0$. Assuming ε to be small, we consider the normal form in the neighborhood of the main parametric resonance. The nonautonomous Hamiltonian corresponding to this equation is expressed as

$$H = \tfrac{1}{2}[p^2 + q^2 + \varepsilon q^2(a + \cos 2t)],$$

where $q \equiv x$ and $p \equiv \dot{x}$. Introduce the new generalized coordinate $\varphi \equiv 2t$ and the corresponding momentum J. Then we obtain the autonomous Hamiltonian of Mathieu's equation in the form

$$H(\varphi, J, q, p) = 2J + \tfrac{1}{2}(q^2 + p^2) + \tfrac{1}{2}\varepsilon q^2(q + \cos\varphi).$$

The Hamiltonian of the degenerate system ($\varepsilon = 0$) is determined by the linear equations

$$\dot{\varphi} = 1, \quad \dot{J} = 0, \quad \dot{q} = p, \quad \dot{p} = -q.$$

The general solution of this system is given by

$$\varphi = 2t + \widetilde{\varphi}, \quad J = \widetilde{J}, \quad q = \widetilde{q}\cos t + \widetilde{p}\sin t, \quad p = -\widetilde{q}\sin t + \widetilde{p}\cos t.$$

Carry out the first approximation of the above normalization algorithm. We have $L_1 \equiv H$ and

$$\int_0^t H\,dt = \int_0^t \left[2\widetilde{J} + \tfrac{1}{2}(\widetilde{q}^2 + \widetilde{p}^2) + \tfrac{1}{2}\varepsilon(\widetilde{q}\cos t + \widetilde{p}\sin t)^2\big(a + \cos(\widetilde{\varphi} + 2t)\big)\right] dt.$$

Integrating yields

$$\begin{aligned}\int_0^t H\,dt = {} & \left[2\widetilde{J} + \tfrac{1}{2}\big(\widetilde{q}^2 + \widetilde{p}^2\big)\big(1 + \tfrac{1}{2}\varepsilon a\big) + \tfrac{1}{8}\varepsilon\big(\widetilde{q}^2 - \widetilde{p}^2\big)\cos\widetilde{\varphi} - \tfrac{1}{4}\varepsilon\widetilde{q}\widetilde{p}\sin\widetilde{\varphi}\right]t + \tfrac{1}{8}\varepsilon\big(\widetilde{q}^2 + \widetilde{p}^2\big) \\ & + \sin(2t + \widetilde{\varphi}) + \tfrac{1}{8}\varepsilon a\big(\widetilde{q}^2 - \widetilde{p}^2\big)\sin 2t + \tfrac{1}{32}\varepsilon\big(\widetilde{q}^2 - \widetilde{p}^2\big)\sin(4t + \widetilde{\varphi}) - \tfrac{1}{16}\varepsilon\widetilde{q}\widetilde{p}\cos(4t + \widetilde{\varphi}) \\ & - \tfrac{1}{4}\varepsilon a\widetilde{q}\widetilde{p}\cos 2t - \tfrac{1}{8}\varepsilon\big(\widetilde{q}^2 + \widetilde{p}^2\big)\sin\widetilde{\varphi} - \tfrac{1}{32}\varepsilon\big(\widetilde{q}^2 - \widetilde{p}^2\big)\sin\widetilde{\varphi} + \tfrac{1}{16}\varepsilon\widetilde{q}\widetilde{p}\cos\widetilde{\varphi} + \tfrac{1}{4}\varepsilon a\widetilde{q}\widetilde{p}.\end{aligned}$$

In accordance with the preceding the coefficient of t in this expression is the first approximation of the normal form for the original Hamiltonian; we have

$$\widetilde{H}_1 = 2\widetilde{J} + \tfrac{1}{2}\big(\widetilde{q}^2 + \widetilde{p}^2\big)\big(1 + \tfrac{1}{2}\varepsilon a\big) + \tfrac{1}{8}\varepsilon\big(\widetilde{q}^2 - \widetilde{p}^2\big)\cos\widetilde{\varphi} - \tfrac{1}{4}\varepsilon\widetilde{q}\widetilde{p}\sin\widetilde{\varphi}.$$

The time-independent coefficient of ε is the generating Hamiltonian of the canonical change of variables which transforms the original Hamiltonian into the normal form; we have

$$Q_0 = -\tfrac{1}{32}\big(5\widetilde{q}^2 + 3\widetilde{p}^2\big)\sin\widetilde{\varphi} + \tfrac{1}{16}\widetilde{q}\widetilde{p}\cos\widetilde{\varphi} + \tfrac{1}{4}a\widetilde{q}\widetilde{p}.$$

The change of variables itself, $(\varphi, J, q, p) \to (\widetilde{\varphi}, \widetilde{J}, \widetilde{q}, \widetilde{p})$, which is defined by the formulas

$$\begin{aligned}\varphi &= \widetilde{\varphi} + \varepsilon\{\widetilde{\varphi}, Q\} + \frac{\varepsilon^2}{2!}\big\{\{\widetilde{\varphi}, Q\}, Q\big\} + \cdots, \\ J &= \widetilde{J} + \varepsilon\{\widetilde{J}, Q\} + \frac{\varepsilon^2}{2!}\big\{\{\widetilde{J}, Q\}, Q\big\} + \cdots, \\ q &= \widetilde{q} + \varepsilon\{\widetilde{q}, Q\} + \frac{\varepsilon^2}{2!}\big\{\{\widetilde{q}, Q\}, Q\big\} + \cdots, \\ p &= \widetilde{p} + \varepsilon\{\widetilde{p}, Q\} + \frac{\varepsilon^2}{2!}\big\{\{\widetilde{p}, Q\}, Q\big\} + \cdots\end{aligned}$$

is obtained, in the first approximation, in the form

$$\begin{aligned}\varphi &= \widetilde{\varphi}, \\ J &= \widetilde{J} - \frac{\varepsilon}{32}\big(5\widetilde{q}^2 + 3\widetilde{p}^2\big)\cos\widetilde{\varphi} - \frac{3\varepsilon}{16}\widetilde{q}\widetilde{p}\sin\widetilde{\varphi}, \\ q &= \widetilde{q} + \frac{3\varepsilon}{16}\widetilde{p}\sin\widetilde{\varphi} - \frac{\varepsilon}{16}\widetilde{q}\cos\widetilde{\varphi} - \frac{\varepsilon a}{4}\widetilde{q}, \\ p &= \widetilde{q} + \frac{\varepsilon a}{4}\widetilde{p} - \frac{5\varepsilon}{16}\widetilde{q}\sin\widetilde{\varphi} + \frac{\varepsilon}{16}\widetilde{p}\cos\widetilde{\varphi}.\end{aligned}$$

By substituting this change of variables into the original Hamiltonian, one can verify that the resulting Hamiltonian coincides with the normal form $\widetilde{H}_1$ obtained.

Return to the nonautonomous representation of the problem. To this end, one should integrate the equation

$$\dot{\widetilde{\varphi}} = \frac{\partial \widetilde{H}_1}{\partial \widetilde{J}}$$

subject to the initial condition $\widetilde{\varphi}(0) = 0$. This yields $\widetilde{\varphi} = 2t$. We substitute this solution into $\widetilde{H}_1$, Q_0, q, and p and omit $\widetilde{J}$ to obtain

$$\begin{aligned}
\widetilde{H}_1 &= \frac{1}{2}(\widetilde{q}^2 + \widetilde{p}^2)\left(1 + \frac{\varepsilon a}{2}\right) + \frac{\varepsilon}{8}(\widetilde{q}^2 - \widetilde{p}^2)\cos 2t - \frac{\varepsilon}{4}\widetilde{q}\widetilde{p}\sin 2t, \\
Q_0 &= -\frac{1}{32}(5\widetilde{q}^2 + 3\widetilde{p}^2)\sin 2t + \frac{1}{16}\widetilde{q}\widetilde{p}\cos 2t + \frac{a}{4}\widetilde{q}\widetilde{p}, \\
q &= \widetilde{q} - \frac{\varepsilon a}{4}\widetilde{q} + \frac{3\varepsilon}{16}\widetilde{p}\sin 2t - \frac{\varepsilon}{16}\widetilde{q}\cos 2t, \\
p &= \widetilde{q} + \frac{\varepsilon a}{4}\widetilde{p} - \frac{5\varepsilon}{16}\widetilde{q}\sin 2t + \frac{\varepsilon}{16}\widetilde{p}\cos 2t.
\end{aligned}$$

To summarize, Mathieu's equation rewritten as the system

$$\dot{q} = p, \quad \dot{p} = -[1 + \varepsilon(a + \cos 2t)]q$$

is reduced by the canonical transformation $(q, p) \to (\widetilde{q}, \widetilde{p})$ specified above to the form

$$\begin{aligned}
\dot{\widetilde{q}} &= \frac{\partial \widetilde{H}_1}{\partial \widetilde{p}} = \widetilde{p}\left(1 + \frac{\varepsilon a}{4}\right) - \frac{\varepsilon}{4}\widetilde{p}\cos 2t - \frac{\varepsilon}{4}\widetilde{q}\sin 2t, \\
\dot{\widetilde{p}} &= -\frac{\partial \widetilde{H}_1}{\partial \widetilde{q}} = -\widetilde{q}\left(1 + \frac{\varepsilon a}{4}\right) - \frac{\varepsilon}{4}\widetilde{q}\cos 2t + \frac{\varepsilon}{4}\widetilde{p}\sin 2t.
\end{aligned}$$

This is just the first approximations of the normal form of Mathieu's equation. It should be emphasized that the normal form is nonautonomous, just as the original system. However, its perturbed and unperturbed components commute and, in accordance with the theorem proved previously, it can be reduced to a system with constant coefficients.

As follows from the mentioned theorem, the solution of the normal form is the composition of the solutions of two systems.

The first system ($\varepsilon = 0$),

$$\dot{\widetilde{q}} = \widetilde{p}, \quad \dot{\widetilde{p}} = -\widetilde{q},$$

has the general solution $\widetilde{q} = q_0 \cos t + p_0 \sin t$, $\widetilde{p} = -q_0 \sin t + p_0 \cos t$.

The second system is determined by the perturbation with frozen time ($t = 0$) and has the form

$$\dot{\widetilde{q}} = \frac{\varepsilon}{4}(2a - 1)\widetilde{p}, \quad \dot{\widetilde{p}} = -\frac{\varepsilon}{4}(2a + 1)\widetilde{q}.$$

It has three sorts of the general solution depending on the value of $4a^2 - 1$. If $4a^2 - 1 > 0$, then all solutions are harmonic with frequency $\sqrt{4a^2 - 1}$ (stability). If $4a^2 - 1 < 0$, then all solutions are exponential (instability). If $4a^2 - 1 = 0$ (the boundary of the stability region), then the general solution is expressed as

$$\begin{aligned}
&\widetilde{q} = C_1, \quad \widetilde{p} = C_2 - \frac{\varepsilon}{2}C_1 t \qquad (2a - 1 = 0), \\
&\widetilde{q} = C_1 - \frac{\varepsilon}{2}C_2 t, \quad \widetilde{q} = C_2 \qquad (2a + 1 = 0).
\end{aligned}$$

For the sake of illustration, we write out the general solution of the normal form for the case $2a = 1$. The mentioned composition of solutions yields

$$\widetilde{q} = C_1 \cos t + \left(C_2 - \frac{\varepsilon}{2} C_1 t\right) \sin t, \quad \widetilde{p} = -C_1 \sin t + \left(C_2 - \frac{\varepsilon}{2} C_1 t\right) \cos t.$$

This is just the general solution of the nonautonomous normal system, which can be verified by straightforward substitution.

Finally, the substitution of this solution into the normalizing canonical change of variables permits one to obtain the solution of the original Mathieu equation.

5.6.2. Averaging of Hamiltonian Systems in Terms of Lie Series

In Section 5.4, it was shown that the normalization procedure for nonlinear systems of general form differs from the procedure of asymptotic averaging by the form of the linear part of the system. If the linear part is first reduced to the Jordan form, then one deals with Poincaré's methods. If the linear part is first reduced to the standard from, then one deals with the Krylov–Bogolyubov method.

If a nonlinear system is Hamiltonian, then its Poincaré normal form is referred to as the Birkhoff normal form. The two ways of normalization which are available for systems of general form are also present in this case. If the Krylov–Bogolyubov method is applied to a Hamiltonian system, then this is equivalent to the construction of the Birkhoff normal form for the system's Hamiltonian, provided that prior to this the linear part of the system has been reduced to the standard form.

In accordance with the definition of the notion of the standard form for nonlinear systems of general form with a small parameter (see Subsection 5.5.1), a Hamiltonian system is in the standard form if the small parameter is a common multiplier of the Hamiltonian, $H = \varepsilon H_*(t, q, p, \varepsilon)$. Then the Hamilton equations

$$\dot{q} = \varepsilon \frac{\partial H_*}{\partial p}, \quad \dot{p} = -\varepsilon \frac{\partial H_*}{\partial q}$$

describe the evolution of the slow variables q and p.

The asymptotic averaging of these equations can be carried out following the scheme presented in Subsection 5.5.1, while disregarding the fact that the system is Hamiltonian. But it is more favorable to perform the asymptotic averaging on the level of the Hamiltonian rather than on the level of the differential equations. In the former case, one has to deal with a scalar object, the Hamiltonian, instead of vectors and manipulate Lie series for Hamiltonians instead of much more complicated Hausdorff series for operators.

The procedure of asymptotic averaging for nonautonomous systems was outlined in the previous subsection and illustrated by Mathieu's equation.

In the current subsection, we consider the Hamiltonian of a system in the standard form. Such a Hamiltonian is a special case of that discussed in the previous subsection.

The above algorithm is applied here to a system in the standard form with no changes.

To this end, the Hamiltonian $\varepsilon H_*(t, q, p)$ must first be rewritten in the autonomous form by treating time as a new generalized coordinate, $t \equiv \varphi$, and introducing its conjugate, the momentum J. The resulting autonomous Hamiltonian of the system in the standard form is expressed as

$$H(\varphi, J, q, p, \varepsilon) = J + \varepsilon H_*(\varphi, q, p, \varepsilon).$$

The asymptotic averaging of this Hamiltonian aims at searching for a change of variables $(\varphi, J, q, p) \to (\widetilde{\varphi}, \widetilde{J}, \widetilde{q}, \widetilde{p})$ that would reduce the Hamiltonian to the form

$$\widetilde{H}(\widetilde{\varphi}, \widetilde{J}, \widetilde{q}, \widetilde{p}, \varepsilon) = \widetilde{J} + \varepsilon \widetilde{H}_*(\widetilde{q}, \widetilde{p}, \varepsilon),$$

in which $\widetilde{H}_*$ is independent of time,

$$\frac{\partial \widetilde{H}_*}{\partial \varphi} \equiv 0 \quad \text{or, equivalently,} \quad \{J, \widetilde{H}_*\} = 0.$$

By definition, the autonomy condition represented in terms of a Poisson bracket is just the condition of the Birkhoff normal form.

As before, the desired canonical change of variables is induced by a generating Hamiltonian $Q(\varphi, q, p)$, so that the solution of the Cauchy problem for the Hamiltonian system,

$$\frac{d\varphi}{d\tau} = \frac{\partial Q}{\partial J} = 0, \quad \frac{dJ}{d\tau} = -\frac{\partial Q}{\partial \varphi}, \quad \frac{dq}{d\tau} = \frac{\partial Q}{\partial p}, \quad \frac{dp}{d\tau} = -\frac{\partial Q}{\partial q},$$
$$\varphi(0) = \widetilde{\varphi}, \quad J(0) = \widetilde{J}, \quad q(0) = \widetilde{q}, \quad p(0) = \widetilde{p},$$

is what determines a one-parameter family of canonical changes of variables (τ is the parameter, which is subsequently identified with ε).

There is a slight difference from the previous subsection. It manifests itself at the stage where the homological system is written out, after the function L_k has been calculated. Since now $H_0 = J$, the homological system acquires the simpler form

$$\widetilde{H}_k - \tau \frac{dQ_{k-1}}{d\varphi} = L_k.$$

To solve the problem in question, it suffices to set

$$\widetilde{H}_k = \langle L_k \rangle, \quad Q_{k-1} = -\frac{1}{\tau} \int \widetilde{L}_k \, d\varphi,$$

where, as before, $\langle L \rangle$ stands for the average of L over φ, and $\widetilde{L}_k$ is the complement of the average: $L_k = \langle L_k \rangle + \widetilde{L}_k$.

As before, the first approximation ($k = 1$) is given by

$$\widetilde{H}_1 = \langle H_1 \rangle, \quad Q_0 = -\frac{1}{\tau} \int (H_1 - \langle H_1 \rangle) \, d\varphi.$$

5.6.3. Artificial Hamiltonization

Due to the asymptotic averaging on the level of the Hamiltonian function, a simplification of the mathematical treatment for Hamiltonian systems is attained. This simplification is so significant that it is sometimes quite profitable first to reduce a non-Hamiltonian system to a Hamiltonian form.

Consider an essentially nonlinear multifrequency system (see Subsection 5.5.2) defined by

$$\frac{d\varphi}{dt} = \omega(x) + \varepsilon \Phi(\varphi, x, \varepsilon),$$
$$\frac{dx}{dt} = \varepsilon X(\varphi, x, \varepsilon), \qquad \varphi = (\varphi_1, \dots, \varphi_m), \quad x = (x_, \dots, x_n).$$

To hamiltonize the system, we introduce the variables ψ and y which are the conjugates of φ and x, respectively, so that the Hamiltonian function is obtained in the form

$$H(\varphi, x, \psi, y) = \omega(x)\psi + \varepsilon[\Phi(\varphi, x, \varepsilon)\psi + X(\varphi, x, \varepsilon)y].$$

By virtue of this function H, the first half of the Hamilton equations coincides with the system under study. The second half of the equations will not be needed in what follows, since ψ and y play a purely auxiliary role (this will be revealed later). These variables do not occur in the original equations, and hence will not occur in the transformed equations either. Thus, the doubling of the dimension of the problem is fictitious; actually, the problem remains $(m+n)$-dimensional.

The objective of the canonical transformations $(\varphi, x, \psi, y) \to (\widetilde{\varphi}, \widetilde{x}, \widetilde{\psi}, \widetilde{y})$ sought below is to reduce the Hamiltonian H to a form in which it (i) would not depend on the fast variables $\widetilde{\varphi}$ and (ii) would be linear in the momenta $\widetilde{\psi}$ and $\widetilde{y}$, i.e.,

$$\widetilde{H}(\widetilde{\varphi}, \widetilde{x}, \widetilde{\psi}, \widetilde{y}) = \omega(\widetilde{x})\widetilde{\psi} + \varepsilon\big[\widetilde{\Phi}(\widetilde{x}, \varepsilon)\widetilde{\psi} + \widetilde{X}(\widetilde{x}, \varepsilon)\big].$$

The independence of the transformed equations on the fast phases is the usual objective of asymptotic averaging. And the linearity of the Hamiltonian in $\widetilde{\psi}$ and $\widetilde{y}$ is required to separate the Hamilton equations for $\widetilde{\varphi}$ and $\widetilde{x}$ from those for $\widetilde{\psi}$ and $\widetilde{y}$.

As usual, the canonical change of variables is sought as the general solution of the following Cauchy problem for a Hamiltonian system with unknown generating Hamiltonian $Q(\varphi, x, \psi, y, \varepsilon)$:

$$\frac{d\varphi}{d\tau} = \frac{\partial Q}{\partial \psi}, \quad \frac{dx}{d\tau} = \frac{\partial Q}{\partial y}, \quad \frac{d\psi}{d\tau} = -\frac{\partial Q}{\partial \varphi}, \quad \frac{dy}{d\tau} = -\frac{\partial Q}{\partial x},$$
$$\varphi(0) = \widetilde{\varphi}, \quad x(0) = \widetilde{x}, \quad \psi(0) = \widetilde{\psi}, \quad y(0) = \widetilde{y}.$$

The further procedure has been already carried out several times. The three Hamiltonians are related by the Lie series

$$\widetilde{H} = H + \tau\{H, Q\} + \frac{\tau^2}{2!}\{\{H, Q\}, Q\} + \cdots.$$

After identifying τ with ε and introducing asymptotics of the Hamiltonians, this series splits into the recurrence chain

$$\begin{aligned}
\widetilde{H}_0 &= H_0,\\
\widetilde{H}_1 &= H_1 + \varepsilon\{H_0, Q_0\},\\
&\dots\dots\dots\dots\\
\widetilde{H}_k &= L_k + \varepsilon\{H_0, Q_{k-1}\},\\
&\dots\dots\dots\dots
\end{aligned}$$

where, as usual,

$$L_k = H_1 \qquad (k = 1),$$
$$L_k = H_k + \varepsilon\{H_{k-1} - H_0, Q_{k-2}\} + \sum_{j=2}^{k} \frac{\varepsilon^j}{j!} \underbrace{\{\dots\{\{}_{j \text{ times}} H_{k-j}, Q_{k-j}\}, Q_{k-j}\}, \dots\} \qquad (k > 1),$$

Proceed to the construction of the first approximation ($k = 1$). We choose $\widetilde{H}_1$ as the time average of H_1,

$$\widetilde{H}_1 = \omega(\widetilde{x})\widetilde{\psi} + \varepsilon \lim_{h\to\infty} \frac{1}{h} \int_0^h \big[\Phi(\omega\theta, \widetilde{x}, \varepsilon)\widetilde{\psi} + X(\omega\theta, \widetilde{x}, \varepsilon)\widetilde{y}\big]\, d\theta.$$

With this average, the Hamiltonian H_1 can be represented in the form

$$H_1 = \widetilde{H}_1 + \varepsilon\big[f^1(\widetilde{\varphi}, \widetilde{x}, \varepsilon)\widetilde{\psi} + g^1(\widetilde{\varphi}, \widetilde{x}, \varepsilon)\widetilde{y}\big].$$

The additional terms must be compensated for by the term $\varepsilon\{H_0, Q_0\}$, which leads to the following equation for Q_0:

$$\{H_0, Q_0\} + f^1\widetilde{\psi} + g^1\widetilde{y} = 0.$$

Since

$$H_0 = \omega(\widetilde{x})\psi,$$

this equation can be rewritten as

$$-\sum_{i=1}^{n}\sum_{j=1}^{m}\widetilde{\psi}_i\frac{\partial\omega_i}{\partial\widetilde{x}_j}\frac{\partial Q_0}{\partial\widetilde{y}_j} + \sum_{i=1}^{n}\left(\omega_i\frac{\partial Q_0}{\partial\widetilde{\varphi}_i} + f^1\widetilde{\psi}_i\right) + \sum_{j=1}^{m} g_j^1\widetilde{y}_j = 0.$$

If the system is quasilinear, then $\partial\omega_i/\partial x_j \equiv 0$, and hence the first term in this equation, which is represented by the double sum, vanishes.

In the general case, a solution of the above equation is given by the following quadrature:

$$\begin{aligned} Q_0 = &-\sum_{i=1}^{n}\widetilde{\psi}_i\int f_i^1(C_1+\omega_1\theta,\ \ldots,\ C_m+\omega_m\theta,\ \widetilde{x},\ \varepsilon)\,d\theta \\ &-\sum_{j=1}^{m} y_j\int g_j^1(C_1+\omega_1\theta,\ \ldots,\ C_m+\omega_m\theta,\ \widetilde{x},\ \varepsilon)\,d\theta \\ &-\sum_{i=1}^{n}\sum_{j=1}^{m}\widetilde{\psi}_i\frac{\partial\omega_i}{\partial\widetilde{x}_j}\iint g_j^1(C_1+\omega_1\theta,\ \ldots,\ C_m+\omega_m\theta,\ \widetilde{x},\ \varepsilon)\,d\theta\,d\theta. \end{aligned}$$

On taking the integrals with respect to θ, one must replace the arbitrary constants $C_1, \ldots, C_m$ in this relation by

$$C_j = \widetilde{\varphi}_j - \omega_j\theta.$$

The above particular solution is linear in $\widetilde{\psi}$ and $\widetilde{y}$. This property is very important for the subsequent analysis. The linearity of Q_0 in $\widetilde{\psi}$ and $\widetilde{y}$ implies the linearity of

$$L_2 = \varepsilon\{H_1 - H_0, Q_0\} + \tfrac{1}{2}\varepsilon^2\{\{H_0, Q_0\}, Q_0\}$$

in $\widetilde{\psi}$ and $\widetilde{y}$ as well, which permits one to repeat this reasoning for the second and higher approximations.

Thus, for the kth approximation ($k = 1, 2, \ldots$), we have

$$\widetilde{H}_k = \langle L_k\rangle$$

and $L_k = \langle L_k\rangle + \varepsilon(f^k\widetilde{\psi} + g^k\widetilde{y})$. Hence, we obtain the following equation for Q_{k-1}:

$$-\widetilde{\psi}\frac{d\omega}{d\widetilde{x}}\frac{\partial Q_{k-1}}{\partial\widetilde{y}} + \omega\frac{\partial Q_{k-1}}{\partial\widetilde{\varphi}} + f^k\widetilde{\psi} + g^k\widetilde{y} = 0.$$

As in the first approximation, a solution of this equation is given by the quadrature

$$\begin{aligned} Q_{k-1} = &-\int\left[f^k(C+\omega\theta,\widetilde{x},\varepsilon)\widetilde{\psi} + g^k(C+\omega\theta,\widetilde{x},\varepsilon)\widetilde{y}\right]d\theta \\ &-\widetilde{\psi}\frac{d\omega}{d\widetilde{x}}\iint g^k(C+\omega\theta,\widetilde{x},\varepsilon)\,d\theta\,d\theta \qquad (C = \widetilde{\varphi} - \omega\theta). \end{aligned}$$

In accordance with the new Hamiltonian H_k, the transformed essentially nonlinear multifrequency system is obtained in the form

$$\frac{d\widetilde{\varphi}}{dt} = \omega(\widetilde{x}) + \varepsilon\widetilde{\Phi}_k(\widetilde{x}, \varepsilon),$$
$$\frac{d\widetilde{x}}{dt} = \varepsilon\widetilde{X}_k(\widetilde{x}, \varepsilon).$$

Thus, the problem of separating the motions is solved.

In the methods just constructed it was implied that a nonresonant case was studied, in which case the time average used for determining $\widetilde{H}_k$ coincided with the space average (see Subsection 5.5.2).

If the system is quasilinear and has a resonance, then, prior to carrying out the asymptotic procedure, one must eliminate the resonance by increasing the dimension of the vector of slow variables, just as in Subsection 5.5.2. Recall that the resonance relation $\Lambda\omega = 0$ was associated with the notion of the frequency detuning, $\Delta = \Lambda\omega$, which was assumed to be small, $|\Delta| \sim \varepsilon$, in the neighborhood of the resonance. If the system is essentially nonlinear, the situation gets more complicated. The vector ω is a function of slow variables, $\omega = \omega(x)$, and the frequency detuning also becomes a function. The notion of smallness of this function requires further formalization.

For example, such a formalization is possible if the evolution of the vector of slow variables, x, is studied in a small neighborhood of the resonance surface $\Lambda\omega(x) = 0$.

The formalization is performed as follows. One introduces a new slow variable $s = \Lambda\omega(x)$ instead of some old variable x_i. After that, the variable s is scaled so that $s = \sqrt{\varepsilon}\, r$. Then, with the vector of slow variables extended and the resonance eliminated, the system remains in the standard form. The introduction of a fractional power of ε in essentially nonlinear systems is necessary, since the ordinary scaling of the form $s = \varepsilon r$ gives no result. After the resonance has been eliminated, one proceeds with the asymptotic averaging procedure.

5.7. Method of Tangent Approximations

The method presented below is convenient for the construction of periodic solutions of nonlinear systems, although other applications of this methods are possible.

Consider a scalar equation of order n of the form

$$\frac{d^n x}{dt^n} = f\left(t, x, x', \ldots, x^{(n-1)}\right).$$

The function $f\left(t, x, x', \ldots, x^{(n-1)}\right)$ is analytic in some domain of its arguments and periodic in time. The case of autonomous systems will be considered separately.

The method of tangent approximations combines the ideas of Ritz's and Taylor's methods.

A periodic solution of the above differential equation is sought in the form

$$x(t) = a_0\varphi_0(t) + a_1\varphi_1(t) + \cdots + a_k\varphi_k(t) + \cdots,$$

where the $\varphi_k(t)$ are basis, or coordinate, functions which are chosen so as to conform with the particular problem. To construct periodic solutions, these function should be taken periodic, which is however unnecessary.

Impose the following condition of the above series. Its any finite segment must have at $t = 0$ all derivatives up to a fixed order and these derivatives must be equal to the derivatives of the exact solution at this point.

In other words, one constructs an approximate periodic function that has the same period as the exact solution and is tangent to the exact solution at $t = 0$ with a prescribed order of tangency.

A Taylor polynomial behaves in the same manner in respect of the function it approximates.

Since the basis functions are assumed to be prescribed, we also assume that their Taylor expansions are know at $t = 0$,

$$\varphi_k = \sum_s \mu_k^s t^s$$

(the functions φ_k are assumed to be analytic in the neighborhood of $t = 0$).

Substituting this expansion into the solution yields

$$x(t) = \sum_{k,s} a_k \mu_k^s t^s.$$

On the other hand, denoting

$$x' = p, \quad x'' = q, \quad \ldots,$$

we rewrite the original equation as a system of first order differential equations. Confining ourselves to a second order system, we have

$$\dot{t} = 1, \quad \dot{x} = p, \quad \dot{p} = f(t, x, p).$$

This system determines a group with infinitesimal generator

$$A = \frac{\partial}{\partial t_0} + p_0 \frac{\partial}{\partial x_0} + f(t_0, x_0, p_0) \frac{\partial}{\partial p_0}.$$

This group transfers the initial values (t_0, x_0, p_0) to the current values (t, x, p). Take advantage of the representation of this group by the Lie series

$$x(t) = x_0 + tAx_0 + \frac{1}{2!} t^2 A^2 x_0 + \cdots.$$

The combination $A^r x_0$ is equal to the rth derivative of the exact solution at $t = 0$. The same derivative at $t = 0$ is expressed as a series expansion in the basis function,

$$\frac{1}{r!} \sum_k a_k \mu_k^r.$$

Setting $t = 0$ and equating the r derivatives of the exact solution with those of its representation, we arrive at the equations

$$\begin{aligned} x_0 &= a_0 \mu_0^0 + a_1 \mu_1^0 + a_2 \mu_2^0 + \cdots, \\ Ax_0 &= a_0 \mu_0^1 + a_1 \mu_1^1 + a_2 \mu_2^1 + \cdots, \\ &\cdots\cdots\cdots\cdots\cdots\cdots\cdots\cdots \\ \frac{1}{r!} A^r x_0 &= a_0 \mu_0^r + a_1 \mu_1^r + a_2 \mu_2^r + \cdots. \end{aligned}$$

Thus, we have obtained a linear system of equations for the unknown coefficients $a_0, a_1, \ldots$ The question arises: How many coefficients can be determined from this system? Note that, apart from a_k's, the initial conditions determining the periodic solution are also unknown. For a second-order system, there are two such initial conditions, x_0 and $p_0 = Ax_0$. Retaining $r - 2$ undetermined coefficients in the system, we obtain

$$\begin{aligned} x_0 &= a_0 \mu_0^0 + \cdots + a_{r-2} \mu_{r-2}^0, \\ &\cdots\cdots\cdots\cdots\cdots\cdots\cdots\cdots \\ \frac{1}{r!} A^r x_0 &= a_0 \mu_0^r + \cdots + a_{r-2} \mu_{r-2}^r. \end{aligned}$$

This is an overdetermined system for a_k's. By the well-known theorem of algebra, for the equations to be consistent it is necessary and sufficient that the rank of the system matrix be equal to the rank of the extended matrix. If the rank is equal to the number of unknowns, then the solution is unique. In our case, without loss of generality, we can consider the rank of the system matrix to be equal to $r-2$ (since the coefficients μ_k^s are chosen quite arbitrarily). The extended matrix of the system has the form

$$\begin{pmatrix} \mu_0^0 & \cdots & \mu_{r-2}^0 & x_0 \\ \vdots & \ddots & \vdots & \vdots \\ \mu_0^r & \cdots & \mu_{r-2}^r & \frac{1}{r!}A^r x_0 \end{pmatrix}.$$

This is an $r \times (r-1)$ rectangular matrix.

To compile the rank correspondence conditions, we take advantage of the theorem on the rank of a matrix: if a matrix has a nonzero minor of order $r-2$ and all minors that contain it are zero, then the rank of the matrix is equal to $r-2$. The matrix has the nonzero minor

$$\begin{vmatrix} \mu_0^0 & \cdots & \mu_{r-2}^0 \\ \vdots & \ddots & \vdots \\ \mu_0^{r-2} & \cdots & \mu_{r-2}^{r-2} \end{vmatrix}.$$

Hence, for the system to be consistent it is necessary and sufficient that the following two minors containing this nonzero minor vanish:

$$\begin{vmatrix} \mu_0^0 & \cdots & \mu_{r-2}^0 & x_0 \\ \vdots & \ddots & \vdots & \vdots \\ \mu_0^{r-2} & \cdots & \mu_{r-2}^{r-2} & \frac{1}{(r-2)!}A^{r-2}x_0 \\ \mu_0^{r-1} & \cdots & \mu_{r-2}^{r-1} & \frac{1}{(r-1)!}A^{r-1}x_0 \end{vmatrix} = 0, \qquad \begin{vmatrix} \mu_0^0 & \cdots & \mu_{r-2}^0 & x_0 \\ \vdots & \ddots & \vdots & \vdots \\ \mu_0^{r-2} & \cdots & \mu_{r-2}^{r-2} & \frac{1}{(r-2)!}A^{r-2}x_0 \\ \mu_0^{r} & \cdots & \mu_{r-2}^{r} & \frac{1}{r!}A^{r}x_0 \end{vmatrix} = 0.$$

These two determinant equations serve to find the two unknowns of the initial conditions, x_0 and p_0. On substituting the expressions of x_0 and p_0 obtained form these equations into the first $r-2$ equations of the system, we arrive at a solvable nonhomogeneous system determining the unique solution for $a_0, \dots, a_{r-2}$.

If the original equation is of order n, then the tangent condition of order r ($r > n$) can be used to find $r-n$ unknown coefficients a_k. The unknown initial conditions can be determined from the zero conditions for n bordering minors.

Consider the autonomous system

$$\dot{x} = p, \quad \dot{p} = f(x,p), \qquad A = p_0\frac{\partial}{\partial x_0} + f(x_0,p_0)\frac{\partial}{\partial p_0}.$$

The above procedure is applicable completely in this case, too. However, there is a difference: the coefficients of the basis functions will depend on a period T unknown a priori.

By virtue of autonomy, one of the initial conditions (e.g., x_0) is considered arbitrary. The above two determinant equations are used to find the period T and the second initial condition in terms of x_0. The other things remain the same as in the nonautonomous cases.

Remark. In solving specific problems, it may happen that the convenient system of basis functions is such that the condition that all principal minors of the system matrix be nonzero is not satisfied. In such cases, the number of derivatives equated at $t = 0$ will be greater than the number of the unknown variables plus the order of the system.

Example. Consider Duffing's equation $\ddot{x} = -x - \varepsilon x^3$. Since the oscillations described by this equation are symmetric, the functions

$$\varphi_0 = \cos\omega t, \quad \varphi_1 = \cos 3\omega t, \quad \varphi_2 = \cos 5\omega t, \quad \ldots$$

can be taken to be the basis functions. In this case, the coefficients μ_k^s are given by

$$\mu_0 = 1, \quad \mu_0^1 = 0, \quad \mu_0^2 = -\tfrac{1}{2}\omega^2, \quad \mu_0^3 = 0, \quad \mu_0^4 = \tfrac{1}{4!}\omega^4, \quad \ldots$$
$$\mu_1 = 1, \quad \mu_1^1 = 0, \quad \mu_1^2 = -\tfrac{9}{2}\omega^2, \quad \mu_1^3 = 0, \quad \mu_1^4 = \tfrac{81}{4!}\omega^4, \quad \ldots$$

Confining ourselves to finding the first two coefficients, we arrive at the following system for them,

$$\begin{aligned} x_0 &= a_0 + a_1, \\ Ax_0 &= 0, \\ A^2x_0 &= -a_0\omega^2 - 9\omega^2 a_1, \\ A^3x_0 &= 0, \\ A^4x_0 &= a_0\omega_4 + 81\omega^4 a_1. \end{aligned}$$

The fact that Ax_0 and A^3x_0 vanish follows from mentioned structure of the solution. Since the second and fourth equations of this system are satisfied identically.

The main system that solves the problem is

$$\begin{aligned} a_0 + a_1 &= x_0, \\ \omega^2 a_0 + 9\omega^2 a_1 &= -A^2x_0, \\ \omega^4 a_0 + 81\omega^4 a_1 &= A^4x_0. \end{aligned}$$

The operator A is given by

$$A = p_0\frac{\partial}{\partial x_0} - (x_0 + \varepsilon x_0^3)\frac{\partial}{\partial p_0}.$$

Calculate the right-hand sides to obtain

$$\begin{aligned} Ax_0 &= p_0, \\ A^2x_0 &= -(x_0 + \varepsilon x_0^3), \\ A^3x_0 &= -(1 + 3\varepsilon x_0^2)p_0, \\ A^4x_0 &= -6\varepsilon x_0 p_0^2 + (1 + 3\varepsilon x_0^2)(x_0 + \varepsilon x_0^3). \end{aligned}$$

Since the solution is symmetric, we have $p_0 = 0$. Thus, we arrive at the following frequency equation for ω:

$$\begin{vmatrix} 1 & 1 & 1 \\ \omega^2 & 9\omega^2 & 1 + \varepsilon x_0^2 \\ \omega^4 & 81\omega^4 & (1 + 3\varepsilon x_0^2)(1 + \varepsilon x_0^2) \end{vmatrix} = 0.$$

Assuming ε to be a small parameter and denoting $\omega = 1 + \varepsilon\omega_1$, we find from the equation that $\omega_1 = \frac{3}{8}x_0^2$. To determine a_0 and a_1, it suffices to solve the system

$$\begin{aligned} a_0 + a_1 &= x_0, \\ a_0 + 9a_1 &= \frac{1}{\omega^2}x_0(1 + \varepsilon x_0^2), \end{aligned}$$

whose solution is given by

$$a_0 = \frac{(9\omega^2 - 1 - \varepsilon x_0^2)x_0}{8\omega^2}, \quad a_1 = -\frac{(\omega^2 - 1 - \varepsilon x_0^2)x_0}{8\omega^2}.$$

Justification of the method. If an exact solution is representable by finitely many basis functions $\varphi_0(t), \dots, \varphi_r(t)$, then the procedure with order of tangency $r+n$, where n is the order of the equation, allows constructing the periodic solution exactly. It follows that, in the case of quasilinear systems, the method in question allows constructing exact expansions of periodic solutions in powers of ε. This follows from the fact that, for quasilinear systems, the higher the number of the harmonic (with sin and cos being the basis functions), the higher the order of smallness of its amplitude in ε. For systems without a small parameter, the problem of justification is much more complicated.

5.8. Classical Examples of Oscillation Theory

5.8.1. Van Der Pol's Equation

Van der Pol's equation

$$\ddot{x} + 2\varepsilon(1 - 4x^2)\dot{x} + x = 0$$

is a simple mathematical example of a self-oscillatory system. This equation approximately describes oscillations in a vacuum-tube generator.

Rewrite this equation as a first order system in the Cauchy normal form:

$$\dot{x} = y, \quad \dot{y} = -x - 2\varepsilon(1 - 4x^2)y.$$

Reduce the linear part of this nonlinear system to a diagonal form by means of the change of variables

$$(x, y) \to (z, \bar{z}): \quad z = x + iy, \quad \bar{z} = x - iy.$$

The inverse change of variables is expressed as

$$x = \frac{1}{2}(z + \bar{z}), \quad y = \frac{1}{2i}(z - \bar{z}).$$

In terms of the new variables, the system becomes

$$\dot{z} = -iz - \varepsilon[1 - (z + \bar{z})^2](z - \bar{z}),$$
$$\dot{\bar{z}} = i\bar{z} - \varepsilon[1 - (z + \bar{z})^2](\bar{z} - z).$$

Construct the first approximation of the normal form (see Section 5.3). In this system, $\lambda_1 = -i$ and $\lambda_2 = i$. Consider the perturbed part of the first equation,

$$[1 - (z + \bar{z})^2](z - \bar{z}) = z - \bar{z} - z^3 + z\bar{z}^2 - z^2\bar{z} + \bar{z}^3.$$

There are two terms, z and $z^2\bar{z}$, satisfying a resonance relation, $\lambda_1 m_1 + \lambda_2 m_2 = \lambda_1$.

Thus, the normal form of the first approximation is given by

$$\dot{z} = -iz - \varepsilon(z - z^2\bar{z}),$$
$$\dot{\bar{z}} = -i\bar{z} - \varepsilon(\bar{z} - \bar{z}^2 z).$$

The operators of the linear and perturbed parts,

$$A_0 = i\left(-z\frac{\partial}{\partial z} + \bar{z}\frac{\partial}{\partial \bar{z}}\right), \quad A_\varepsilon = -\varepsilon(1 - z\bar{z})\left(z\frac{\partial}{\partial z} - \bar{z}\frac{\partial}{\partial \bar{z}}\right),$$

commute,

$$[A_0, A_\varepsilon] = 0.$$

Therefore, the general solution of the normal form is the composition of two systems

$$\begin{array}{ll} \text{I} & \dot{z} = -iz, \quad \dot{\bar{z}} = i\bar{z}, \\ \text{II} & \dot{z} = -\varepsilon z(1 - z\bar{z}), \quad \dot{\bar{z}} = -\varepsilon\bar{z}(1 - z\bar{z}). \end{array}$$

The general solution of the first system is expressed as

$$z = z_0 e^{-it}, \quad \bar{z} = \bar{z}_0 e^{it}.$$

To construct the general solution of the second system, we multiply its first equation by $\bar{z}$ and add to the second equation multiplied by z. We obtain

$$\frac{d}{dt}(z\bar{z}) = -2\varepsilon z\bar{z}(1 - z\bar{z}).$$

The general solution of this equation is given by

$$z\bar{z} = \frac{z_0\bar{z}_0}{z_0\bar{z}_0 + (1 - z_0\bar{z}_0)e^{2\varepsilon t}}.$$

On substituting this solution into the first (or second) equation of system II and integrating the resulting separable equation, we find

$$z = \frac{z_0}{\sqrt{z_0\bar{z}_0 + (1 - z_0\bar{z}_0)e^{2\varepsilon t}}}.$$

For $\bar{z}$ the solution is the complex conjugate.

The general solution of the complete system, or the normal form, results from substituting the general solution of the first system for the arbitrary constants in the general solution of the second system, or vice versa. In either case, we obtain

$$z = \frac{z_0 e^{-it}}{\sqrt{z_0\bar{z}_0 + (1 - z_0\bar{z}_0)e^{2\varepsilon t}}}.$$

In terms of the original variables, the general solution is

$$x = \frac{x_0 \cos t + \dot{x}_0 \sin t}{\sqrt{x_0^2 + \dot{x}_0^2 + (1 - x_0^2 - \dot{x}_0^2)e^{2\varepsilon t}}}.$$

This solution shows that Van der Pol's equation has a single periodic solution, which satisfies the initial conditions $x_0^2 + \dot{x}_0^2 = 1$ and has unit amplitude. The other solutions tend to it asymptotically, provided that $\varepsilon < 0$.

5.8.2. Mathieu's Equation

This equation was studied as an example of constructing the Birkhoff normal form in Subsection 5.6.1.

Now we consider it to construct the second zone of instability. Rewrite the equation in the form

$$\ddot{x} + (1 + \Delta + \varepsilon \cos t)x = 0,$$

where Δ is a small quantity of the order of ε. Unlike the example considered previously, where we sought the first, main zone of instability, here the period of the nonautonomous term is twice as much.

In this example we will demonstrate how the procedure of asymptotic normalization, which is presented in Section 5.3 for autonomous systems, can be applied to nonautonomous systems.

Rewrite Mathieu's equation as the system

$$\begin{aligned} \dot{x}_1 &= x_2, \\ \dot{x}_2 &= -x_1 - \Delta x_1 - \varepsilon x_1 x_3, \\ \dot{x}_3 &= x_4, \\ \dot{x}_4 &= -x_3, \qquad x_3(0) = 1, \quad x_4(0) = 0. \end{aligned}$$

The system is represented in polynomial form, which is attained by means of doubling the dimension of the system. As in Subsection 5.6.3, this increase in the dimension is fictitious—the last two equations will not be subjected to transformations, they are independent of the first two equations, and the variables x_3 and x_4 do not occur in the final result.

Just as in the previous example (Van der Pol's equation), we reduce the linear part of the system to a diagonal form by passing to complex variables,

$$y_1 = x_2 + ix_1, \quad \bar{y}_1 = x_2 - ix_1, \quad y_2 = x_4 + ix_3, \quad \bar{y}_2 = x_4 - ix_3.$$

The inverse change of variables is expressed as

$$x_1 = \frac{1}{2i}(y_1 - \bar{y}_1), \quad x_2 = \frac{1}{2}(y_1 + \bar{y}_1), \quad x_3 = \frac{1}{2i}(y_2 - \bar{y}_2), \quad x_4 = \frac{1}{2}(y_2 + \bar{y}_2).$$

In terms of the new variables the system acquires the form

$$\begin{aligned} \dot{y}_1 &= iy_1 + \frac{i\Delta}{2}(y_1 - \bar{y}_1) + \frac{\varepsilon}{4}(y_1y_2 - y_1\bar{y}_2 - \bar{y}_1y_2 + \bar{y}_1\bar{y}_2), \quad \dot{y}_2 = iy_2, \\ \dot{\bar{y}}_1 &= -i\bar{y}_1 - \frac{i\Delta}{2}(\bar{y}_1 - y_1) + \frac{\varepsilon}{4}(\bar{y}_1\bar{y}_2 - \bar{y}_1y_2 - y_1\bar{y}_2 + y_1y_2), \quad \dot{\bar{y}}_2 = -i\bar{y}_2. \end{aligned}$$

The operators of the linear and nonlinear parts of the system are given by

$$\begin{aligned} A_0 &= iy_1\frac{\partial}{\partial y_1} + iy_2\frac{\partial}{\partial y_2} + \cdots, \\ A_1 &= \left[iy_1 + \frac{i\Delta}{2}(y_1 - \bar{y}_1) + \frac{\varepsilon}{4}(y_1y_2 - y_1\bar{y}_2 - \bar{y}_1y_2 + \bar{y}_1\bar{y}_2)\right]\frac{\partial}{\partial y_1} + iy_2\frac{\partial}{\partial y_2} + \cdots. \end{aligned}$$

Here and in what follows, the dots stand for the second half of the operators which correspond to the complex conjugate variables $\bar{y}_1$ and $\bar{y}_2$.

Let us construct the first approximation of the method. Following the procedure of Section 5.3, we perform a change of variables $(y_1, y_2) \to (z_1, z_2)$ that transforms A_1 into B_1,

$$B_1 = A_1 + \varepsilon[A_0, U_0].$$

The resonant part of the operator A_1 is

$$(A_1)_{\mathrm{R}} = \left(iz_1 + \frac{i\Delta}{2}z_1\right)\frac{\partial}{\partial z_1} + iz_2\frac{\partial}{\partial z_2} + \cdots = B_1.$$

The nonresonant part of A_1 is

$$(A_1)_N = \left[-i\frac{\Delta}{2}\bar{z}_1 + \frac{\varepsilon}{4}(z_1z_2 - z_1\bar{z}_2 - \bar{z}_1z_2 + \bar{z}_1\bar{z}_2)\right]\frac{\partial}{\partial z_1} + \cdots .$$

The operator U_0 is determined by the equation

$$[A_0, U_0] = -\frac{1}{\varepsilon}(A_1)_N.$$

To find U_0 we use the formula $h_m = g_m/[\lambda_l - (\lambda, m)]$ derived in Section 5.4, where h_m and g_m are the coefficients of the identical terms in the desired operator U_0 and $(A_1)_N$, respectively. Thus,

$$U_0 = \left(-\frac{\Delta}{4\varepsilon}\bar{z}_1 + \frac{i}{4}z_1z_2 + \frac{i}{4}z_1\bar{z}_2 + \frac{i}{4}\bar{z}_1z_2 - \frac{i}{12}\bar{z}_1\bar{z}_2\right)\frac{\partial}{\partial z_1} + \cdots .$$

Proceed to the second approximation:

$$B_2 = A_1 + \varepsilon[A_0, U_1] + \varepsilon[A_1 - A_0, U_0] + \frac{\varepsilon^2}{2}\left[[A_0, U_0], U_0\right].$$

Since $A_1 - A_0 = \frac{i}{2}\Delta z_1\frac{\partial}{\partial z_1} + (A_1)_N$ and $\varepsilon[A_0, U_0] = -(A_1)_N$, the expression for B_2 can be rewritten as

$$B_2 = A_1 + \varepsilon[A_0, U_1] + \frac{\varepsilon}{2}[(A_1)_N, U_0] + \varepsilon\left[i\frac{\Delta}{2}z_1\frac{\partial}{\partial z_1} + \cdots , U_0\right].$$

Since the further approximations do not interest us, we will not seek the operator U_1.

The operator of the normal form of the second approximation, B_2, can be evaluated as

$$B_2 = (A_1)_R + \frac{\varepsilon}{2}\left([(A_1)_N, U_0]\right)_R,$$

since the commutator

$$\left[i\frac{\Delta}{2}z_1\frac{\partial}{\partial z_1} + \cdots , U_0\right]$$

does not contain resonant terms in view of the fact that U_0 does not contain resonant terms.

Write out the expression of the commutator $[(A_1)_N, U_0]$:

$$\begin{aligned}
[(A_1)_N, U_0] = \Bigg\{ & \frac{i}{4}(z_2+\bar{z}_2)\left[\frac{\varepsilon}{4}z_1(z_2-\bar{z}_2) + \bar{z}_1\left(-i\frac{\Delta}{2} - \frac{\varepsilon}{4}z_2 + \frac{\varepsilon}{4}\bar{z}_2\right)\right] \\
& + \left(-\frac{\Delta}{4\varepsilon} + \frac{i}{4}z_2 - \frac{i}{12}\bar{z}_2\right)\left[\frac{\varepsilon}{4}\bar{z}_1(\bar{z}_2 - z_2) + \bar{z}_1\left(i\frac{\Delta}{2} - \frac{\varepsilon}{4}\bar{z}_2 + \frac{\varepsilon}{4}z_2\right)\right] \\
& - \frac{\varepsilon}{4}(z_2-\bar{z}_2)\left[\frac{i}{4}z_1(z_2+\bar{z}_2) + \bar{z}_1\left(-\frac{\Delta}{4\varepsilon} + \frac{i}{4}z_2 - \frac{i}{12}\bar{z}_2\right)\right] \\
& + \left(\frac{i\Delta}{2} + \frac{\varepsilon}{4}z_2 - \frac{\varepsilon}{4}\bar{z}_2\right)\left[-\frac{i}{4}z_1(z_2+\bar{z}_2) + z_1\left(-\frac{\Delta}{4\varepsilon} - \frac{i}{4}\bar{z}_2 + \frac{i}{12}z_2\right)\right]\Bigg\}\frac{\partial}{\partial z_1} + \overline{\{\cdots\}}\frac{\partial}{\partial \bar{z}_1}.
\end{aligned}$$

The resonant terms in this expression are z_1, $\bar{z}_1z_2^2$, and $z_1z_2\bar{z}_2$. It follows that

$$\left([(A_1)_N, U_0]\right)_R = \left(-i\frac{\Delta^2}{4\varepsilon}z_1 - i\frac{\varepsilon}{4}\bar{z}_1z_2^2 - i\frac{\varepsilon}{6}z_1z_2\bar{z}_2\right)\frac{\partial}{\partial z_1} + \overline{(\cdots)}\frac{\partial}{\partial \bar{z}_1}.$$

Hence, for B_2 we finally obtain

$$B_2 = \left[iz_1\left(1 + \frac{\Delta}{2} - \frac{\Delta^2}{8} - \frac{\varepsilon^2}{8} - \frac{\varepsilon^2}{12}z_2\bar{z}_2\right) - i\frac{\varepsilon^2}{8}\bar{z}_1 z_2^2\right]\frac{\partial}{\partial z_1} + iz_2\frac{\partial}{\partial z_2} + \overline{\left[\cdots\right]}\frac{\partial}{\partial \bar{z}_1} - i\bar{z}_2\frac{\partial}{\partial \bar{z}_2}.$$

Thus, up to the third order terms, the normal form of Mathieu's equation is

$$\dot{z}_1 = iz_1 + iz_1\left(\frac{\Delta}{2} - \frac{\Delta^2}{8} - \frac{\varepsilon^2}{12}z_2\bar{z}_2\right) - i\frac{\varepsilon^2}{8}\bar{z}_1 z_2^2,$$
$$\dot{z}_2 = iz_2.$$

Take advantage of the principle of superposition, which holds for systems in the normal form, and write out an equivalent system, from the viewpoint of stability (i.e., the terms corresponding to A_0, which is the generator of a symmetry group of the normal form, must be omitted). We have

$$\dot{z}_1 = iz_1\left(\frac{\Delta}{2} - \frac{\Delta^2}{8} - \frac{\varepsilon^2}{12}z_2\bar{z}_2\right) - i\frac{\varepsilon^2}{8}\bar{z}_1 z_2^2, \quad \dot{z}_2 = 0, \qquad z_2(0) = 1.$$

Or, taking into account the solution for z_2 ($z_2 \equiv 1$),

$$\dot{z}_1 = iz_1\left(\frac{\Delta}{2} - \frac{\Delta^2}{8} - \frac{\varepsilon^2}{12}\right) - i\frac{\varepsilon^2}{8}\bar{z}_1,$$
$$\dot{\bar{z}}_1 = i\bar{z}_1\left(\frac{\Delta}{2} - \frac{\Delta^2}{8} - \frac{\varepsilon^2}{12}\right) + i\frac{\varepsilon^2}{8}z_1.$$

The stability boundary is determined by the equation

$$\begin{vmatrix} \dfrac{\Delta}{2} - \dfrac{\Delta^2}{8} - \dfrac{\varepsilon^2}{12} & -\dfrac{\varepsilon^2}{8} \\ \dfrac{\varepsilon^2}{8} & -\left(\dfrac{\Delta}{2} - \dfrac{\Delta^2}{8} - \dfrac{\varepsilon^2}{12}\right) \end{vmatrix} = 0.$$

Whence it follows that

$$\left(\frac{\Delta}{2} - \frac{\Delta^2}{8} - \frac{\varepsilon^2}{12}\right)^2 - \frac{\varepsilon^4}{64} = 0.$$

Solving this equation for Δ, we obtain two boundaries,

$$\Delta = \frac{5}{12}\varepsilon^2 \quad \text{and} \quad \Delta = -\frac{1}{12}\varepsilon^2.$$

The trivial solution is stable in the neighborhood of the resonance in question if $\Delta < -\frac{1}{12}\varepsilon^2$ or $\Delta > \frac{5}{12}\varepsilon^2$.

5.8.3. Forced Oscillations of Duffing's Oscillator

We will study forced oscillations of a one-dimensional nonlinear oscillator governed by the differential equation

$$\ddot{x} + a\dot{x} + bx + cx^3 = d\sin\omega t.$$

The aim of the analysis that follows is to find out what new quality is added by the nonlinear term cx^3. For convenience, we denote the parameters that occur in the equation as follows: $a = 2h$,

$b = 1 - 2\Delta$, $c = \frac{8}{3}\varepsilon$, and $d = 2\mu$. The quantity Δ is the frequency detuning. If the unit of time is chosen so that $\omega = 1$, then a resonance occurs in the linear system at $\Delta = 0$. The dependence of the amplitude of steady-state oscillations, whose period coincides with that of the external force, on the detuning Δ is what determines the amplitude-frequency characteristic.

We will solve the nonlinear equation by the method of the Poincaré normal form. To this end, we first rewrite the original equation in the Cauchy normal form

$$\dot{x} = y,$$
$$\dot{y} = -x(1 - 2\Delta) - 2hy - \tfrac{8}{3}\varepsilon x^3 + 2\mu \cos t.$$

Just as in the case of Mathieu's equation, we reduce the nonautonomous system to autonomous form in order to apply the method of normal form. To this end, we add an auxiliary oscillator $\ddot{u} + u = 0$, whose solution for the initial conditions $u(0) = 1$ and $\dot{u}(0) = 0$ is given by $u = \cos t$, which formally coincides with the external force applied to Duffing's oscillator. Thus, we have the equivalent system

$$\dot{x} = y, \quad \dot{y} = -x(1 - 2\Delta) - 2hy - \tfrac{8}{3}\varepsilon x^3 + 2\mu u,$$
$$\dot{u} = v, \quad \dot{v} = -u, \qquad u(0) = 1, \quad v(0) = 0.$$

This system is autonomous and Poincaré's method is applicable to it. As before, it is convenient to use complex variables. We introduce $z = x + iy$ and $w = u + iv$. In terms of the new variables the system becomes

$$\dot{z} = -i(1 - \Delta)z + i\Delta\bar{z} - h(z - \bar{z}) - \tfrac{1}{3}i\varepsilon(z + \bar{z})^3 + i\mu(w + \overline{w}),$$
$$\dot{w} = -iw, \qquad w(0) = 1.$$

The equations for the complex conjugate variables are not written out but implied.

The unperturbed system (i.e., the system with $\Delta = h = \varepsilon = \mu = 0$) has the eigenvalues $\lambda_1 = \lambda_2 = -i$ and $\lambda_3 = \lambda_4 = i$. The first two pertain to the system just written out and the other two to the conjugate system.

In accordance with this, the normal form of the first approximation is given by

$$\dot{z} = -i(1 - \Delta)z - hz - i\varepsilon z^2\bar{z} - i\mu w,$$
$$\dot{w} = -iw, \qquad w(0) = 1.$$

By seeking the solution of this system in the form

$$z = z_0 e^{-it}, \quad w = w_0 e^{-it},$$

we obtain the following equations for $z_0(t)$ and $w_0(t)$:

$$\dot{z}_0 = i\Delta z_0 - hz_0 - i\varepsilon z_0^2\bar{z}_0 + i\mu,$$
$$\dot{w}_0 = 0.$$

The solution is periodic if $z_0 = \text{const}$. This leads to the following equation for z_0:

$$i\Delta z_0 - hz_0 - i\varepsilon z_0^2\bar{z}_0 + i\mu = 0.$$

Denoting the square of the amplitude by $A^2 = z_0\bar{z}_0 = x_0^2 + y_0^2$, we find

$$z_0 = \frac{-i\mu}{i\Delta - h - i\varepsilon A^2}, \quad \bar{z}_0 = \frac{-i\mu}{i\Delta + h - i\varepsilon A^2}.$$

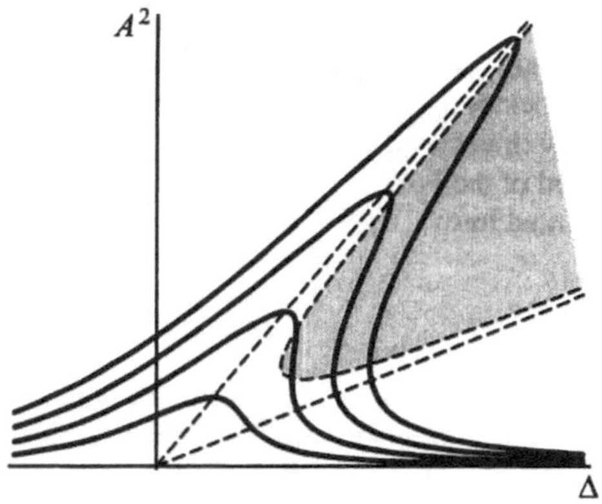

Figure 5.3

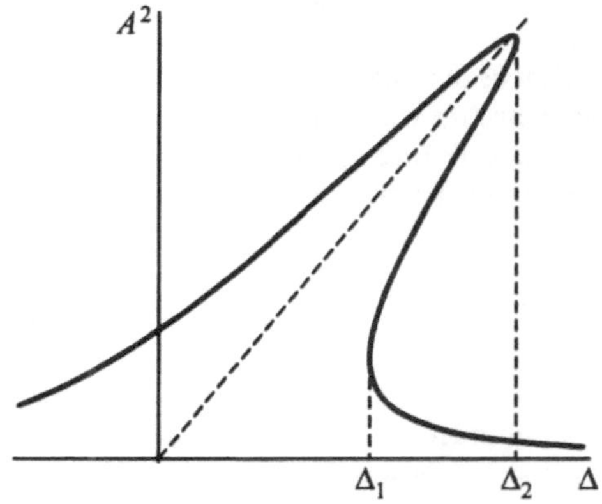

Figure 5.4

Multiplying the first relation by the second yields

$$A^2 = \frac{\mu^2}{h^2 + (\varepsilon A^2 - \Delta)^2} \quad \Longrightarrow \quad A^2[h^2 + (\varepsilon A^2 - \Delta)^2] = \mu^2.$$

This equation determines the amplitude-frequency characteristic $A(\Delta)$ in implicit form. The equation can be solved for Δ; we have

$$\Delta = \varepsilon A^2 \pm \sqrt{\frac{\mu^2}{A^2} - h^2}.$$

Figure 5.3 depicts the family of amplitude-frequency characteristics based on this formula (the parameter of the family is μ).

The main difference of this characteristic from that in the linear case ($\varepsilon = 0$) is that the resonance peak is tilted to the right (for $\varepsilon > 0$) or left (for $\varepsilon < 0$). As a result, a range of frequencies (from Δ_1 to Δ_2) arises where the amplitude of steady-state oscillations of the oscillator can have three different values. The values Δ_1 and Δ_2 possess the property that if Δ deviates from them by an arbitrarily small quantity, the state of the system can change a finite value. This phenomenon is referred to as a *bifurcation*, and the values Δ_1 and Δ_2 are said to be bifurcation values of the parameter Δ (see Fig. 5.4).

Determine the points of the amplitude-frequency characteristic where the tangents to the curve are vertical. Obviously, these vertical tangents must intersect the frequency axis at Δ_1 and Δ_2. The condition determining these points is that the derivative of the equation for the amplitude-frequency characteristic, $A^2[h^2 + (\varepsilon A^2 - \Delta)^2] = \mu^2$, with respect to A^2 is zero. Differentiating yields

$$h^2 + (\varepsilon A^2 - \Delta)(3\varepsilon A^2 - \Delta) = 0.$$

This equation just defines the locus of the points at which the amplitude-frequency characteristics of the family in question have vertical tangents. The curve is a hyperbola with asymptotes $\Delta = \varepsilon A^2$ and $\Delta = 3\varepsilon A^2$. It is apparent that for sufficiently small μ (the amplitude of the external force) and fixed h or sufficiently large h and fixed μ, there are no vertical tangents, and hence the amplitude is always single-valued.

Let us analyze the stability of the found modes of steady-state oscillations. Introduce a small deviation δz_0 from the stationary value z_0 and, using the differential equation $\dot{z}_0 = i\Delta z_0 - hz_0 - i\varepsilon z_0^2 \bar{z}_0 + i\mu$, write out a system of equations in variations. We have

$$\delta\dot{z}_0 = (i\Delta - h - 2i\varepsilon A^2)\delta z_0 - i\varepsilon z_0^2\,\delta\bar{z}_0,$$
$$\delta\dot{\bar{z}}_0 = (-i\Delta - h + 2i\varepsilon A^2)\delta\bar{z}_0 + i\varepsilon \bar{z}_0^2\,\delta z_0.$$

The characteristic equation of this system,

$$\begin{vmatrix} i\Delta - h - 2i\varepsilon A^2 - \lambda & -i\varepsilon z_0^2 \\ i\varepsilon \bar{z}_0^2 & -i\Delta - h + 2i\varepsilon A^2 - \lambda \end{vmatrix} = \lambda^2 + 2h\lambda + h^2 + (\varepsilon A^2 - \Delta)(3\varepsilon A^2 - \Delta) = 0,$$

makes it possible to write out the conditions of asymptotic stability of steady-state oscillations in the form

$$h > 0, \quad h^2 + (\varepsilon A^2 - \Delta)(3\varepsilon A^2 - \Delta) > 0.$$

The second condition is the most important. It does not depend on μ and defines, in terms of (Δ, A^2), the boundary of the stability region. This boundary exactly coincides with the locus of the points determining the bifurcation. The instability region is shaded in Figure 5.3. Thus, in the region where the amplitude is not single-valued, the oscillations with maximum and minimum amplitudes are stable and the oscillations with intermediate amplitudes are unstable.

5.8.4. Forced Oscillations of Van Der Pol's Oscillator

In oscillation theory the phenomenon to be studied is referred to as *harmonic capture of self-oscillations*. Whenever a self-oscillatory system is acted upon by an external periodic force, oscillations can occur which are similar to both self-oscillations and forced oscillations. This manifests itself in the fact that, under certain conditions, a periodic process arises whose amplitude is close to that of self-oscillations, while frequency coincides with that of the external force.

Consider the equation

$$\ddot{x} + x = 2\varepsilon(1 - 4x^2)\dot{x} + 2\Delta x + 2\mu \cos t.$$

The system

$$\begin{aligned} \dot{x} &= y, \\ \dot{y} &= -x + 2\varepsilon(1 - 4x^2)y + 2\Delta x + 2\mu u, \\ \dot{u} &= v, \\ \dot{v} &= -u, \qquad u(0) = 1, \quad v(0) = 0, \end{aligned}$$

is equivalent to this equation. Introduce the complex variables

$$z = x + iy, \quad w = u + iv$$

and reduce the system to the form

$$\begin{aligned} \dot{z} &= -iz + i\Delta(z + \bar{z}) + \varepsilon[1 - (z + \bar{z})^2](z - \bar{z}) + i\mu(w + \overline{w}), \\ \dot{w} &= -iw, \qquad w(0) = 1. \end{aligned}$$

The equations for the conjugate variables are the complex conjugate of these equations.

Apart from the three resonant terms that were mentioned in Subsection 5.8.1, one more resonant term, $i\mu w$, have appeared in this system. Thus, the normal form of the system is

$$\begin{aligned} \dot{z} &= -iz + i\Delta z + \varepsilon z(1 - z\bar{z}) + i\mu w, \\ \dot{w} &= -iw. \end{aligned}$$

Just as in the previous subsections, the perturbed part of the normal form is of most interest. Thus, we seek a solution of the form $z = z_0 e^{-it}$, $w = w_0 e^{-it}$. Omitting the zero subscript in what follows, we obtain

$$\dot{z} = i\Delta z + \varepsilon z(1 - z\bar{z}) + i\mu.$$

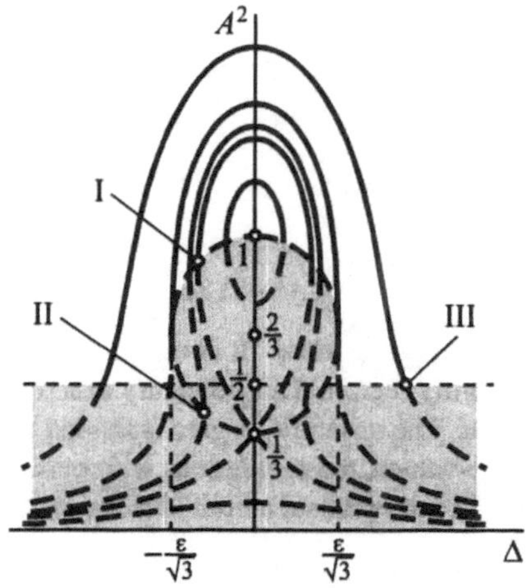

Figure 5.5

To the steady-state oscillations there corresponds $\dot{z} = 0$. This leads to the equation

$$i\Delta z + \varepsilon z(1 - z\bar{z}) + i\mu = 0;$$

whence,

$$z = \frac{-i\mu}{i\Delta + \varepsilon(1 - A^2)}, \quad \bar{z} = \frac{i\mu}{-i\Delta + \varepsilon(1 - A^2)}, \quad \text{where} \quad A^2 = z\bar{z}.$$

We multiply the first relation by the second to obtain

$$A^2 = \frac{\mu^2}{\Delta^2 + \varepsilon^2(1 - A^2)^2}.$$

This is the equation of the amplitude-frequency characteristic. Solving this equation for Δ yields

$$\Delta = \pm\sqrt{\frac{\mu^2}{A^2} - \varepsilon^2(1 - A^2)^2}.$$

To gain a correct qualitative understanding of the behavior of amplitude-frequency characteristic, it is necessary, just as in the previous subsection, to find the locus of the points at which the amplitude-frequency characteristic have vertical tangents. To this end, one should differentiate the function expressing the dependence of A^2 on Δ with respect to A^2 and equate the derivative to zero. We have

$$F(\Delta, A^2) = A^2[\Delta^2 + \varepsilon^2(1 - A^2)^2] - \mu^2 = 0,$$
$$\frac{\partial F}{\partial A^2} = \Delta^2 + \varepsilon^2(1 - A^2)(1 - 3A^2) = 0.$$

The curve represented by this equation in the (Δ, A^2) plane is an ellipse with center at the point $(0, 2/3)$, minor semiaxis $\varepsilon/\sqrt{3}$ (along the Δ-axis), and major semiaxis $1/3$ (along the A^2-axis). If an amplitude-frequency characteristic intersects this ellipse, then the tangents to the characteristic at the points of intersection are vertical (see Fig. 5.5).

To perform a stability analysis for steady-state modes, it is necessary to compile an equation in variations with respect to the stationary value z_0. Varying the differential equation $\dot{z} = i\Delta z + \varepsilon z(1 - z\bar{z}) + i\mu$, we arrive at the linear system of equations

$$\delta\dot{z} = [i\Delta + \varepsilon(1 - 2A^2)]\,\delta z - \varepsilon z^2\,\delta\bar{z},$$
$$\delta\dot{\bar{z}} = -\varepsilon\bar{z}^2\,\delta z + [-i\Delta + \varepsilon(1 - 2A^2)]\,\delta\bar{z}.$$

The characteristic equation of this system,

$$\begin{vmatrix} i\Delta + \varepsilon(1-2A^2) - \lambda & -\varepsilon z^2 \\ -\varepsilon \bar{z}^2 & -i\Delta + \varepsilon(1-2A^2) - \lambda \end{vmatrix} = \lambda^2 - 2\varepsilon(1-2A^2)\lambda + \Delta^2 + \varepsilon^2(1-A^2)(1-3A^2) = 0,$$

leads to the following conditions of asymptotic stability:

$$(1)\quad 2A^2 - 1 > 0,$$
$$(2)\quad \Delta^2 + \varepsilon^2(1-A^2)(1-3A^2) > 0.$$

(It is assumed that Van der Pol's oscillator, when not acted upon by an external force, is in the stable mode of oscillations, i.e., $\varepsilon > 0$.)

Part of the stability boundary coincides with the above ellipse.

The results obtained permit us to display the family of amplitude-frequency characteristics graphically (Fig. 5.5). The instability region is shaded. The parameter of the family is μ, the amplitude of the external force.

The analysis performed permits us to make an important observation. In the case of forced oscillations of Duffing's oscillator, the bifurcation points, at which the number of steady-state modes alters, are simultaneously the points at which the stability/instability change takes place. Contrastingly, the example of forced oscillations of Van der Pol's oscillator shows that this coincidence is not a rule. Points with any combinations of these properties are possible. In Fig. 5.5, point I is a point of bifurcation and also a point at which the stability/instability change takes place. At point II a bifurcation occurs but the system remains unstable. Point III is a point of stability change but there is no bifurcation.

BRIEF HISTORICAL SKETCH

The term "group" was introduced into science by Evariste Galois in 1831 in connection with the solvability of algebraic equations in radicals. Galois understood a group as a group of substitutions. The general abstract definition of a group was provided by Arthur Cayley in his works between 1849 and 1854.

Meanwhile, it was not only Galois' works that brought group theory to life. Implicitly, the notion of a group was used by Lagrange, Cauchy, and Abel. At that time, ideas of group theory could be encountered not only in studies of mathematicians. Auguste Bravais, a French scientist in the field of crystallography, should be mentioned here; in 1848, he studied rotations and translations that take a crystal to itself.

The most important role in spreading the ideas of group theory was played by the fundamental monograph "Traité des substitutions et des équations algébriques" (1870) by Camille Jordan [1]. That was the first systematic course of group theory and Galois' theory, in which Jordan not only generalized but also considerably extended the results obtained by Galois and Bravais.

In one of his early works, Jordan applies finite groups to studying differential equations. However, works by Bravais suggested him the idea of studying *infinite* (*continuous*) groups of transformations. It is continuous groups that were destined to play subsequently the basic role in the analysis of differential equations.

Sophus Lie [2], a pupil of Jordan, became the founder of continuous groups. He understood continuous groups as groups of transformations. Subsequently, such groups were named Lie groups, and the term "continuous group" was retained to denote abstract groups, in which relations of limit passage were established. Also such groups are called *topological* (see Pontryagin [3]). Today, a Lie group is understood as a continuous group which is also a manifold.

The application of group theory to differential equations has an uneven history. Having introduced the notion of a continuous group, Lie also provided examples of group analysis of differential equations [4]. Ordinary differential equations were first dealt with; the knowledge of a symmetry group for them permits one to lower their order. However, already in [2], Lie considered also partial differential equations; in particular, he calculated the symmetry group for the heat equation.

It is worthwhile mentioning Noether's theorem [5], which links a reduction in the order of a system of ordinary differential equations to the existence of a symmetry group of the Hamiltonian action, provided that the system is Lagrangian. Noether's theorem permits one to calculate a first integral for such a system.

Subsequently, this direction had been practically forgotten for a long time. This is accounted for, on the one hand, by the lack of constructive methods for calculating symmetry groups and, on the other hand, the realization of great potentialities of using groups in other fields and for other purposes, as well as by the growing interest in purely internal problems of group theory.

It was not until the 1950s that a new wave of interest in group analysis of differential equations arose after Birkhoff's work [6], where he paid attention to the possibility of using group analysis in hydrodynamics.

The further significant development of group analysis methods for studying differential equations was achieved in the 1960s and succeding years by Ovsiannikov and his school [7, 8]. In the books by Ibragimov [9, 10], a wide range of problems concerning the usage of Lie groups in the theory of differential equations (symmetries, exact solutions, conservation laws and some physical applications) are considered. These books contain an extensive bibliography on the subject. Note also the paper [11], where the theory of approximate groups of transformations is presented; this theory allows constructing approximate symmetries of equations with a small parameter.

Today, we have a lot of wonderful textbooks on the methods of group analysis of differential equations; among them it is worthwhile noting the texts by Bluman and Cole, Miller, and Olver [12, 13, 14].

Although group theory was first created to meet the needs of studying algebraic and differential equations, it practically immediately had found application in other fields of science. Felix Klein [15], another pupil of Jordan, writes about the three sources of group theory in mathematics: *the theory of equations*, *geometry*, and *the theory of functions*.

The utilization of groups in geometry radically influenced the analysis of the foundations of mechanics and physics. In this respect, the application of groups to relativistic mechanics is most significant. This is a unique example of how a natural science paradigm can be synthesized on the basis of experimental data with the decisive role of group-theoretic analysis.

This happened as follows. The gradual accumulation of observational data in the field of electromagnetic phenomena had led James Maxwell to his famous equations, which had made it possible to predict the existence of electromagnetic waves. These waves were later experimentally discovered by Heinrich Hertz. However, having allowed predicting a lot of phenomena, Maxwell's equations led unexpectedly to a rather acute problem. The wave equation that follows from Maxwell's equations and describes the propagation of electromagnetic waves is no different from the wave equation that describes the simplest wave processes in various media such as gas, liquid, and elastic bodies. However, for these media, the following question does nor arise: Relative to what do the waves in them propagate? The waves propagate relative to the same medium in its unperturbed state. Therefore, if the wave displacements of the medium are treated in a reference frame that moves relative to the medium, then the velocity of propagation of the wave in this reference frame is added to the velocity of the frame. Hence, the wave velocity will be different in different directions. This perfectly well agrees with the fact that the wave equation has a different form in the moving reference frame.

However, there is no any special medium for electromagnetic waves. The equations for the waves are written in *vacuum*. Therefore, the question about relative to what electromagnetic waves propagate becomes senseless. Suppositions were discussed that there is a hypothetical medium—ether, but these did not find experimental confirmation.

It follows from the aforesaid that Maxwell's equations are *obliged* to have the same form in all inertial reference frames (a reference frame is inertial if a point particle which is not acted upon by any forces moves in this frame in a straight line).

Hendrik Lorentz found the transformations of the spatial coordinates and time which leave Maxwell's equations unchanged [16].

Henri Poincaré [17, 18] postulated the Lorentz transformations, thus making this the main axiom of new mechanics. In development of the theory, Poincaré established the group nature of the Lorentz transformations for a particular value of some coefficient, which Lorentz left arbitrary. After that, the construction of the theory reduced to purely technical issues. In particular, Poincaré determined the law of transformation of velocities, constructed the relativistic Lagrangian function, introduced the notion a four-dimensional spacetime with one imaginary coordinate and indefinite metric, showed that the Lorentz transformations define a rotation in this space, introduced the notion of the force and velocity four-vectors, resolved the problem of the modification of Newton's law of universal gravitation with the aim to obtain its covariant representation within the framework of relativistic mechanics, introduced the concept of gravitational waves, and constructed relativistic electrodynamics.

In principle, there was no need to postulate the Lorentz transformations; these could be calculated if the invariance of Maxwell's equations over the ensemble of inertial frames was postulated. To this end, it suffices to find the intersection of the symmetry algebra for Maxwell's equations with the algebra of the projective group. The latter links the inertial reference frames to each other, since it takes any straight line to a straight line again. This resolves the previously discussed question on nonuniqueness of transformations that leave Maxwell's equation, or the fundamental quadratic form, unchanged.

Although the utilization of group theory in analytical mechanics is partly aimed at the integration of differential equations, it also goes far beyond this aim. Noether's theorem may not already be treated within the framework of this aim. The theorem establishes the connection of first integrals of Lagrangian systems with groups of translations in the generalized coordinates and time, and also with rotation groups. And this connection reveals the group nature of conservation laws. This fact is remarkable by itself, without using it to lower the order of the system, the more so as the presence of appropriate first integrals is obvious without any group analysis.

The application of group theory to analytical mechanics was stimulated by Elie Cartan's works. He introduced the definition of a manifold and gave first examples of using new geometric ideas in analytical mechanics [19, 20]. Subsequently, this field had been actively developed (e.g., see [21]).

As early as 1901, Poincaré published his paper [22] in which he derived new equations of mechanics by using infinitesimal generators of a transitive group acting in the space of generalized coordinates to write out variables that define the phase state of the system. For example, special cases of these equations were Lagrange's equations and the equations of motion of a rigid body (Euler's equations).

It is not difficult to notice that the chief merit of applying the group methods in mechanics is in the formation of models and construction of the corresponding equations rather than the possibility to integrate the equations in quadrature. The kinematics of a rigid body is a demonstrative example.

The first kinematic and dynamic equations were written by Leonard Euler in 1760. Euler did not know about groups and, to describe the orientation of a rigid body, he introduced angles of finite rotation—the first example of what is now called local coordinates of a rotation group. In local coordinates, the kinematic equations are nonlinear and have singularities, which presents serious practical inconveniences.

The kinematic equations in terms of orthogonal matrices (the SO(3) group)—written by Siméon Poisson in 1833—are linear and, in particular, have no singularities.

Note that the replacement of the kinematic equations in terms of Eulerian angles by the kinematic equations in terms of elements of a rotation group leads to an increase in the dimensionality of these equations. Contrastingly, the initial idea of using groups in equations was aimed at lowering the dimensionality of the equations.

It has long been realized that the desire to lower the order of equations—brought by a fairly old-fashioned idea to necessarily integrate them in quadrature—is not always reasonable. Often, at the expense of increasing the order, one succeeds in attaining a substantially better quality of the model employed.

Another example of this sort are the Kustaanheimo–Stiefel equations in celestial mechanics. These equations are written in terms of unit norm quaternions, i.e., the elements of a simply connected covering group of rotations, and are also linear.

Quaternions were introduced into science by Willian Hamilton [23, 24] in 1843. He attempted to find an $\mathbb{R}^3$-analogue of complex numbers so that the geometric interpretation of the hypercomplex numbers in space be, if possible, as effective as that of the complex numbers in the plane.

The multiplication of a complex number by another complex number of unit norm is a transformation of rotation in the plane. Hamilton discovered that a transformation of rotation in $\mathbb{R}^3$ can be specified by means of quaternions of unit norm. The kinematic equations for a rigid body written in terms of quaternions are linear, just as the equations in terms of orthogonal matrices. Moreover, the former are of smaller order that the latter—four versus six.

J. J. Moreau found an unexpected application of rotation groups. He showed in [25] that the motion of an incompressible fluid in a fixed domain can be identified with the rotation of an infinite-dimensional rigid body about a fixed point. Both ideal and viscous fluids are allowed, which corresponds to the motion with or without friction.

The above ideas were subsequently taken up by numerous researchers. For example, see [26].

Recently, the initial idea of using groups to integrate differential equations found a new expression in connection with attempts to improve asymptotic methods.

The need for this improvement was felt when the algorithms for constructing asymptotic expansions (we mean Poincaré's method of normal form and the Krylov–Bogolyubov method) had become to be effectively used in practice. Note that these algorithms were formulated, mainly, in the language of the existence of such expansions. Although the construction of the first approximation is fairly easy, it turned out that the second and higher approximations are extremely cumbersome.

The idea to invoke group theory to construct asymptotic expansions is quite natural. Indeed, if the formation of an algorithm is subordinated to some algebraic structure, then all operations will be reduced to operations from the corresponding algebra, which promises significant simplifications.

Ideas of this sort were implemented for constructing the Poincaré normal form [27] and reducing a nonautonomous system to an autonomous form in the spirit of the Krylov–Bogolyubov method [28].

In analytical mechanics, asymptotic expansions are known in the problem of normalization of Hamiltonian functions for conservative mechanical systems; this problem was first formulated by George Birkhoff. Here also, restating the algorithm for carrying out the normalization in the spirit of group theory leads to radical simplifications [29].

References

1. Jordan, C., *Traité des Substitutions et des Équations Algébriques*, Paris, 1870.
2. Lie, S., *Theorie der Transformationsgruppen*, Vol. I (1888), Vol. II (1890), Vol. III (1893), Leipzig.
3. Pontryagin, L. S., *Topological Groups*, 2nd ed., Gordon & Breach, New York, 1966.
4. Lie, S., *Klassifikation und Integration von gewohnlichen Differetialgleichungen zwischen x, y, die eine Gruppe von Transformationen gestatten*, Math. Ann., Vol. 32, pp. 213–281, 1888.
5. Noether, E., *Invariante Variationsprobleme*, Nachr. Konig. Gesell. Wissen. Göttingen, Math.-Phys. Kl., pp. 235–257, 1918.
6. Birkhoff, G. D., *Hydrodynamics—A Study in Logic, Fact and Similitude*, Revised edition, A Greenwood Press Reprint, New York, 1960.
7. Ovsiannikov, L. S., *Group Analysis of Differential Equations*, Academic Press, New York, 1982.
8. Ibragimov, N. H., *Transformation Groups Applied to Mathematical Physics*, Reidel, Boston, 1985.
9. Ibragimov, N. H. (editor), *CRC Handbook of Lie Group to Differential Equations, Vol. 1, Symmetries, Exact Solutions and Conservation Laws*, CRC Press, Boca Raton, 1994.
10. Ibragimov, N. H. (editor), *CRC Handbook of Lie Group to Differential Equations, Vol. 2, Applications in Engineering and Physical Scienses*, CRC Press, Boca Raton, 1995.
11. Baikov, V. A., Gazizov, R. K., and Ibragimov, N. H., *Perturbation methods in group analysis*, J. Sov. Math., Vol. 55, No. 1, p. 1450, 1991.
12. Bluman, G. W. and Cole, J. D., *Similarity Methods for Differential Equations*, Appl. Math. Sci., No. 13, Springer-Verlag, New York, 1974.
13. Miller, W., Jr., *Symmetry and Separation of Variables*, Addison-Wesley, Reading, Mass., 1977.
14. Olver, P. J., *Applications of Lie Groups to Differential Equations*, Springer-Verlag, New York, 1986.
15. Klein, F., *Vorlesungen uber die Entwicklung der Mathematik in 19 Jahrhundert*, Verlag Springer, Berlin, 1926.
16. Lorentz, H., *Electromagnetic phenomena in a system moving with any velocity smaller than of light*, Proc. Acad. Sci., Amsterdam, Vol. 6, p. 809, 1904.

17. Poincaré, H., *Sur la dynamique de l'électron*, Comptes Rendues, Vol. 140, p. 1504, 1905.

18. Poincaré, H., *Sur la dynamique de l'électron*, Rendiconti del Circolo Matematico di Palermo, Vol. XXI, p. 129, 1906.

19. Cartan, E., *Leçons sur les Invariants Intégraux*, Hermann, Paris, 1922.

20. Cartan, E., *La Théorie des Groupes Finis et Continus et l'Analysis Situs*, Mem. Sci. Vath., No. 42, Gauthier-Villars, Paris, 1930.

21. Kozlov, V. V., *Symmetries, Topology and Resonances in Hamiltonian Mechanics*, Izd-vo UGU, Izhevsk, 1995 [in Russian].

22. Poincaré, H., *Sur une forme nouvelle des équations de la mécanique*, Comptes Rendues, 1er Semestre, Vol. CXXXII, No. 7, p. 369, 1901.

23. Hamilton, W. R., *Lectures on Quaternions*, Hodges and Smith, Dublin, 1853.

24. Hamilton, W. R., *Elements of Quaternions*, Vols. 1–2, 2nd ed., Longmans, Green & Co., London, 1899–1901.

25. Moreau, J. J., *Une méthode de "cinématique fonctionnelle" en hydrodynamique*, Comptes Rendues, Vol. 249, No. 21, p. 2156, 1959.

26. Arnold, V. I., *Sur la géometrie différentielle des groupes de Lie de dimension infinie et des applications a l'hydrodynamique des fluides parfaits*, Annales de l'Institut Fourier, XVI, No. 1, pp. 319–361, 1966.

27. Zhuravlev, V. Ph. and Klimov, D. M., *Applied Methods in Oscillation Theory*, Nauka, Moscow, 1988 [in Russian].

28. Zhuravlev, V. Ph., *The application of monomial Lie groups to the problem of asymptotically integrating equations of mechanics*, J. Appl. Math. Mech. (PMM), Vol. 50, No. 3, pp. 260–265, 1986.

29. Zhuravlev, V. Ph., *A new algorithm for Birkhoff normalization of Hamiltonian systems*, J. Appl. Math. Mech. (PMM), Vol. 61, No. 1, pp. 9–14, 1997.

INDEX

A

B

C

H

I

N

O

P

Q

R